CHARGED BEAM INTERACTION WITH SOLIDS

CHARGED BEAM INTERACTION WITH SOLIDS

Yoshi-Hiko OHTSUKI

Waseda University, Tokyo, Japan

Taylor & Francis Inc.
New York
1983

First published 1983 by Taylor & Francis Inc.,
114 East 32nd Street, New York, NY 10016

Typeset by Thomson Press (India) Limited New Delhi
Printed in Great Britain by
Taylor & Francis (Printers) Ltd, Rankine Road,
Basingstoke, Hampshire RG24 0PR

Library of Congress Cataloging in Publication Data

Ótsuki, Yoshihiko, 1936–
Charged beam interaction with solids.
Bibliography: p.
Includes index.
1. Solids—Effect of radiation on. 2. Channeling (Physics). 3. Particle beams. 4. Scattering (Physics).
I. Title.
QC176.8.R3087 1983 539.7′54 83-12596
ISBN 0-8002-3083-3

PREFACE

In the past, scientists interested in the tracks seen in the Wilson Cloud Chamber made by charged particles passing through it related the tracks to the interaction of charged particles with gases or solids. Further, with the development of the emulsion chamber, the nuclear film, and the solid state detector, the study of the interactions between charged particles and solids, or atomic collisions in solids, became very important. This study is not only of interest because of the physics involved, but also because of its application to problems in the fields of ion scattering in solids, structure analysis of solids, ion implantation, and plasma-wall interaction.

In this book I present the fundamental subject of the interactions of (1) electron, (2) positron, (3) proton, (4) light ion and (5) heavy ion beams with solids and solid surfaces. However, I want to limit discussion to the first stages of the interactions, and I do not therefore touch on the secondary phenomena, such as secondary particle emission.

Chapter 1 reviews atomic physics and atomic scattering, followed by a discussion of the fundamental characteristics of atomic collisions in solids. Chapters 2 and 3 explain electron and ion channelling in detail, including electron diffraction and electron and positron channelling; special emphasis is given to Lindhard's theory of channelling and dynamical diffraction theory.

Chapter 4 discusses inelastic scattering effects in channelling. It covers ordinary stopping as described by the Bethe–Bloch and Firsov formulae; the inelastic scattering function, which depends on the beam's position in relation to the channelling wall; inelastic scattering with the Umklapp process, as described by Kikuchi patterns; and channelling radiation.

Chapter 5 presents the exact theory of dechannelling in terms of the inelastic scattering function and the Fokker–Planck equation. An explanation of Lindhard's dechannelling theory in terms of exact channelling theory, and the applications of dechannelling theory to crystal analysis and the interpretation of the diffusion of diffraction spots are given.

Chapter 6 discusses the response of a solid to the impact of a charged particle and the resulting modification of the charge state of that particle due to polarization of the medium. The chapter covers equilibrium charge, effective charge, the Okorokov effect and the screening of the interaction potential, including the wake potential. Special emphasis is given at the end of the chapter to the study of the wake potential in terms of canonical transformation due to plasmon, exciton and single electron excitations.

Chapter 7, the final chapter, reviews charged particle interactions with solid surfaces. Surface structure analysis by electron diffraction and channelling, and proton neutralization and H^- production at the solid's surface are discussed. Surface dynamical potential is introduced and the skipping motion of light ions, asymmetrical alignment effects, transition radiation and circular polarizations of the emitted radiation at the surface are then discussed. The energy loss of ions at the surface is shown to be due to surface plasmons.

I am grateful for the many discussions with young colleagues during the preparation of this manuscript, especially those in my seminar classes at Waseda University in 1972 and 1981. The manuscript was partly prepared at the University of Strasbourg (1967, 1970, 1975), at Harwell (1975), at the University of Munich (1975), and at the Institute of Plasma Physics at Nagoya University (1980, 1981). Therefore, I would like to thank the late Professor S. Goldztaub, Professor C. Burggraf, Dr. A. Mosser, Dr. G. Dearnaley, Professor R. Sizmann, Dr. C. Rau and Dr. M. Kitagawa for their hospitality and helpful discussions during my stays in the above places. I have also benefited from many discussions with Professor T. Ichinokawa, Professor M. Namiki, Professor K. Kohra, Dr. R.H. Ritchie, Dr. N. Eckstein, Professor N. Itoh, Professor A.F. Tulinov and Dr. M.A. Kumakhov. I am particularly indebted to Mr. H.C. Marciniak for reading and criticizing much of the manuscript and for making many valuable suggestions.

Tokyo
December 1982

Y.H. Ohtsuki

CONTENTS

Preface v

Chapter 1. INTRODUCTION 1
1.1. Atomic collisions in solids 1
1.2. Atomic structure 2
1.3. Scattering of charged particles 5

Chapter 2. CHANNELLING AND BLOCKING 9
2.1. Introduction 9
2.2. Continuum potential and critical angle 10
2.3. Rutherford shadow (blocking) 15
2.4. Modification of the Lindhard theory 18

Chapter 3. QUANTUM EFFECTS IN ELECTRON AND POSITRON CHANNELLING 23
3.1. Introduction 23
3.2. One-dimensional periodic field. 23
3.3. Quantum corrections to the Lindhard theory . . 24
3.4. Electron channelling 26
3.5. Electron diffraction and the scattering factor . . 30
3.6. Multiple scattering model 33
3.7. Bethe's dynamical theory of electron diffraction . 36
3.8. Correction to the two-wave theory 44

Chapter 4. INELASTIC SCATTERING OF ELECTRONS AND IONS IN CRYSTALS 54
4.1. Introduction 54
4.2. Anomalous absorption 54
4.3. Calculation of absorption coefficients 58
4.4. Complex potential (Yoshioka term) 67
4.5. Comparison with experiments 70
4.6. Diffuse patterns 71
4.7. Kikuchi patterns 75
4.8. Stopping power, stopping power function and inelastic scattering function 83
4.9. Energy loss due to photon emission 108

Chapter 5. DECHANNELLING 125
5.1. Phenomenological theory. 125
5.2. Calculation of diffusion coefficients by the inelastic scattering function 129
5.3. Planar dechannelling 142
5.4. Axial dechannelling 148
5.5. Diffusion of electron beams 149
5.6. Applications of dechannelling 151

Chapter 6. CHARGE STATES, SCREENING AND WAKE . 155
6.1. Introduction 155
6.2. Equilibrium charge states and effective charge . . 155
6.3. Charged states of channelled ions and the Okorokov effect 167
6.4. Screening and wake 174

Chapter 7. PARTICLES AND SOLID SURFACES 186
7.1. Introduction 186
7.2. LEED 187
7.3. Backscattering, surface scattering and structure analysis 200
7.4. Neutralization of ions at surfaces 206
7.5. Dynamical surface potential 221
7.6. Skipping motion, surface energy loss and photon emission 233
7.7. Interaction between two ions near the surface . . 239

Index 247

CHAPTER 1

introduction

1.1. Atomic collisions in solids

Charged particle penetration in matter is a typical phenomenon in physics in which elementary processes are treated using atomic collision or ion collision methods. The atomic collision phenomena in gas targets have been well established in the last few decades. As an indication of this, almost a thousand scientists attend the international conference in this field every two years. On the other hand, atomic collisions in solids is a new field, which started only about ten years ago as the use of small accelerators became more common.

What is the difference between gaseous targets and solid targets? It is obvious that the solid target is more dense than the gaseous one; therefore, sometimes we can understand atomic collisions in solids in terms of atomic collisions in gases by merely changing the atomic density (or electron density) in the appropriate formula. For example, the mean charge of He^+ ions with an energy of 500 keV in a solid is derived by changing the atomic density in the expression of the mean charge of He^+ ions in a gaseous target.

However, these simple modifications between collisions in gaseous targets and solid targets are valid only in rather exceptional cases. The above mean charge may not be derived by changing only the atomic density when the charge-exchange mean free path L_{ce} is smaller than the atomic spacing d in the solid. In this case, a few atoms within the length of order L_{ce} contribute to the charge-exchange process as a whole, and the binary encounter model is not adequate.

In solid targets we have to at least take into account the valence electrons (conduction electrons), the periodic lattice orientation and surface effects. We note that the valence electrons in the solid form the solid's band structure and collective oscillations. Plasmon excitation and exciton excitation are very typical collective motions originating from the valence electrons. The energy transfer to the collective oscillation from the ion beam is very small compared to the energy of the ion moving through the solid. However, there are a lot of cases in which these collective oscillations play an important role.

The most interesting and exciting phenomena which have been discovered are those which depend on crystal orientation and lattice periodicity. Among them electron diffraction and channelling have been given much attention. When the de Broglie wavelength of the particle in the crystal is not smaller than the lattice spacing, quantum mechanical interference, or diffraction, appears. The electron wavelength is comparable with the lattice spacing ($\sim 0{\cdot}1$ nm) when the electron energy is about 100 eV. We can find electron diffraction phenomena for electrons with energies up to 1 MeV. Channelling is another orientational effect in crystals, and will be understood by the transverse (to the atomic row) energy space. The periodicity along the atomic axis plays an important role in the penetration of the charged particles into the crystal. Lattice orientation and lattice vibration (or phonon excitation) should be taken into account at the same time. The energy of the phonon is very small ($\lesssim 0{\cdot}1$ eV), but its momentum is comparable with the scattering momentum in the atomic collision.

There are many kinds of phenomena in atomic collisions with the strictly defined solid surface. For example, the electron density in a metal surface decreases to zero within a distance of approximately 0·01 nm. Due to the clear-cut surfaces of solids there are effects which are brought about as a result of the presence of these surfaces. For example, the lattice constant and the lattice vibration are modified in the region of the surface. The band structure and the collective motion of the valence electrons near the surface are also changed from those in the bulk.

Understanding atomic collisions in solids by taking into account the characteristics of the solid state is very complicated, but this understanding was increased very much recently by the increasing interest in its application to many other fields, such as nuclear fusion (plasma–wall interaction), surface structure analysis, ion implantation, and so on. We start our study with a fundamental review of atomic physics and scattering phenomena.

1.2. Atomic structure

The hydrogen model of the atom assumes a single electron revolving about a nucleus of charge Ze. It is represented by the binding energy

$$E_n = -(me^4/2\hbar^2)(Z^2/n^2) \tag{1.2.1}$$

and the wave function

$$\varphi_{nlms} = R_{nl}(r)\,Y_{lm}(\theta,\varphi)\chi_s(\sigma) \tag{1.2.2}$$

where n, l, m and s are the principal quantum number, the orbital quantum number, the magnetic quantum number and the spin quantum number, respectively. For $n=1$, $l=0$ (s orbit) and $m=0$, we obtain

$$R_{1,0}(r) = 2(Z/a_B)^{3/2}\exp(-Zr/a_B)$$
$$Y_{0,0}(\theta,\varphi) = 1/\sqrt{4\pi}$$

and for $n=2$, $l=0, 1$, $m=0, \pm 1$, we get

$$R_{2,0}(r)=\frac{1}{2\sqrt{2}}\left(\frac{Z}{a_B}\right)^{3/2}\left(2-\frac{Z}{a_B}r\right)\exp(-Zr/a_B)$$

$$Y_{1,0}(\theta,\varphi)=(\sqrt{3}/\sqrt{4\pi})\cos\theta$$

$$Y_{1,\pm 1}(\theta,\varphi)=\left(1/\sqrt{4\pi}\right)\sqrt{3/2}\sin\theta\exp(\pm i\varphi)$$

where $a_B=\hbar^2/me^2$ is the Bohr radius (0·053 nm). The velocity v_O of the orbital electron with quantum number n is obtained by

$$v_O=Zv_B/n$$

where v_B is the Bohr velocity, $e^2/\hbar=2{\cdot}188\times 10^8\,\mathrm{cm\,s^{-1}}$.

When we consider multi-electron atoms, the hydrogenic model is still useful. In this approximation, the many electron interactions are included in the single electron Hamiltonian, which takes the form

$$H_i=(-\hbar^2/2m)\nabla_i^2-(Ze^2/r_i)+\sum_{\substack{j\\(i\neq j)}}\int(e^2/r_{ij})|\phi_j(\mathbf{r}_j)|^2\,\mathrm{d}\mathbf{r}_j \qquad (1.2.3)$$

where the second term describes the Coulomb potential due to the nucleus and the third term the interaction between the ith electron and the rest of the electrons whose density is given by $\phi_j(\mathbf{r}_j)\,\mathrm{d}\mathbf{r}_j$. The wave function ϕ_i for the above Hamiltonian should be found using the 'self consistent field method'

$$H_i\phi_i=\varepsilon_i\phi_i \qquad (1.2.4)$$

The above method of obtaining the wave function ϕ_i and the energy ε_i is called the 'Hartree approximation', or, if we include the exchange interaction, the 'Hartree–Fock approximation'.

Another model for the energy levels of the multi-electron atom is the screening model: the Coulomb potential of the nucleus is modified by the electron cloud within the orbit of the electron to be considered, whence the modified potential takes the form

$$U(r)=-\frac{Z_{\mathrm{eff}}e^2}{r}+U_0 \qquad (1.2.5)$$

where $Z_{\mathrm{eff}}e$ is the effective charge of the nucleus screened by the electron cloud within the sphere of radius r and U_0 is the sum of the Coulomb potential of the electrons outside the sphere. Solving the Schrödinger equation, we obtain energy levels, E_n

$$E_n=\frac{Z_{\mathrm{eff}}^2}{n^2}E_r-U_0 \qquad (1.2.6)$$

where E_r is the Rydberg energy equal to 13·59 eV, or

$$E_n=\theta_s\frac{Z_{\mathrm{eff}}^2}{n^2}E_r \qquad (1.2.7)$$

where θ_s is defined instead of the parameter U_0. θ_s takes values from 0 to 1 and is called the 'screening parameter'. The effective charge number $Z_{eff}e$ is determined from the wave function in the Hartree–Fock approximation. For the K shell in He we obtain $Z_{eff} = Z - 5/16$, and for other atoms $Z_{eff} = Z - 0{\cdot}3$. On the other hand, the screening parameter θ_s is determined from experimental energy levels. For large Z, θ_s takes values from 0·8 to 0·85 for the K shell, and from 0·6 to 0·65 for L shell.

The simple Thomas–Fermi model is very useful in understanding atomic collisions in an analytic manner. We consider the atomic electrons as the 'electron gas'. The electron density $n(r)$ at position r measured from the nucleus is given by

$$n(r) = \frac{2}{(2\pi\hbar)^3}\frac{4\pi}{3}[p_F(r)]^3 = \frac{1}{3\pi^2}\left(\frac{p_F(r)}{\hbar}\right)^3 \tag{1.2.8}$$

where $p_F(r)$ is the maximum momentum of the electrons at position r.

We assume the potential energy for the electron in the Thomas–Fermi model to be $U(r)$. In a stationary state the maximum energy of the electrons takes a constant value U_0

$$\frac{[p_F(r)]^2}{2m} - U(r) = U_0 \tag{1.2.9}$$

Inserting equation 1.2.9 into equation 1.2.8 we obtain

$$n(r) = \frac{1}{3\pi^2\hbar^3}[2m(U(r) - U_0)]^{3/2} \tag{1.2.10}$$

If we insert the potential of the form

$$U(r) = \frac{Ze}{r}\chi(r)$$

(where $\chi(0) = 1$, $\chi(\infty) = 0$) into the Poisson equation $\nabla^2 U = 4\pi ne^2$, we obtain the following differential equation for $\chi(x)$

$$x^{1/2}\frac{d^2\chi}{dx^2} = \chi^{3/2} \tag{1.2.11}$$

where x is defined by

$$r = \tfrac{1}{2}\left(\frac{3\pi}{4}\right)^{2/3} a_B Z^{-1/3}x = 0{\cdot}88534 a_B Z^{-1/3}x$$

The Thomas–Fermi radius a_{TF} is defined by $\chi(x) = \frac{1}{2}$, which means that $x = 1$ or

$$a_{TF} = 0{\cdot}88534 a_B Z^{-1/3} \tag{1.2.12}$$

The number of electrons with momentum intervals p and $p + \mathrm{d}p$ is obtained by equation 1.2.8

$$w(p)\mathrm{d}p = \frac{4r_\mathrm{p}^3}{3\pi^2\hbar^3} p^2 \,\mathrm{d}p$$

where r_p is defined by $p^2 = 2mU(r_\mathrm{p})$. The mean velocity of electrons is easily derived

$$v_0 = 0{\cdot}688 Z^{2/3} v_\mathrm{B} \tag{1.2.13}$$

If the exchange interaction is included in the Thomas–Fermi model, equations 1.2.10 and 1.2.11 are modified to the following forms

$$n(r) = \frac{1}{3\pi^2\hbar^3}(2m)^{3/2}\{[U(r) + b^2]^{1/2} + b\}^3$$

and

$$\frac{\mathrm{d}^2\chi}{\mathrm{d}x^2} = x\left[\frac{6^{1/3}}{4(\pi Z)^{2/3}} + \left(\frac{\chi}{x}\right)^{1/2}\right]^3 \tag{1.2.14}$$

where $b = \sqrt{2m}/2\pi\hbar$ and $U(r) + b^2 = \dfrac{Ze^2}{r}\chi$.

1.3. Scattering of charged particles

If we consider collisions of charged particles of mass M_1 with target atoms of mass M_2 in the centre-of-mass system, we may solve for the scattering of the charged particles by a fixed scatterer employing the reduced mass defined by

$$\frac{1}{M} = \frac{1}{M_1} + \frac{1}{M_2} \tag{1.3.1}$$

Classically, the impact parameter b determines the scattering angle uniquely. For example, the scattering angle is determined from the potential $U(r)$ to be

$$\theta = \pi - 2b\int_{r_0}^{\infty} \frac{1}{\left[1 - \dfrac{b^2}{r^2} - \dfrac{U(r)}{E_\mathrm{R}}\right]^{1/2}} \frac{\mathrm{d}r}{r^2} \tag{1.3.2}$$

where E_R is the energy of the particle in the centre-of-mass system

$$E_\mathrm{R} = \tfrac{1}{2}Mv^2 = \frac{M_2}{M_1 + M_2}E$$

E being the energy of the particle in the laboratory system. In equation 1.3.2 r_0 is the minimum distance to the scatterer. In the case of Rutherford scattering,

$U(r) = (Z_1 Z_2 e^2)/r$, we obtain

$$\cot \tfrac{1}{2}\theta = \frac{mv^2}{Z_1 Z_2 e^2} b \tag{1.3.3}$$

In fundamental physics, the differential scattering cross-section $\sigma(\theta)$ is calculated using the above relation between impact parameter and scattering angle

$$\sigma(\theta)\,\mathrm{d}\theta = 2\pi b \left(\frac{\mathrm{d}\theta}{\mathrm{d}b}\right)^{-1} \mathrm{d}\theta \tag{1.3.4}$$

or

$$\sigma(\theta)\,\mathrm{d}\Omega = b \left(\frac{\mathrm{d}\theta}{\mathrm{d}b}\right)^{-1} \operatorname{cosec}\theta\,\mathrm{d}\Omega \tag{1.3.4$'$}$$

Inserting equation 1.3.3 into equation 1.3.4, we obtain

$$\begin{gathered}\sigma(\theta)\,\mathrm{d}\Omega = (\tfrac{1}{4}B)^2 \operatorname{cosec}^4 \tfrac{1}{2}\theta\,\mathrm{d}\Omega \\ (B = 2Z_1 Z_2 e^2/Mv^2)\end{gathered} \tag{1.3.5}$$

which is a very famous expression derived by Rutherford.

On the other hand, the differential cross-section $\sigma(T)$ for the energy transfer between T and $T + \mathrm{d}T$ takes the form (see equation 1.3.4)

$$\begin{aligned}\sigma(T)\,\mathrm{d}T &= 2\pi b \left(\frac{\mathrm{d}T}{\mathrm{d}b}\right)^{-1} \mathrm{d}T \\ &= \pi \frac{\mathrm{d}(b^2)}{\mathrm{d}T} \mathrm{d}T\end{aligned} \tag{1.3.6}$$

The maximum energy transfer T_m is simply calculated by momentum and energy conservation in classical mechanics

$$\begin{aligned}T_\mathrm{m} &= \frac{4M_1 M_2}{(M_1 + M_2)^2} E \\ &= \left[1 + 4\left(\frac{2Z_1 Z_2 e^2}{Mv^2}\right)^{-2} b^2\right] T\end{aligned} \tag{1.3.7}$$

Inserting equation 1.3.7 into equation 1.3.6, we obtain

$$\sigma(T)\,\mathrm{d}T = \frac{\pi}{4}\left(\frac{2Z_1 Z_2 e^2}{Mv^2}\right)^2 T_\mathrm{m} \frac{\mathrm{d}T}{T^2} \tag{1.3.8}$$

In quantum mechanics, we can also derive the differential cross-section for the charged particle

$$\sigma(\theta)\,\mathrm{d}\Omega = \frac{|\mathbf{k}'|}{|\mathbf{k}|} |f(\theta)|^2 \,\mathrm{d}\Omega \tag{1.3.9}$$

where $\hbar\mathbf{k}$ and $\hbar\mathbf{k}'$ are the incident and scattered momentum vectors, respectively. The scattering amplitude $f(\theta)$, calculated quantum mechanically from the interaction potential, is

$$f(\theta) = \frac{1}{2ik}\sum_{l=0}^{\infty}(2l+1)[\exp(2i\delta_l) - 1]P_l(\cos\theta) \tag{1.3.10}$$

where δ_l is the phase shift. The first term of the expansion of $[\exp(2i\delta_l) - 1]$ in equation 1.3.10 corresponds to the Born approximation.

The Born approximation is represented by the golden rule for transitions from an initial state $|i\rangle$ with energy E_i to the final state $|f\rangle$ with energy E_f

$$W_{fi} = \frac{2\pi}{\hbar}|\langle f|H'|i\rangle|^2\delta(E_f - E_i)\rho_f \tag{1.3.11}$$

where H' is the interaction Hamiltonian and ρ_f is the number of the final states. In the case of Rutherford scattering H' is the Coulomb interaction

$$H' = \frac{Z_1 Z_2 e^2}{r}$$

and

$$|i\rangle = \exp(i\mathbf{k}\cdot\mathbf{r})|0\rangle$$
$$|f\rangle = \exp(i\mathbf{k}'\cdot\mathbf{r})|0\rangle \tag{1.3.12}$$

where $|0\rangle$ is the target state. Inserting equation 1.3.12 into equation 1.3.11 we obtain

$$W_{fi} = \frac{2\pi}{\hbar}|H'(q)_{00}|^2\delta\left(\frac{\hbar^2k^2}{2M} - \frac{\hbar^2k'^2}{2M}\right)\rho_f \tag{1.3.13}$$

where

$$H'(q) = \int H'(r)\exp(i\mathbf{q}\cdot\mathbf{r})\,\mathrm{d}\mathbf{r} = \frac{4\pi Z_1 Z_2 e^2}{q^2} \tag{1.3.14}$$

$$\hbar\mathbf{q} = \hbar(\mathbf{k} - \mathbf{k}')(\equiv \mathbf{p} - \mathbf{p}') \tag{1.3.15}$$

$$\rho_f = \frac{Vp'\mathrm{d}(p')^2}{16\pi^3\hbar^3}\mathrm{d}\Omega \tag{1.3.16}$$

It is easy to show from equation 1.3.13 that the differential cross-section for Rutherford scattering derived from quantum mechanics coincides with the classical one, equation 1.3.5.

Inelastic scattering can also be analyzed with the Born approximation, making use of the golden rule where the final state is an excited state $|n\rangle$

$$|f\rangle = \exp(i\mathbf{k}'\cdot\mathbf{r})|n\rangle$$

The nth excited state of the target is formed by a single electron excitation (core excitation, inter-band excitation of the valence electron, and so on), plasmon excitation, phonon excitation, and so on.

We shall now discuss the interaction potential $U(r)$ for atomic collisions. The most widely used potential is the screening potential in the form

$$U(r) = \frac{Z_1 Z_2 e^2}{r} \exp\left(-\frac{r}{a}\right) \quad (1.3.17)$$

where the screening length a is given by

$$a = (Z_1^{2/3} + Z_2^{2/3})^{-1/2} a_B \quad (1.3.18)$$

and the Thomas–Fermi potential (or Firsov potential[1,2]) is in the form

$$U(r) = \frac{Z_1 Z_2 e^2}{r} \chi\left(\frac{r}{a_{TF}}\right) \quad (1.3.19)$$

which originated from equations 1.2.10 and 1.2.12. However, a_{TF} in equation 1.3.19 should be modified according to the complicated structure of the incident atom with Z_1 electrons

$$a_{TF} = 0{\cdot}88534 \frac{a_B}{(Z_1^{1/2} + Z_2^{1/2})^{2/3}} \quad (1.3.20)$$

or

$$a_{TF} = 0{\cdot}88534 \frac{a_B}{(Z_1^{2/3} + Z_2^{2/3})^{1/2}} \quad (1.3.21)$$

The screening potential differs very much from the Firsov potential for large r.

Other empirical potentials are the Born–Mayer potential and the Molière potential. The Born–Mayer potential takes the form

$$U(r) = U_0 \exp(-r/r_{BM}) \quad (1.3.22)$$

where r_{BM} is a parameter which corresponds to the atomic radius in the crystal. Abrahamson[3] derived the parameters U_0 and r_{BM} in order to fit the Thomas–Fermi–Dirac potential. However, the Born–Mayer potential with Abrahamson's parameters is correct only for a region of r less than 0·1 nm.

We call the Thomas–Fermi potential the 'Moliere potential' when we replace the function χ by

$$\chi_M(x) = 0{\cdot}35 \exp(-0{\cdot}3x) + 0{\cdot}55 \exp(-1{\cdot}2x) + 0{\cdot}10 \exp(-6{\cdot}0x) \quad (1.3.23)$$

where $x = r/a_{TF}$. The Molière potential is useful in studying atomic collisions analytically because $\chi_M(r/a_{TF})$ is expressed only by exponential terms.

References

1. Firsov, O.B., *Dokl. Akad. Nauk* **91**, 515 (1953).
2. Firsov, O.B., *J. Exp. Theor. Physics* **6**, 534 (1958).
3. Abrahamson, A.A., *Phys. Rev.* **178**, 76 (1969).

CHAPTER 2

channelling and blocking

2.1. Introduction

Until 1962 only a few differences between the phenomena of ions incident on crystalline materials and ions incident on amorphous materials were known. The experiments of Davies *et al.*[1] (1960) which measured the penetration depth of ions in crystalline materials gave one example of such differences. These authors showed that, unlike the case of ions incident on amorphous materials, the distribution of ions in crystalline materials as a function of penetration depth is not Gaussian, but has a long tail for large penetration depths.

Referring to these experiments, Robinson and Oen[2] (1963) tried to make a computer simulation of the penetration of ions in the crystal, and showed that a few ions incident at a small angle to the atomic plane or the atomic row are localized within a channel as shown in figure 2.1. On the other hand, ions incident at a large angle are not localized, and are scattered. This effect is called 'channelling'. It should be noted that channelling phenomena had already been proposed by Stark[3] (1912), who wrote "charged particles will be able to penetrate much greater distances along certain particular axis...".

Lindhard[4] developed a theory of channelling in 1965, and defined the critical angle for channelling based on classical mechanics. Quantum mechanical

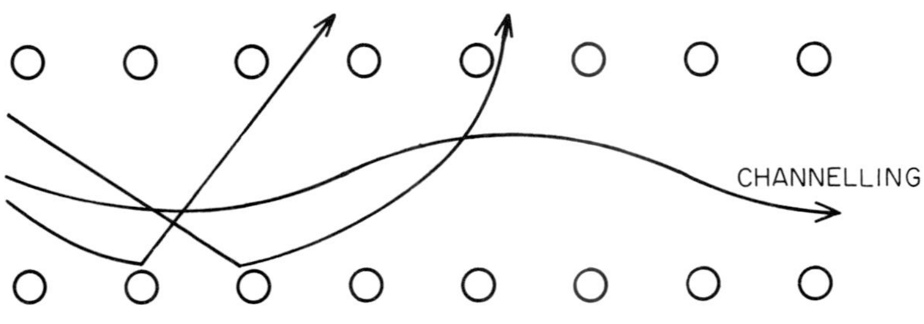

Figure 2.1. Ion collision with atoms in the crystal. Ions incident at a small angle to the atomic plane (or atomic axis) are localized within a channel.

treatments of channelling for positrons or electrons were also developed by many authors, connecting it to the electron diffraction theory.

Since the Lindhard theory many experimenters in nuclear physics, accelerator physics and radiation physics have moved to this field, and almost all phenomena related to channelling have been investigated both theoretically and experimentally. In this chapter we shall study the fundamental aspects of channelling.

2.2. Continuum potential and critical angle

Firstly we shall consider ion beams passing through the atomic row (z axis). As is shown in figure 2.2(a) we take d and r to be the inter-atomic distance in the atomic row and the distance between the atomic row and the ion, respectively. The static potential due to the nth atom (its position being z_n) in the atomic row is represented by

$$v\sqrt{(z-z_n)^2+r^2}$$

Then we obtain the total potential $V(\mathbf{R})$

$$V(\mathbf{R})=\sum_n v\sqrt{(z-z_n)^2+r^2} \tag{2.2.1}$$

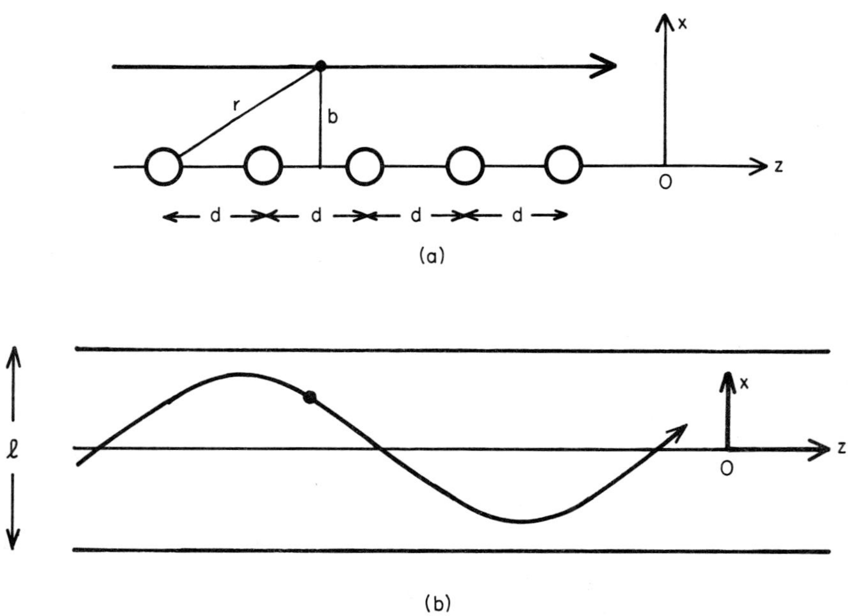

Figure 2.2. (a) Moving ion with the impact parameter b to the atomic string. (b) Trajectory of ions in the (x, z) plane.

where $\mathbf{R} = (\mathbf{r}, z)$. Now we consider the average potential over a straight long path $\left[-\frac{L}{2}, \frac{L}{2}\right]$ of ions

$$U(r) = \frac{1}{L}\int_{-L/2}^{L/2} v\sqrt{(z - z_n)^2 + r^2}\,\mathrm{d}z$$

$$= \frac{1}{Nd}\sum_n \int_{-L/2-z_n}^{L/2-z_n} v\sqrt{z'^2 + r^2}\,\mathrm{d}z' \simeq \frac{1}{Nd}\sum_n \int_{-\infty}^{\infty} v\sqrt{z'^2 + r^2}\,\mathrm{d}z'$$

$$= \int_{-\infty}^{\infty} \frac{\mathrm{d}z}{d} v\sqrt{z^2 + r^2} \qquad (2.2.2)$$

where N is the number of atoms on the atomic row with length L.

The averaged potential $U(r)$ obtained in equation 2.2.2 is related to the Fourier component of the potential. The potential $V(r)$ may be expanded in Fourier series with the one-dimensional reciprocal lattice vector $\mathbf{g}$, since the position of atoms in the atomic row is periodic. The Fourier component $V_g(r)$ is obtained as follows

$$V_g = \frac{1}{L}\int_{-\infty}^{\infty} V(R)\exp(-2\pi i g z)\,\mathrm{d}z$$

$$= \frac{1}{L}\sum_n \exp(-2\pi i g z_n)\int_{-\infty}^{\infty} v\sqrt{z^2 + r^2}\exp(-2\pi i g z)\,\mathrm{d}z$$

$$= \frac{N}{L}\int_{-\infty}^{\infty} v\sqrt{z^2 + r^2}\exp(-2\pi i g z)\,\mathrm{d}z \qquad (2.2.2)'$$

Comparing equation 2.2.2 with equation 2.2.2′ we have the following relation

$$U(r) = V_0(r) \qquad (2.2.3)$$

Lindhard[4] called the average potential $U(r)$ 'the string potential', which we frequently call the 'continuum potential'.

Similarly, we may obtain the average potential for ions moving parallel to the atomic plane (y–z plane in figure 2.2(*a*)). The average potential $U^{\mathrm{pl}}(x)$ is a function of the distance between the ion and the atomic plane. We obtain

$$U^{\mathrm{pl}}(x) = \frac{1}{s}\int_0^{\infty} 2\pi r\,\mathrm{d}r\, V\sqrt{x^2 + r^2} \qquad (2.2.4)$$

where s is the area per atom on the atomic plane.

Figure 2.3 shows two examples of continuum potential (string potential) for Cu according to Lehmann and Leibfried[5], in which the mark Q denotes the lowest potential or the stable position for ions. The mark P in the figure shows the saddle point.

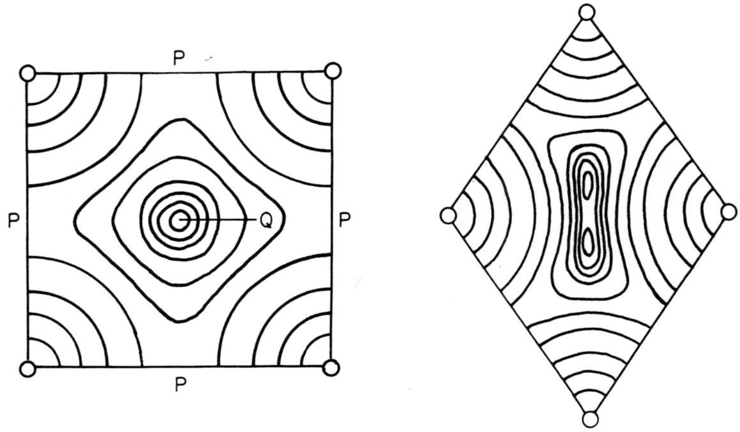

Figure 2.3. Examples of continuum potentials for Cu, according to Lehmann and Leibfried.[5] The points Q and P show the lowest potential position and the saddle points respectively.

We now turn to study ion motion under these continuum potentials. We start from the ordinary Newton equation, since the de Broglie wavelength of ions is much smaller than the atomic distance. Some quantum mechanical effects will be discussed in Chapter 3. Taking the co-ordinates as shown in figure 2.2(*b*), the Newton equations for the x and the z direction are obtained

$$M\frac{\mathrm{d}^2x}{\mathrm{d}t^2}+\frac{\mathrm{d}V_1(x)}{\mathrm{d}x}=0 \tag{2.2.5}$$

and

$$\frac{\mathrm{d}^2z}{\mathrm{d}t^2}=0 \tag{2.2.6}$$

In the above

$$V_1(x)=U\left(\frac{l}{2}-x\right)+U\left(\frac{l}{2}+x\right)$$

where l means the distance between atomic planes.

When the angle between the ion beam direction and the atomic plane is θ, the energy conservation law is obtained from equation 2.2.5

$$\tfrac{1}{2}M\dot{x}^2+[V_1(x)-V_1(0)]=E_{\text{in}}\sin^2\theta \tag{2.2.7}$$

where E_{in} means the ion beam energy. It is worth noting that the ion moves only in the region given by

$$E_{\text{in}}\sin^2\theta-[V_1(x)-V_1(0)]\geq 0$$

If we assume the minimum distance of approach of the ion from the atomic row, $x = x_m$, we obtain

$$\sin\theta \leq \sin\theta_c \equiv \sqrt{\frac{V_1(x_m) - V_1(0)}{E_{in}}}$$

which corresponds to a periodic motion of the ion in x space. This is the channelling condition. If the ion impinges on the atomic plane with an angle greater than θ_c, then the ion path deviates from the atomic plane. The angle θ_c, the Lindhard critical angle, describes the angular condition of a periodic motion along a channelling plane. In figures 2.4(*a*) and 2.4(*b*) we show the results of computer simulation by Robinson and Oen[6] for Cu ⟨110⟩ and Cu ⟨100⟩ channelling. Starting from the points S, the beam trajectories are confined within the inter-axis or inter-plane positions.

The next step is to calculate the continuum potentials in more detail. Lindhard[4] and Erginsoy[7] took the Thomas–Fermi type potential, i.e.

$$V(R) = \frac{Z_1 Z_2 e^2}{R}\chi\left(\frac{R}{a_{TF}}\right) \tag{2.2.8}$$

for the atomic potential $V(R)$, where Z_1 and Z_2 are the atomic numbers of the ion beam and the atoms comprising the crystal, respectively. The function $\chi(x)$ represents the Fermi function[8] and a_{TF} the screening length of Thomas–Fermi

$$a_{TF} = 0{\cdot}46(Z_1^{2/3} + Z_2^{2/3})^{-1/2}\,\text{Å} \tag{2.2.9}$$

(see equation 1.3.21, p. 8).

Inserting equation 2.2.8 into equation 2.2.2 we obtain

$$U(r) = \frac{Z_1 Z_2 e^2}{d}\ln\left[\left(\frac{a_{TF}\sqrt{3}}{r}\right)^2 + 1\right] \tag{2.2.10}$$

Similarly, for the planar case we obtain

$$U^{pl}(y) = 2\pi Z_1 Z_2 e^2 N_p[(y^2 + 3a_{TF}^2)^{1/2} - y] \tag{2.2.11}$$

where N_p is the number density of atoms in the atomic plane. Equations 2.2.10 and 2.2.11 are valid when R is not much larger than a_{TF}.

The planar continuum potential calculated by the Molière potential, equation 1.3.23, can then be obtained

$$U^{pl}(y) = \frac{4\pi\rho Z_1 Z_2 e^2 l}{b}\{0{\cdot}35\exp(-by) + 0{\cdot}1375\exp(-4by) + 0{\cdot}005\exp(-20by)\} \tag{2.2.12}$$

where $b = 0{\cdot}3/a_{TF}$ and ρ and l mean the atomic density and inter-planar distance respectively.†

---

†In the axial case, $U(r) = \frac{2Z_1 Z_2 e^2}{d}[0{\cdot}35K_0(0{\cdot}3r/a_{TF}) + 0{\cdot}55K_0(1{\cdot}2r/a_{TF}) + 0{\cdot}1K_0(6r/a_{TF})]$

Finally, we can determine the minimum approach introduced in deriving the critical angle θ_c. If ions approach the core with a Thomas–Fermi length of a_{TF}, they are scattered with a large angle (Rutherford scattering) and deviate from the channelling path. So, we may take x_m as

$$x_m = \frac{l}{2} - a_{TF} \tag{2.2.13}$$

We then obtain

$$\begin{aligned} V_1(x_m) - V_1(0) &= U\left(\frac{l}{2} - x_m\right) + U\left(\frac{l}{2} + x_m\right) - V_1(0) \\ &\simeq U\left(\frac{l}{2} - x_m\right) = U(a_{TF}) \\ &\simeq \frac{Z_1 Z_2 e^2}{d} 2\log\left(\frac{\sqrt{3}a_{TF}}{a_{TF}}\right) \\ &= \frac{2\log\sqrt{3}Z_1 Z_2 e^2}{d} \end{aligned}$$

whence the critical angle θ_c for axial cases is given by

$$\theta_c \simeq \sqrt{\frac{2Z_1 Z_2 e^2 \log\sqrt{3}}{E_{in} d}} \tag{2.2.14}$$

Similarly, we can obtain the critical angle for the planar case

$$\theta_c \simeq \sqrt{\frac{2\pi N_p Z_1 Z_2 e^2 a_{TF}}{E_{in}}}$$

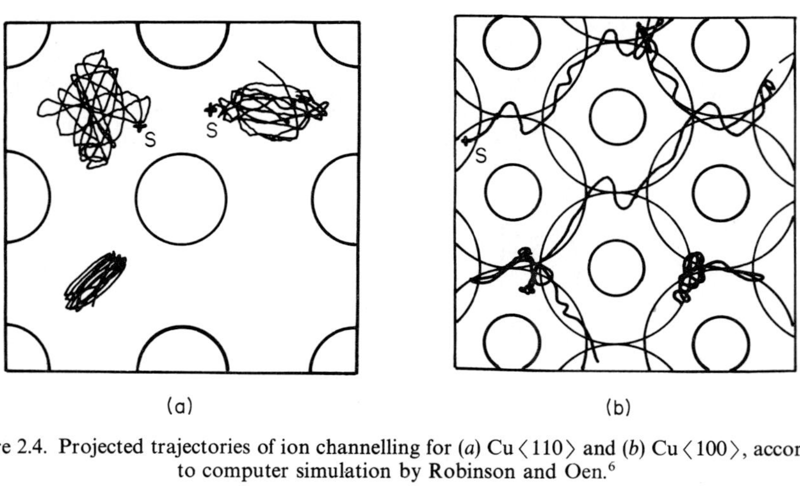

Figure 2.4. Projected trajectories of ion channelling for (*a*) Cu ⟨110⟩ and (*b*) Cu ⟨100⟩, according to computer simulation by Robinson and Oen.[6]

Therefore, we may conclude that the ratio of the planar critical angle to the axial critical angle is about $\sqrt{\pi a_{TF}/d}$. In general, the planar critical angle is smaller than the axial critical angle.

It is noted in figure 2.4 that there are different types of trajectories of axial channelling: those confined only to one channel, and those not confined (wandering freely between channels during passage through the crystal). If we consider the case where the channelling is confined to one channel, the critical angle is smaller than that given by equation 2.2.14, because the potential barrier is not given by equation 2.2.8 but by the boundary potential between channels, marked as P in figure 2.3. This kind of channelling, occurring when the incident angle to the axis is very small, is called 'hyperchannelling'.

2.3. Rutherford shadow (blocking)

We may consider the blocking of ion beams by the atomic row or the atomic plane as being contrary to channelling. In this case a shadow of the atomic row or atomic plane casts a shadow on the ion beams. This shadowing effect was discovered experimentally by Gemmell and Holland[9] and by Domeij

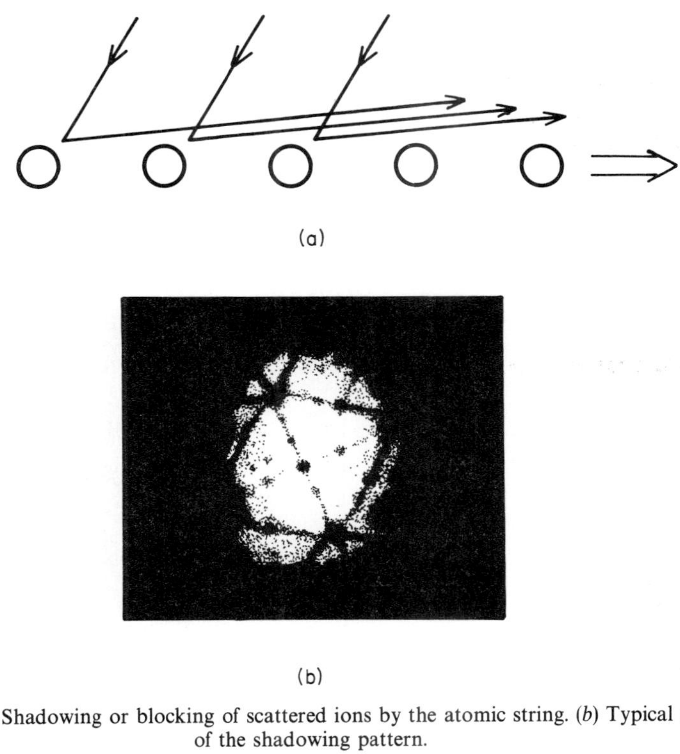

(a)

(b)

Figure 2.5. (*a*) Shadowing or blocking of scattered ions by the atomic string. (*b*) Typical example of the shadowing pattern.

and Björqvist.[10] Gemmell and Holland studied the large angle (60°) scattering of 4 MeV protons for the silicon single crystal, in which they found a shadow in the direction of ⟨110⟩ where proton beams scarcely penetrate. As is shown in figure 2.5(*a*), protons scattered through larger angles avoid the atomic strings. Figure 2.5(*b*) shows a typical example of the shadowing pattern.

Quite similar patterns were obtained in the experiment by Domeij and Björqvist, who detected α particles emitted from the radioactive isotope ^{222}Rn embedded in a single tungsten crystal. Figure 2.6 shows that the intensity of α particles emitted has a minimum in the ⟨111⟩ direction. We call these shadowing effects 'blocking'.

Lindhard derived the critical angle for blocking as follows. First we consider ions emitted from an atomic position in the atomic row with the angle ϕ. Assuming that an ion is scattered with the angle ψ, we obtain the energy conservation equation at the point $z=\frac{1}{2}$ (see figure 2.7)

$$E_{\text{in}}\psi^2 = E_{\text{in}}\phi^2 + U\left(\frac{\phi d}{2}\right)$$

Utilizing the expression 2.2.10 in the above equation, we obtain

$$\psi^2 = \phi^2 + \frac{\theta_c^2}{2}\log\left(\frac{12\,a_{\text{TF}}^2}{d^2\phi^2}+1\right) \tag{2.3.1}$$

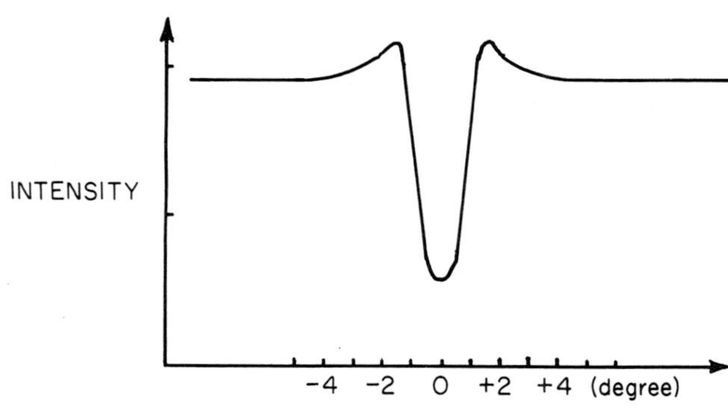

Figure 2.6. Intensity profile of the blocking of α particles in W crystal (Domeij and Björqvist[10]). The minimum corresponds to the ⟨111⟩ direction.

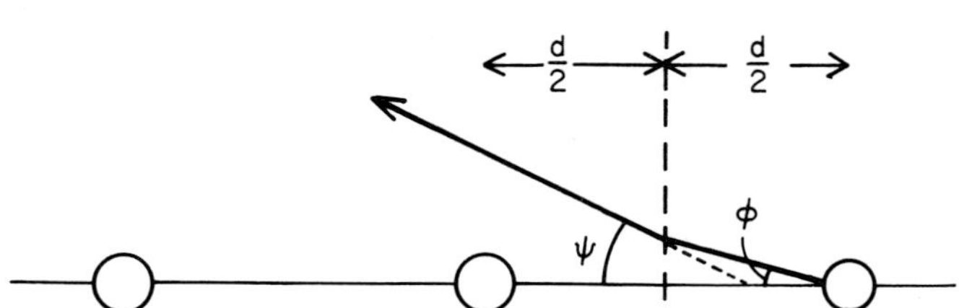

Figure 27. Ions emitted from an atomic position with angle ϕ are scattered by the second atom with angle ψ.

When

$$\phi^2 = -\frac{6a_{TF}^2}{d^2} + \left[\left(\frac{6a_{TF}^2}{d^2}\right)^2 + \theta_c^2\left(\frac{6a_{TF}^2}{d^2}\right)\right]^{1/2}$$

ψ^2 is at a minimum.

For higher energy, i.e., $\theta_c < (a_{TF}/d)$, ψ^2 is at a minimum when $\phi^2 = \theta_c^2/2$. The minimum value of ψ^2 is obtained by

$$\psi_{min}^2 = \frac{\theta_c^2}{2} + \frac{\theta_c^2}{2}\log\left(\frac{12a_{TF}^2}{d^2}\frac{2}{\theta_c^2} + 1\right)$$

$$\simeq \theta_c^2\left[\frac{1}{2} + \log\left(\frac{2\sqrt{6}\,a_{TF}}{\theta_c d}\right)\right]$$

or

$$\psi_{min} = \theta_c\sqrt{\log\left(\frac{\alpha a_{TF}}{\theta_c d}\right)} \tag{2.3.2}$$

where $\alpha = 2\sqrt{6e}$. In the above the value of the root, $\sqrt{\log(\;\;)}$, is $1{\cdot}5 \sim 2$. The value ψ_{min} is the critical angle for blocking; ions emitted from an atom in the atomic row rarely propagate within the angle ψ_{min}.

However, in real crystals there are lattice vibrations and impurities which increase the intensity of ions in the direction within ψ_{min}. We call this intensity the 'minimum yield'. Thus, the minimum yield increases with increasing temperature.

When lattice vibrations and impurities are neglected, the ions blocked become random beams (background beams) whose intensity is the minimum yield. So, ions contributing to the minimum yield in this case are ones which touch the atomic cores with radius a_{TF} (the cross-section being πa_{TF}^2). Since the number of atoms in a cylinder with height d and area 1 is $N(d \times 1)$, N being the density of atoms in the crystal, then the probability of transition to random beams is

$$\chi_{min} = \pi a_{TF}^2 N d \tag{2.3.3}$$

which is the minimum yield for axial blocking. In a similar way, we obtain the minimum yield for planar cases

$$\chi_{min} = 2a_{TF}/d_p \tag{2.3.4}$$

where d_p is the inter-plane distance.

Taking into account the lattice vibrations, we should modify the Thomas–Fermi length a_{TF} by averaged thermal amplitude of the lattice vibration of the atomic row or the atomic plane. Thus, a_{TF}^2 in equation 2.3.3 is replaced by $a_{TF}^2 + \rho_\perp^2$, and a_{TF} in equation 2.3.4 by $a_{TF} + \rho_y$, where $\rho_\perp$ is the root-mean-square amplitude of deviation of the atomic row and ρ_y the root-mean-square amplitude of deviation of the atomic plane, which will be calculated in the next section.

2.4. Modification of the Lindhard theory

It may seem that the Lindhard theory of channelling and blocking is a rough phenomenological one. However, for heavy particles such as ions the Lindhard theory agrees well with experiment. We can now make some modifications to the Lindhard theory in order to obtain a better description of channelling and blocking.

First, we obtain the contribution of the thermal vibration to the continuum potential. Since the ion velocity is much larger than the velocity of the lattice vibration of the atoms, the time of the interaction between ions and atoms is considerably shorter than the period of the lattice vibration. Therefore atoms which are a distance $\delta\mathbf{R}$ from their equilibrium position result in an atomic potential that is 'frozen' during the interaction time of the ions with the atomic row. The distance $\delta\mathbf{R}$ is a two-dimensional vector $\delta\mathbf{R}_\perp$ perpendicular to the atomic row for the axial channelling case, and a one-dimensional vector $\delta\mathbf{R}_y$ perpendicular to the atomic plane for planar channelling. The potential to be derived here is an average potential due to the thermal vibrational displacements[11] $\delta\mathbf{R}$, whose probability distribution is given by $f(\delta\mathbf{R})$.

For axial channelling, we obtain

$$U_{\mathrm{T}}(r) = \int_0^\infty \mathrm{d}(\delta R_\perp) \int_0^{2\pi} \mathrm{d}\theta\, f^{\mathrm{AX}}(\delta R_\perp)\, U\sqrt{r^2 + \delta R_\perp^2 - 2r\delta R_\perp \cos\theta} \tag{2.4.1}$$

and for planar channelling

$$U_{\mathrm{T}}^{\mathrm{pl}}(y) = \int_{-\infty}^{\infty} \mathrm{d}(\delta R_y)\, f^{\mathrm{pl}}(\delta R_y)\, U^{\mathrm{pl}}(|y - \delta R_y|) \tag{2.4.2}$$

In the above, $f^{\mathrm{AX}}(\delta R_\perp)$ and $f^{\mathrm{pl}}(\delta R_y)$ are calculated by assuming the Einstein model for lattice vibration, in which the deformation probability is of Gaussian form. With this assumption we obtain

$$f^{\mathrm{AX}}(\delta R_\perp) = \frac{\delta R_\perp}{\pi\langle \delta R_\perp^2\rangle} \exp(-\delta R_\perp^2/\langle \delta R_\perp^2\rangle) \tag{2.4.3}$$

and

$$f^{\mathrm{pl}}(\delta R_y) = \frac{1}{\sqrt{2\pi\langle \delta R_y^2\rangle}} \exp(-\delta R_y^2/2\langle \delta R_y^2\rangle) \tag{2.4.4}$$

A detailed derivation of equations 2.4.3 and 2.4.4 using the Debye–Waller factor is given in Chapter 3. Figure 2.8 shows the average potential of Appleton *et al.*[12] for Si⟨110⟩ axial channelling and Si(111) planar channelling as a function of temperature.

We can derive the detailed continuum potential, from which we obtain the channelling orbit and the channelling intensity. Here we discuss the anharmonic oscillation of the channelling orbit and the intensity profile for the blocking pattern.

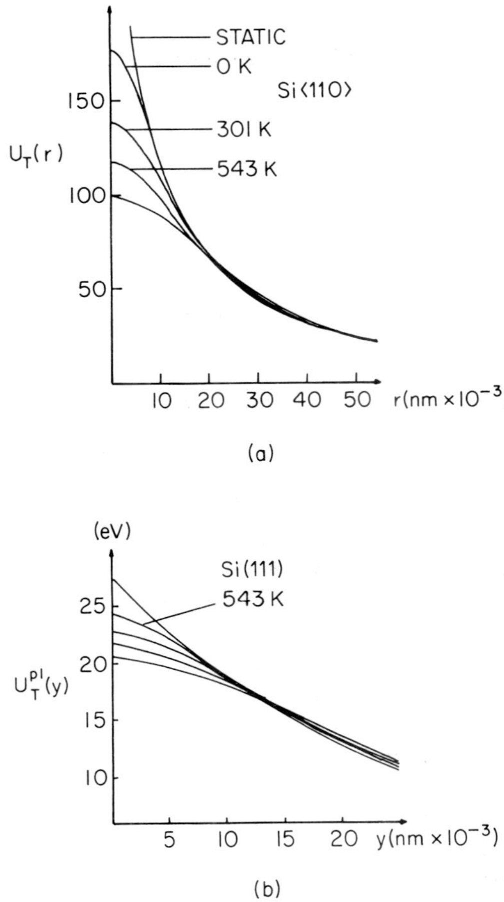

Figure 2.8. Average (continuum) potential obtained by Appleton *et al.*[12] for (*a*) Si⟨110⟩ and (*b*) Si(111) for various temperatures.

Robinson[13] solved for the anharmonic motion by taking the planar potential $V_1(y)$ of Molière

$$V_1(y) = U^{\text{pl}}(\tfrac{1}{2}l - y) + U^{\text{pl}}(\tfrac{1}{2}l + y) \tag{2.4.5}$$

$$= V_0' \cosh by \tag{2.4.6}$$

$$V_0' = 0{\cdot}35(8\pi\rho Z_1 Z_2 e^2 l/2b)\exp(\gamma^2 - \tfrac{1}{2}bl) \tag{2.4.7}$$

where $\gamma^2 = \frac{1}{2}b^2\langle \delta U_y^2 \rangle$. The equation of motion is

$$\frac{\mathrm{d}y}{\mathrm{d}t} = \left(\frac{2V_0'}{M}\right)^{1/2} (\cosh bx_{\text{m}} - \cosh by)^{1/2}$$

from which $\sin\psi(=\dot{y}\sqrt{M/2E})$ becomes Jacobi's elliptic function with period

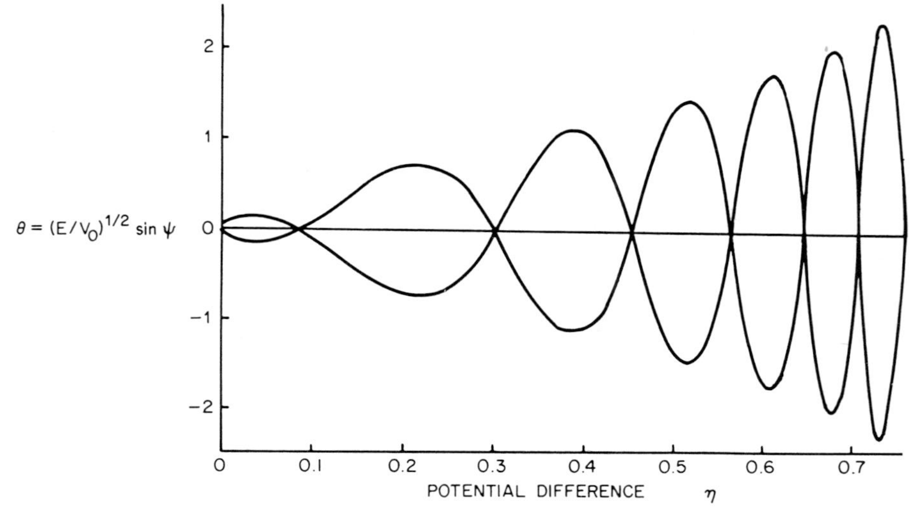

Figure 2.9. Relation of $\sin\psi$ to the potential difference $V(x_m) - V(0)$.

(with respect to the variable $\xi = bt\sqrt{V_0'/M}$)

$$\tau(\eta) = 4(1-\eta)^{1/2}K(\eta) \tag{2.4.8}$$

$$\left(\eta = \tanh^2 \tfrac{1}{2}bx_m = \frac{V(x_m) - V(0)}{V(x_m) + V(0)}\right)$$

where $K(\eta)$ is the elliptic integral of the first kind. In figure 2.9 we show $\theta = (E/V_0)^{1/2} \sin\psi$ as a function of η.

Now we discuss the intensity of the channelled or blocking ions. Assuming the transverse energy to be $E_{in}\psi_e^2$ above the crystal surface, we want to obtain the transition probability from $E_{in}\psi_e^2$ to the transverse energy $E_\perp$ in the crystal, which corresponds to the probability distribution of ions at position r, determined by the energy conservation relation

$$E_\perp - U(r) = E_{in}\psi_e^2$$

We note that the probability of ions in the region between r and $r + dr$ is roughly† $(2\pi r\,dr)/\pi r_0^2$, where πr_0^2 is the area of the unit cell in the transverse plane. Thus, the probability of ions with energy between $E_\perp$ and $E_\perp + dE_\perp$ is

$$T(E_\perp, E_{in}\psi_e^2)\,dE_\perp = \int_0^{r_0} \frac{2\pi r\,dr}{\pi r_0^2}\,\delta[E_\perp - E_{in}\psi_e^2 - U(r)] \tag{2.4.9}$$

†Here we assume a homogeneous equilibrium distribution of ions in the region πr_0^2 after sufficient ion penetration. If we consider a minimum distance r_{min} to the atomic row, which depends on $E_\perp$, then the area πr_0^2 should be replaced by the accessible area $A(E_\perp) = \pi(r_0^2 - r_{min}^2)$.

Assuming the initial distribution of ions is given as

$$P(E_{\text{in}}\psi_e^2)\,\mathrm{d}(E_{\text{in}}\psi_e^2)$$

we obtain the final distribution

$$g(E_\perp, z=0) = \int P(E_{\text{in}}\psi_e^2)\, T(E_\perp, E_{\text{in}}\psi_e^2)\,\mathrm{d}(E_{\text{in}}\psi_e^2) \tag{2.4.10}$$

The distribution $g(E_\perp)$ changes greatly due to the complicated multiple scattering, which will be discussed in Chapter 5.

Next, we study the intensity of the blocking pattern of ions scattered from an atom which has deviated a small distance ε from the atomic row. Taking the distance between the source atom and the blocking atom as d, and the emission angle to the atomic plane as (φ, θ), the initial distribution function $P_{\text{out}}(\psi)$ may be obtained as before. Thus we have

$$P_{\text{out}}(\psi) = \int^{r_0} dP(\varepsilon)\,\mathrm{d}(E_{\text{in}}\varphi^2) \int_0^{2\pi} \frac{d\theta}{2\pi}\,\delta(E_{\text{in}}\psi^2 - U(r^*) - E_{\text{in}}\varphi^2) \tag{2.4.11}$$

where $r^{*2} = \varepsilon^2 + (\varphi d/2)^2 + \varepsilon\varphi d$ and $dP(\varepsilon)$ is the probability that the atom has deviated a distance ε from its equilibrium position in the atomic row. The ions with the initial distribution $P_{\text{out}}(\psi)$ inside the crystal are emitted from the exit surface. Thus the final distribution outside the crystal is represented by

$$P_e(E_{\text{in}}\psi_e^2) = \int dE_\perp P_{\text{out}}(E_\perp)\, T(E_{\text{in}}\psi_e^2, E_\perp) \tag{2.4.12}$$

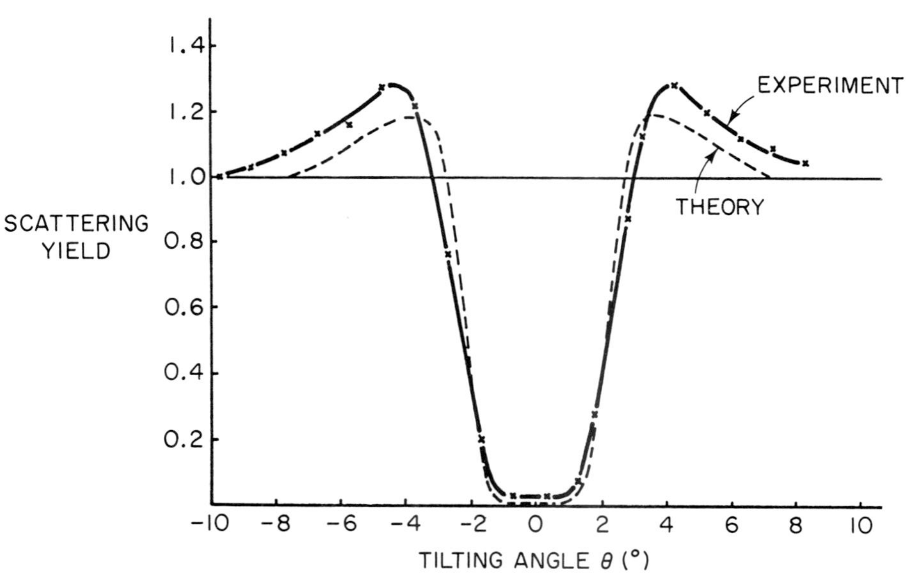

Figure 2.10. Comparison of experimental and theoretical results of the blocking pattern for W⟨100⟩, according to Andersen.[14]

where $T(E_{in}\psi_e^2, E_\perp)$ is the probability of emission at the exit surface from the transverse energy $E_\perp$ to $E_{in}\psi_e^2$. In figure 2.10 the experimental intensity distribution[14] of the blocking pattern for W⟨100⟩ is compared with the theoretical one.

References

1. Davies, J.A., Friesen, J. and McIntyre, J.D., *Can. J. Chem.* **38**, 1526 (1960).
2. Robinson, M.T. and Oen, O.S., *Appl. Phys. Letters* **2**, 30 (1963).
3. Stark, J., Z. *Physik* **13**, 973 (1912).
4. Lindhard, J., *Kgl. Dan. Viden. Sels. Mat-Fys. Medd.* **34**, No. 14 (1965).
5. Lehmann, C. and Leibfried, G., *J. Appl. Phys.* **34**, 2821 (1963).
6. Robinson, M.T. and Oen, O.S., *Phys. Rev.* **132**, 2385 (1963).
7. Erginsoy, C., *Phys. Rev. Letters* **15**, 360 (1965).
8. Gombas, P., *Die statistische Theorie des Atoms* (Vienna: Springer-Verlag, 1949).
9. Gemmell, D.S. and Holland, R.E., *Phys. Rev. Letters* **14**, 945 (1965).
10. Domeij, B. and Björqvist, K., *Phys. Letters* **14**, 127 (1965).
11. Kitagawa, M. and Ohtsuki Y.H., *Phys. Rev.* **B8**, 3117 (1973).
12. Appleton, B.R., Erginsoy, C. and Gibson, W.M., *Phys. Rev.* **161**, 330 (1967).
13. Robinson, M.T., *Phys. Rev.* **179**, 327 (1969).
14. Andersen, J.U., *Kgl. Dan. Viden. Sels. Mat-Fys. Medd.* **36**, No. 7 (1967).

CHAPTER 3

quantum effects in electron and positron channelling

3.1. Introduction

The Lindhard theory, introduced in the previous chapter, is based on classical mechanics and agrees well with the experimental results on ion channelling due to elastic scattering obtained thus far. However, the Lindhard theory is not adequate for particles of small mass such as electrons and positrons and, therefore, a theory based on quantum mechanics is needed to describe electron and positron channelling in crystals. Electrons and positrons are diffracted in a crystal when their energies are in a range where quantum effects are not negligible for such particles. So, electron and positron channelling have a close relationship with diffraction theory. This relationship is discussed in this chapter.

3.2. One-dimensional periodic field

The Schrödinger equation for a charged particle of mass M moving in a crystal field $V(\mathbf{r})$ is represented by

$$-\nabla^2\psi + \frac{2M}{\hbar^2}V(\mathbf{r})\psi = \frac{2ME}{\hbar^2}\psi \tag{3.2.1}$$

We assume that the dependence of ψ on z (the channelling direction) can be represented by plane waves. Thus we may represent the wave function ψ by

$$\psi = \varphi(x, y)\exp(ik_z z) \tag{3.2.2}$$

Inserting equation 3.2.2 into equation 3.2.1, multiplying $\exp(-ik_z(z/d))$ and integrating over z from $-\infty$ and $+\infty$, we obtain

$$-\nabla^2\varphi(x, y) + (2M/\hbar^2)V(x, y)\varphi(x, y) = (2M/\hbar^2)E_\perp\varphi(x, y)$$

where $E_\perp = E - \hbar^2k_z^2/2M \simeq E_{\text{in}}\theta^2$ is the energy component transverse to the channelling direction, and

$$V(x, y) = \frac{1}{d}\int_{-\infty}^{\infty} V(x, y, z)\,\mathrm{d}z \tag{3.2.3}$$

is simply the Lindhard potential with the periodic atom spacing d. In axial channelling, $V(x,y)$ is the axially symmetric potential, i.e.

$$V(x, y) = V(r) = \frac{1}{d}\int_{-\infty}^{\infty} V\sqrt{r^2 + z^2}\,\mathrm{d}z \tag{3.2.4}$$

which is $U(r)$ in equation 2.2.2.

In planar channelling, taking the y axis perpendicular to the atomic plane we obtain the Schrödinger equation

$$-\frac{\partial^2 \varphi(y)}{\partial y^2} + 2M/\hbar^2\, U^{\mathrm{pl}}(y)\varphi(y) = (2M/\hbar^2)\, E_\perp \varphi(y) \tag{3.2.5}$$

where $U^{\mathrm{pl}}(y)$ is the Lindhard planar potential.

For now we shall confine our discussion to planar channelling. The planar potential $U^{\mathrm{pl}}(y)$ can be expanded in a Fourier series over reciprocal lattice vectors, since the potential is periodic

$$U^{\mathrm{pl}}(y) = \sum_{g_\perp} V_{g_\perp} \exp(2\pi i g_\perp \cdot y) \tag{3.2.6}$$

where $g_\perp$ is a one-dimensional reciprocal lattice vector perpendicular to the periodic crystal planes.

Our problem is nothing but that of solving for one-dimensional motion in the periodic field, which is well known as the one-dimensional band theory for electrons in a solid.

3.3. Quantum corrections to the Lindhard theory

We can now derive, using quantum mechanics, the condition for which the Lindhard classical theory is applicable for electron and positron channelling. The condition depends on the continuum potential $U^{\mathrm{pl}}(y)$ and the mass of the particle. Taking the planar potential to be of the harmonic type

$$U^{\mathrm{pl}}(y) = \tfrac{1}{2} U^{\mathrm{pl}''}(0) y^2$$

and neglecting tunnelling effects, the difference between the two neighbouring eigen-energies of the harmonic oscillation is $\hbar\sqrt{U^{\mathrm{pl}}(0)''/M}$, which corresponds to the difference in transverse energy

$$\Delta(E_{\mathrm{in}}\theta^2) = \hbar\sqrt{U^{\mathrm{pl}''}(0)/M}$$

or

$$\Delta\theta = \frac{\hbar}{2\theta E_{\text{in}}} \sqrt{U^{\text{pl}}(0)''/M}$$

If $\Delta\theta$ is much smaller than the critical angle θ_c

$$\Delta\theta = \frac{\hbar}{2\theta_c E_{\text{in}}} \sqrt{U^{\text{pl}}(0)''/M} \ll \theta_c \tag{3.3.1}$$

our problem is the classical one. For the case of a particle of mass M smaller than M_c defined by

$$M_c = \Gamma m \tag{3.3.2}$$

(m electron mass) where

$$\Gamma = \frac{\hbar^2 U^{\text{pl}''}(0)}{4m\theta_c^4 E_{\text{in}}^2} \tag{3.3.3}$$

we have to take quantum effects into account. For the electrons, $\Gamma \gtrsim 1$ for energy $\lesssim 1$ MeV, and $M = m \lesssim M_c$, hence quantum effects become important. For the protons, with energy ~ 1 MeV and $M \gg M_c$, the classical treatment is adequate (see, for example, references 1 and 2).

Thus, for electron beams with energy ~ 1 MeV, the Lindhard theory is not adequate, and a theory based on quantum mechanics has to be applied.

In general, the quantum effect is larger for the planar case than for the axial case because the number of energy levels for the axial case is much larger than for the planar case. Figure 3.1 shows[3,4] the number of quantum levels

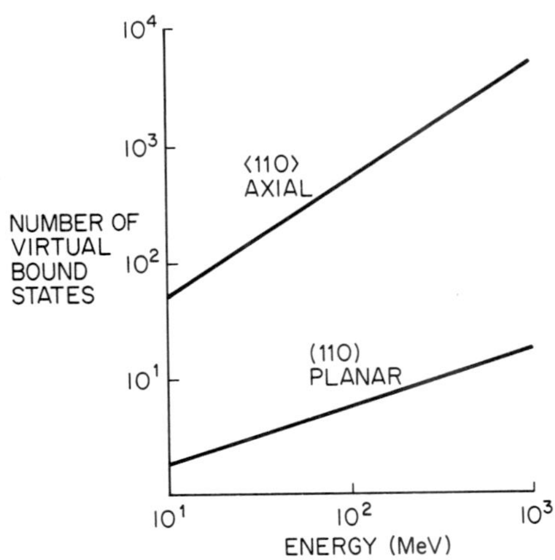

Figure 3.1. Number of quantum levels of electrons with $\langle 110 \rangle$ axial channelling and (110) planar channelling in Cu as a function electron energy.

of electrons with ⟨110⟩ axial channelling and (110) planar channelling in Cu. We see in this figure that for electrons with energy of 100 MeV the number of quantum levels for planar channelling is less than 10; here the classical theory is not adequate. However, the number of levels for axial channelling is a few hundred; here we may neglect quantum effect and the Lindhard theory is good enough.

3.4. Electron channelling

Since the electron has a negative charge, we cannot see the channelling effect; rather we see the blocking effect along the atomic direction in the crystal. Electrons in the crystal are attracted to the positive ions in the lattice of the crystal and scattered through large angles (Rutherford scatterings). Thus, we can not see electron beams in the atomic direction (channelling axis). This is nothing but the blocking effect. The channelling–blocking profile for electrons is the inverse of that for protons (or, more accurately, positrons).[5,6,7]

For the quantum mechanical treatment of electron (and positron) channelling, the electrons penetrating the crystal can be represented by Bloch wave functions. The electron probability density in the crystal is determined by those Bloch wave functions and depends on the beam energy and the initial conditions. The electron beams penetrate by channelling under special conditions; this was first found and discussed in more detail by Humphreys *et al.*[8,9] and Fujimoto *et al.*[10] Here we show the mechanism of electron channelling according to Kubota and Ohtsuki.[11]

We note that the planar potential $U^{\mathrm{pl}}(y)$ in equation 3.2.5 is a periodic function with period d. This is shown in figure 3.2 and can be represented by

$$U^{\mathrm{pl}}(y) = V_0 + V' \cos\left(\frac{2\pi}{d} y\right) \tag{3.4.1}$$

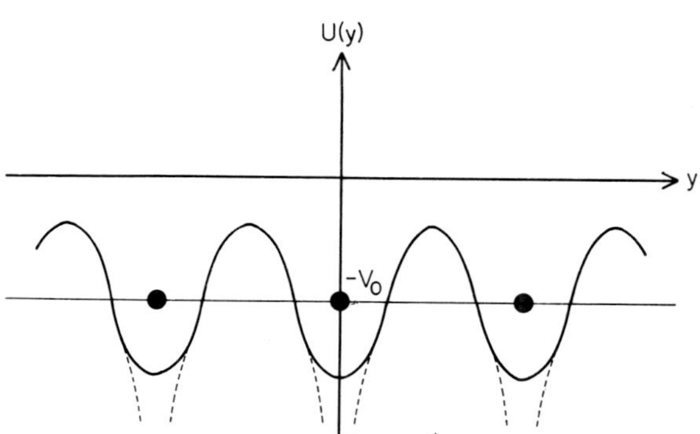

Figure 3.2. Periodic planar potential represented by $\cos(2\pi y/d)$.

assuming that V_0 and V', which are determined by the experimental conditions, are constant. Inserting equation 3.4.1 into equation 3.2.5 we obtain

$$\frac{d^2\varphi}{d\xi^2} + (\lambda + \lambda' \cos 2\xi)\varphi = 0 \tag{3.4.2}$$

where

$$\lambda = \frac{2m}{\hbar^2}\left(\frac{d}{\pi}\right)^2 (E_\perp - V_0) \quad \left[\equiv \left(\frac{d}{\pi}\right)^2 k_\perp^2 \right]$$

$$\lambda' = -\frac{2m}{\hbar^2}\left(\frac{d}{\pi}\right)^2 V'$$

$$\xi = \frac{\pi}{d} y$$

We note that equation 3.4.2 is Mathieu's equation, whose general solution, as obtained by Floquet, is

$$\varphi(\xi) = me_\nu(\xi) = ce_\nu(\xi) + ise_\nu(\xi)$$

where $ce_\nu(\xi)$ and $se_\nu(\xi)$ are Mathieu function of even and odd order, respectively. In the above, $\nu = \sqrt{\lambda} = 0, 1, 2,$ and so on.† Mathieu functions are periodic functions. For example, if $\nu = 1$ we have $\lambda = 1$ and $\left(\frac{d}{\pi}\right)^2 k_\perp^2 = 1$, which represents the first Brillouin boundary. So, replacing ν by $2-\nu$, $2+\nu$, $4-\nu$, and so on, we obtain the reduced band formalism‡ in the normal band theory. We distinguish branches $2-\nu$, $2+\nu$, $4-\nu$, and so on by the suffix (j). Thus we have the general solution of the Schrödinger equation

$$\psi(y) = \sum_j \psi_j \exp(ik_z^{(j)} z) me_\nu^{(j)}\left(y\frac{\pi}{d}\right) \tag{3.4.3}$$

where $\nu \leq 1$ and ψ_j should be determined by the boundary condition

$$\exp(iK_{0x} y) = \exp(iK_0 y \sin\theta) = \sum_j \psi_j me_\nu^{(j)}\left(y\frac{\pi}{d}\right) \tag{3.4.4}$$

Figure 3.3 shows the probability density $|\psi(y)|^2$ for different incidence angles $\theta = 0, \theta_B, 2\theta_B, 3\theta_B$, and $4\theta_B$. θ_B is the Bragg angle for 1 MeV electrons incident on the Au (111) plane. In this figure we used the values $V_0 = -3{\cdot}2 \times 10^{-12}\,\mu J$, $V' = -1{\cdot}6 \times 10^{-12}\,\mu J$. It is worth noting that the electron probability is higher at the atomic plane for $\theta = \theta_B$ and $2\theta_B$ than for $\theta = 0$ and $4\theta_B$,

---

† $se_0(\xi) = 0$.
‡ If we assume $k'_\perp = k_\perp - (\pi/d)$, we may consider the momentum region $-(\pi/d) < k'_\perp \leq (\pi/d)$ only.

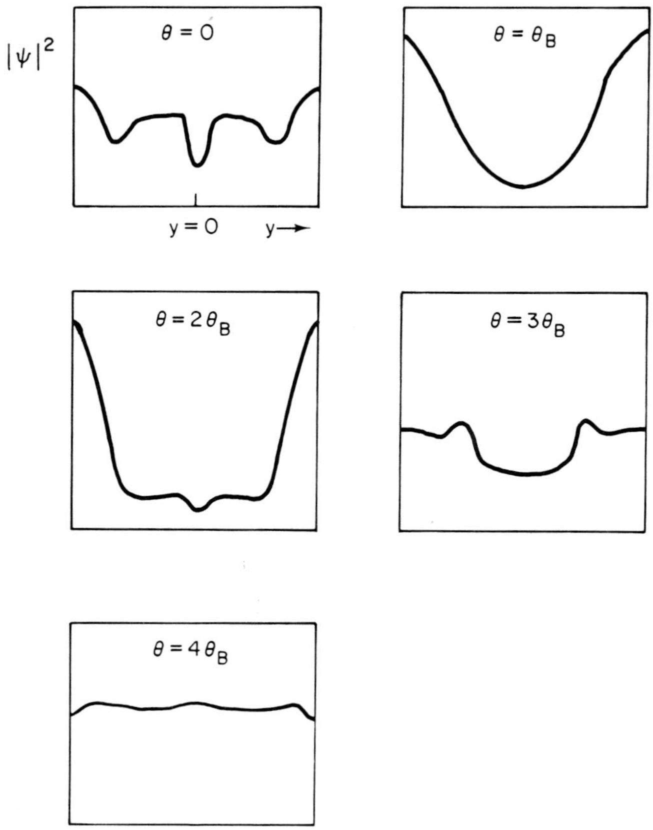

Figure 3.3. Probability density $|\psi|^2$ for different incidence angle, $\theta = m\theta_B$ ($m = 0, 1, \ldots 4$), θ_B being the Bragg angle for 1 MeV electrons on an Au (111) plane.

which means that the absorption becomes smaller for normal incidence than for Bragg angle incidence. This is nothing but electron channelling.

We now consider once more the classical limit of our problem. Increasing the electron mass m, we can not distinguish branches v, $v + 2$, and so on. Thus, k_z and v (i.e. $k_\perp$) are related by energy conservation

$$E_{\text{in}}\theta^2 = E_\perp - U^{\text{pl}}(y)$$

or

$$k_\perp = \left[\frac{2m}{\hbar^2}(E_{\text{in}}\theta^2 + U^{\text{pl}}(y))\right]^{1/2}$$

In this case, we obtain

$$\begin{aligned}\psi(y) &\propto \exp(ik_z z) me_v(y) \\ &= \exp(ik_z z)\exp(ik_\perp y)\end{aligned}$$

However, when $|U^{pl}(y)| > E_{in}\theta^2, k_\perp$ becomes imaginary, which we represent by $i\kappa_\perp$, for example. So, for $y > y_{max}$, where $|U^{pl}(y_{max})| = E_{in}\theta^2$, we obtain

$$\exp(ik_\perp y) = \exp(-\kappa_\perp y)$$

and the wave function is confined within $\kappa_\perp^{-1}$, which represents the channelling condition.

Since the electron trajectories with non-zero angular momentum are at a distance from an atomic axis, the electrons move in an axial channelling path. The calculation of axial channelling of electrons has been performed by Tamura and Ohtsuki,[12] who assumed only one atomic axis (hyper-channelling) and a continuum potential. With these assumptions the Schrödinger equation was solved in the plane perpendicular to the atomic axis with the centre of symmetry. This calculation shows that 3 MeV electrons in W⟨111⟩ move in an axial channelling path with $m = \pm 3$ and ± 4 and agrees well with the experiments of Fujimoto *et al.*[10†]

However we cannot assume hyper-channelling for weakly bound electron paths, for which we have to develop another quantum mechanical theory which is strongly related to electron diffraction. This theory will be presented in Section 3.8. However, for very high energy electrons we can make a computer simulation using classical mechanics.

Finally, we move on to the case of very high energy electron beams (with energies greater than 10 MeV). Here a classical beam trajectory is modified by relativity. This type of channelling is essentially the same as the positive ion channelling described in the previous chapter if we replace the positive potential by the negative one and apply the Lindhard theory using the continuum potential. The electron channelling refers to electron oscillations centred about the atomic rows. The electron beams which are exactly in the atomic rows are scattered through large angles by the atomic rows. Thus, we may conclude that the channelled electrons, which are concentrated by the negative continuum potential, move at a distance from the atomic rows.

Figure 3.4 shows the results of computer simulation for very high energy electron channelling as found by Yamamura and Ohtsuki,[13,14] where 10 MeV electrons impinge on Cu⟨110⟩ at the position marked X. The trajectories are projected on the plane perpendicular to the ⟨110⟩ axis. The two trajectories on the left are obtained for the electron beams whose trajectories have small angles to the axis (i.e., small angular momentum with respect to the axis), and those on the right are for electron beams with large angles to the axis (i.e., with large angular momentum). In addition, the top three are results at very low temperature and the bottom ones are at room temperature. The trajectories obtained here correspond well to the experimental ones, as will be seen in subsection 4.9.2.

† In this article they also calculated the energy of the bound states of channelled electrons by the multi-beam method.

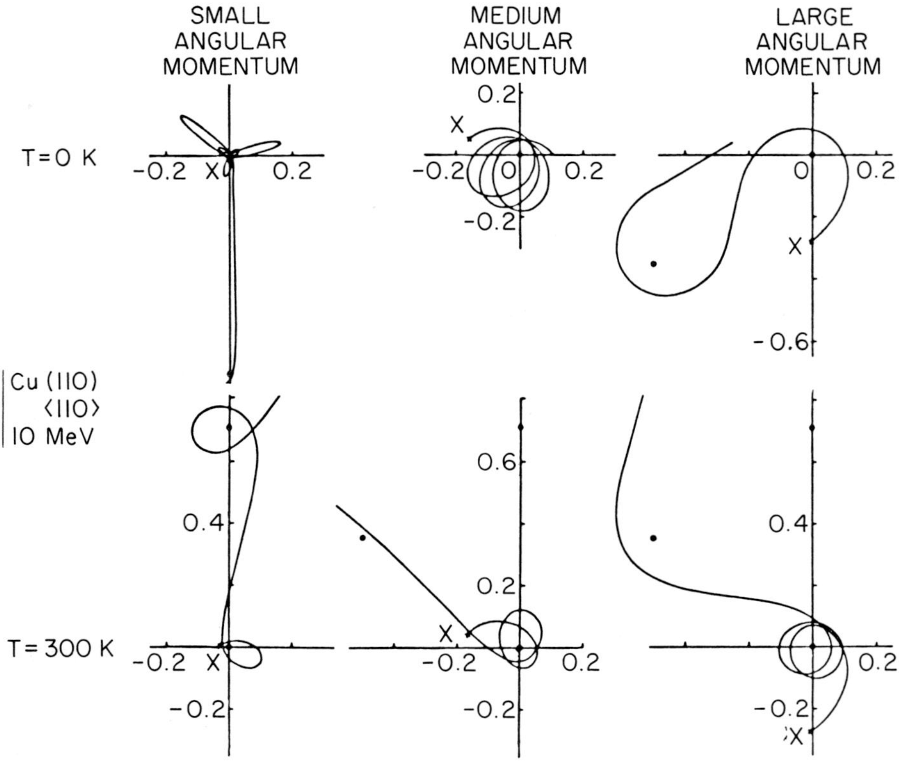

Figure 3.4. Electron channelling trajectories for 10 MeV electrons on a Cu⟨110⟩ axis found by computer simulation.

3.5. Electron diffraction and the scattering factor

For the electrons and positrons of energy $\lesssim 1\,\text{MeV}$ the penetration phenomena of electrons and positrons in crystals become very specialized because of the quantum effect called electron diffraction. The theory of electron diffraction has been studied thoroughly, and agrees well with experiments. Since electron diffraction is closely related to electron and positron channelling, we want to discuss in detail the dynamical theory of electron diffraction.

Assuming the position of the electron from an atomic position to be $\mathbf{r}$, the elastic scattering amplitude $f(\mathbf{Q})$ is obtained from the well known Born approximation

$$f(\mathbf{Q}) = \frac{-m}{2\pi\hbar^2}\int \exp(i\mathbf{Q}\cdot\mathbf{r})\, V(\mathbf{r})\, \mathrm{d}\mathbf{r} \tag{3.5.1}$$

where $\hbar\mathbf{Q}$ is the momentum transferred in the scattering and $V(\mathbf{r})$ is the atomic potential. Taking the initial wave number and the scattered wave number to be $\mathbf{K}$ and $\mathbf{K}'$, respectively, we obtain

$$\mathbf{Q} = \mathbf{K}' - \mathbf{K} \tag{3.5.2}$$

When the potential has spherical symmetry $V(\mathbf{r}) = V(|\mathbf{r}|) = V(r)$, we can easily integrate the integral of equation 3.5.1 over the solid angle, obtaining

$$f(\theta) = -\frac{2m}{\hbar^2}\int_0^\infty \frac{\sin Qr}{Qr} V(r) r^2 \, \mathrm{d}r \tag{3.5.3}$$

where 2θ is the scattering angle given by

$$\cos 2\theta = (\mathbf{K}\cdot\mathbf{K}')/K^2 \tag{3.5.4}$$

We can now divide the atomic potential into two parts, the electron cloud part and the nuclear part, i.e.

$$V(\mathbf{r}) = -\frac{Ze^2}{r} + \sum_j \frac{e^2}{|\mathbf{r} - \mathbf{r}_j|} \tag{3.5.5}$$

where $\mathbf{r}_j$ is the position vector of the jth electron. Inserting equation 3.5.5 into equation 3.5.1, we obtain

$$f(\mathbf{Q}) = \frac{m}{2\pi\hbar^2}\frac{4\pi e^2}{Q^2}[Z - F(\mathbf{Q})] \tag{3.5.6}$$

where

$$F(\mathbf{Q}) = \sum_j \exp(i\mathbf{Q}\cdot\mathbf{r}_j) \tag{3.5.7}$$

Equation 3.5.7 can be rewritten as

$$F(\mathbf{Q}) = \sum_j \exp(i\mathbf{Q}\cdot\mathbf{r}_j) = \int \exp(i\mathbf{Q}\cdot\mathbf{r}) \sum_j \delta(\mathbf{r} - \mathbf{r}_j) \, \mathrm{d}\mathbf{r}$$
$$= \int n(\mathbf{r}) \exp(i\mathbf{Q}\cdot\mathbf{r}) \, \mathrm{d}\mathbf{r}$$

where $n(\mathbf{r})$ is the electron density of the electron cloud. So, in other words, $F(\mathbf{Q})$ is the Fourier component of the electron density, which we call the scattering factor for X-rays. On the other hand, the first term of equation 3.5.6 is the Rutherford scattering factor of the nucleus of atomic number Z.

Rewriting equation 3.5.6 by using $Q = 2K \sin\theta$, we obtain

$$f(\theta) = \frac{me^2}{2h^2}\left(\frac{\lambda}{\sin\theta}\right)^2 [Z - F(\theta)] \tag{3.5.8}$$

where λ is the electron wavelength and $f(\theta)$ is the atomic scattering factor which represents the scattering amplitude of electrons due to the atom.

Many authors have calculated the atomic scattering factor in detail for various materials.[15] A few analytical formulae for the atomic scattering factor have been proposed. For example, Forsyth and Wells[16] gave

$$f(S)=\frac{2e^2m}{h^2}\frac{1}{S^2}\{Z-A\exp[-(a/4)S^2]-B\exp[-(b/4)S^2]-C\} \quad (3.5.9)$$

where

$$S=\frac{2\sin\theta}{\lambda}$$

and A, B, C, a and b are the constants given in table 3.1.

In the above we assumed only one atom at the origin. We can also obtain the scattering amplitude due to the lth atom in the position $\mathbf{R}_l$ from the origin by replacing $\mathbf{r}$ by $\mathbf{r}-\mathbf{R}_l$ in equation 3.5.1. Substituting $\mathbf{r}-\mathbf{R}_l$ for $\mathbf{r}'$ in equation 3.5.1, we obtain the scattering factor, equation 3.5.8, multiplied by the phase factor $\exp(i\mathbf{Q}\cdot\mathbf{R}_l)$.

If we take into account all atoms in a crystal in which there is only one atom per unit cell, then the scattering factor, equation 3.5.8, is multiplied by the factor

$$N\Delta(\mathbf{Q})=\sum_l \exp(i\mathbf{Q}\cdot\mathbf{R}_l) \quad (3.5.10)$$

where N is the atomic number. In the above, the Δ function is represented[17] for large N by

$$\Delta(\mathbf{Q})=\sum_{\mathbf{G}}\delta_{\mathbf{Q},2\pi\mathbf{G}}$$

where $\mathbf{G}$ is the reciprocal lattice vector. Thus, the scattering factor becomes very large when $\mathbf{Q}=0$ or when

$$\mathbf{Q}=\mathbf{K}'-\mathbf{K}=2\pi\mathbf{G} \quad (3.5.11)$$

The condition given by equation 3.5.11 is called the 'Laue–Bragg condition', where the scattering intensity $I_G \propto |f(2\pi G)|^2$ becomes infinite for an infinite crystal.

Table 3.1. Constants used in Forsyth and Wells's analytical formula for the atomic scattering factor.

	Constant				
Element	A	B	C	a	b
Li	1·731	1·072	0·1767	2·652	110·3
Be	1·621	2·096	0·2631	1·531	53·27
Al	7·675	3·411	1·772	2·834	67·03
Si	7·741	4·275	1·862	2·395	69·49
Fe	11·04	8·630	5·441	2·788	24·38
Ni	12·76	8·638	5·650	2·637	19·88
Cu	12·87	9·932	5·583	2·288	13·92
Ge	19·11	5·968	6·637	2·887	38·88
Ag	21·23	14·67	10·50	1·857	19·20

In a real crystal the factor given by equation 3.5.9 is not represented by equation 3.5.10 but increases in intensity at $\mathbf{Q} \simeq 2\pi\mathbf{G}$ with a band width $\sim 1/$(crystal size). We can easily obtain this result by making a summation over l in equation 3.5.10 within a finite crystal region. This explains the diffuseness of the Laue–Bragg spot for finite crystals.

When there is more than one atom in the unit cell, we obtain the resultant scattering factor

$$F(\mathbf{Q}) = \sum_{i}^{\text{unit cell}} f_i(\mathbf{Q}) \exp(i\mathbf{Q}\cdot\mathbf{R}_i)$$

where $f_i(\mathbf{Q})$ is the scattering factor for the ith atom at position $\mathbf{R}_i$ in the unit cell. We call $F(\mathbf{Q})$ the structure factor.

3.6. Multiple scattering model

First we shall consider the case where there is only one reciprocal lattice point satisfying the Laue–Bragg condition. In this case multiple waves, some reflected and some not reflected, interact and form a kind of standing wave, whose wave number is fixed, depending on the electron energy and the value of the crystal potential. The sharp intensity peaks of the Laue–Bragg condition calculated by the diffraction theory introduced in the previous section are then smeared by such multiple scattered wave interference.

We assume the wave number of the standing wave formed by multiple scatterings to be $\mathbf{K}$. In the crystal there is another wave with wave number $\mathbf{K}_G\,(=\mathbf{K}+2\pi\mathbf{G})$, which is reflected by the reciprocal lattice vector $\mathbf{G}$ at the same time. Our starting Hamiltonian of electron propagation is

$$H_0 = \sum_G \frac{\hbar^2 K_G^2}{2m} a^+_{\mathbf{K}G} a_{\mathbf{K}G} \tag{3.6.1}$$

The interaction Hamiltonian H' between electrons and the crystal is obtained assuming the scattering potential $V_{\mathbf{G}-\mathbf{G}'}$

$$H' = \sum_{\substack{\mathbf{G},\mathbf{G}' \\ (\mathbf{G}\neq\mathbf{G}')}} V_{\mathbf{G}-\mathbf{G}'} a^+_{\mathbf{K}G} a_{\mathbf{K}G'} \tag{3.6.2}$$

In equations 3.6.1 and 3.6.2 $a^+_{\mathbf{Q}}$ and $a_{\mathbf{Q}}$ are the creation and annihilation operators of electrons, respectively.

If we want to know the properties of the propagation of the electron in the crystal it is very useful to study the one-body Green function (see, for example, reference 18) given by

$$G(\mathbf{K}; t-t') = -i\langle T a_{\mathbf{K}}(t)\, a^+_{\mathbf{K}}(t')\rangle \tag{3.6.3}$$

where $a_{\mathbf{K}}(t')$ is the Heisenberg representation defined by

$$a_{\mathbf{K}}(t) = \exp(iHt/\hbar)\, a_{\mathbf{K}} \exp(-iHt/\hbar) \tag{3.6.4}$$

with the total Hamiltonian $H = H_0 + H'$.

We can now define the free Green function for the case of no interaction (i.e., $H' = 0$), denoting it as $G_0(\mathbf{K}; t-t')$. The free Green function $G(\mathbf{K}; t-t')$ is represented by

$$G_0(\mathbf{K}; t-t') = -i \exp[-i(E/\hbar)(t-t')]$$

when $t > t'$, and

$$G_0(\mathbf{K}; t-t') = 0 \tag{3.6.5}$$

when $t < t'$.

Performing the Fourier transformation we obtain

$$\begin{aligned} G_0(\mathbf{K}; \Omega) &= \int_{-\infty}^{\infty} \exp(+i\Omega\tau - \delta|\tau|)\, G_0(\mathbf{K}; \tau)\, \mathrm{d}\tau \\ &= -i \int_0^{\infty} \exp(-iE\tau/\hbar) \exp(+i\Omega\tau - \delta\tau)\, \mathrm{d}\tau = \frac{1}{\Omega - (E/\hbar) + i\delta} \end{aligned} \tag{3.6.6}$$

Next we must obtain $G(\mathbf{K}; \Omega)$, assuming only one reciprocal vector $\mathbf{G}$ (two-wave approximation). According to Wick's theorem[18] $G(\mathbf{K}; \Omega)$ can be represented in terms of $G_0(\mathbf{K}; \Omega)$, for example as

$$G_0(\mathbf{K}; \Omega) \frac{V_{-\mathbf{G}}}{\hbar} G_0(\mathbf{K}_G; \Omega) \frac{V_{\mathbf{G}}}{\hbar} G_0(\mathbf{K}; \Omega) \tag{3.6.7}$$

It is useful to consider this topologically. We let G_0 be represented by a solid line and $V_{\mathbf{G}}/\hbar$ be represented by a dotted line. The term is represented in figure 3.5. Within the two-wave approximation there are no terms represented by figure 3.6.

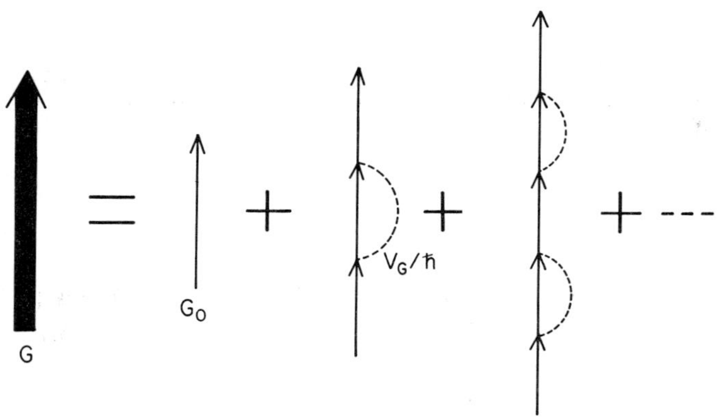

Figure 3.5. Diagrammatic representation of multiple scattering. The solid and dotted lines show the free Green function G_0 and the potential $V_{\mathbf{G}}/\hbar$ respectively.

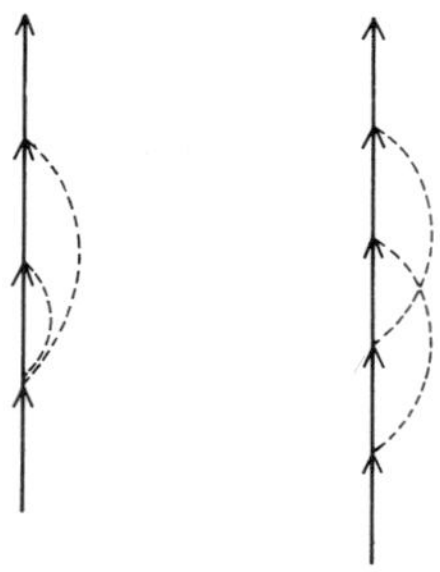

Figure 3.6. Complicated higher order graph.

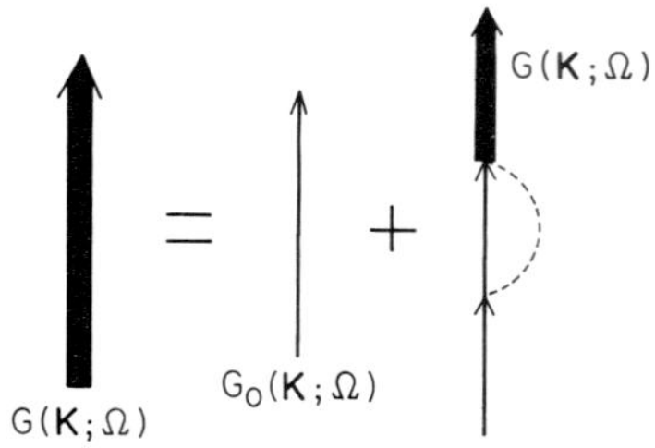

Figure 3.7. Showing renormalization of the graphs. The bold solid line corresponds to the Green function $G(K;\Omega)$.

So, the terms to be considered for the two-wave approximation are shown in figure 3.5, which represents

$$G(\mathbf{K};\Omega) = G_0(\mathbf{K};\Omega) + G_0(\mathbf{K};\Omega)\frac{V_{-\mathbf{G}}}{\hbar}G_0(\mathbf{K}_G;\Omega)\frac{V_{\mathbf{G}}}{\hbar}G_0(\mathbf{K};\Omega) + \ldots \quad (3.6.8)$$

Denoting $G(\mathbf{K};\Omega)$ by the bold solid line, figure 3.5 is redrawn in figure 3.7, which represents

$$G(\mathbf{K};\Omega) = G_0(\mathbf{K};\Omega) + G_0(\mathbf{K};\Omega)\frac{V_{-\mathbf{G}}}{\hbar}G_0(\mathbf{K}_G;\Omega)\frac{V_{\mathbf{G}}}{\hbar}G(\mathbf{K};\Omega) \quad (3.6.9)$$

From equation 3.6.9 we obtain

$$\begin{aligned} G(\mathbf{K};\Omega) &= \frac{G_0(\mathbf{K};\Omega)}{1-(V_{\mathbf{G}}V_{-\mathbf{G}}/\hbar^2)\,G_0(\mathbf{K};\Omega)\,G_0(\mathbf{K}_G;\Omega)} \\ &= \frac{1}{G_0^{-1}(\mathbf{K};\Omega)-(V_{\mathbf{G}}V_{-\mathbf{G}}/\hbar^2)\,G_0(\mathbf{K}_G;\Omega)} \end{aligned} \quad (3.6.10)$$

Now, it is well known[18] that the pole of the Green function determines the wave number **K**. Using $G(\mathbf{K};\Omega)^{-1}=0$, or (from equation 3.6.10)

$$G_0^{-1}(\mathbf{K};\Omega)-(V_{\mathbf{G}}V_{-\mathbf{G}}/\hbar^2)\,G_0(\mathbf{K}_G;\Omega)=0 \tag{3.6.11}$$

we can determine the wave number **K**, thus giving the dispersion equation. Inserting equation 3.6.6 into equation 3.6.11, we obtain

$$\left(\Omega-\frac{\hbar K^2}{2m}\right)\cdot\left(\Omega-\frac{\hbar K_G^2}{2m}\right)=V_{\mathbf{G}}V_{-\mathbf{G}}/\hbar^2 \tag{3.6.12}$$

The above method of determining the dispersion equation 3.6.12 from the pole of the Green function was developed by Ohtsuki and Yanagawa.[19] It should be noted that the dispersion equation 3.6.12 could not be obtained by truncating the graphs at a finite number of series of graphs. The pole of the Green function obtained by truncating the graphs at a finite point results in the dispersion relation for a single scattering

$$\begin{aligned}\Omega-\frac{\hbar K^2}{2m}&=0\\ \Omega-\frac{\hbar K_G^2}{2m}&=0\end{aligned} \tag{3.6.13}$$

which gives the ordinary (simple) Laue–Bragg condition 3.5.11.

The standing wave, and its wave number in X-ray diffraction, was introduced and called the polarization wave by Ewald;[20–25] and was introduced into electron diffraction by Bethe.[26,27] This will be studied in the following section.

3.7. Bethe's dynamical theory of electron diffraction

The Schrödinger equation of electrons in a periodic potential $V(\mathbf{r})$ is given by

$$\left[-\frac{\hbar^2}{2m}\nabla_+^2+V(\mathbf{r})\right]\psi(\mathbf{r})=E\,\psi(\mathbf{r}) \tag{3.7.1}$$

The periodic potential $V(\mathbf{r})$ can be expanded† in a Fourier series as

$$V(\mathbf{r})=\sum_{\mathbf{G}}V_{\mathbf{G}}\exp(2\pi i\mathbf{G}\cdot\mathbf{r}) \tag{3.7.2}$$

---

†From $\langle V(\mathbf{r})\rangle=(1/L^3)\int_{L^3}V(\mathbf{r})\,d\mathbf{r}=(1/L^3)\sum_{\mathbf{G}}V_{\mathbf{G}}\int\exp(2\pi i\mathbf{G}\cdot\mathbf{r})\,d\mathbf{r}=V_0$, it is shown that V_0 is the mean inner potential.

and also[5] as

$$\psi(\mathbf{r}) = \sum_{\mathbf{G}} \psi_{\mathbf{G}} \exp(i\mathbf{K}_G \cdot \mathbf{r}) \tag{3.7.3}$$

Inserting equations 3.7.2 and 3.7.3 into equation 3.7.1, we obtain

$$(K_{\mathbf{G}}^2 - k_0^2)\psi_{\mathbf{G}} + \sum_{\mathbf{H}} \frac{2m}{\hbar^2} V_{\mathbf{H}} \psi_{\mathbf{G}-\mathbf{H}} = 0 \tag{3.7.4}$$

In the above we defined $E = \hbar^2 k_0^2/2m$, where $\mathbf{k}_0$ is the wave number of electrons *in vacuo*.

First, we shall consider the case of no Laue–Bragg reflection. Equation 3.7.4 then becomes

$$(K^2 - k_0^2)\psi_0 + (2m/\hbar^2)V_0\psi_0 = 0$$

and K is determined by

$$K^2 - k_0^2 + (2m/\hbar^2)V_0 = 0$$

from which we obtain

$$K \approx k_0 - \frac{m}{\hbar^2 k_0} V_0 \tag{3.7.5}$$

since V_0 is much smaller than the energy of the electron beam. It is worth noting that the refractive index $n = K/k_0$ is greater than 1 since V_0 is negative.

Next, we shall consider the cases of $K_G^2 - k_0^2 \approx 0$ and $K^2 - k_0^2 \approx 0$ simultaneously, which correspond to condition 3.6.13. If we consider the reciprocal lattice vector $\mathbf{H}$ ($\neq 0, \mathbf{G}$) we have the relation from equation 3.7.4

$$(K_H^2 - k_0^2)\psi_H + \left(\frac{2m}{\hbar^2}\right)(V_0\psi_{\mathbf{G}} + V_{\mathbf{G}}\psi_0 + V_{\mathbf{H}}\psi_{\mathbf{G}-\mathbf{H}}) = 0$$

which leads to

$$\psi_{\mathbf{H}} \sim \frac{(2m/\hbar^2)V_{\mathbf{G}}\psi_0}{K_{\mathbf{H}}^2 - k_0^2} \ll \psi_0$$

Then we can see that only two components, ψ_0 and $\psi_{\mathbf{G}}$ are large in equation 3.7.4

$$[(K^2 - k_0^2) + (2m/\hbar^2)V_0]\psi_0 + (2m/\hbar^2)V_{-\mathbf{G}}\psi_{\mathbf{G}} = 0 \tag{3.7.6}$$

$$(2m/\hbar^2)V_{\mathbf{G}}\psi_0 + [K_{\mathbf{G}}^2 - k_0^2 + (2m/\hbar^2)V_0]\psi_{\mathbf{G}} = 0 \tag{3.7.7}$$

To get non-zero values for ψ_0 and $\psi_{\mathbf{G}}$ from equations 3.7.6 and 3.7.7 it is necessary that

$$[K^2 - k_0^2 + (2m/\hbar^2)V_0][K_{\mathbf{G}}^2 - k_0^2 + (2m/\hbar^2)V_0] = (2m/\hbar^2)^2\, V_{\mathbf{G}}V_{-\mathbf{G}} \tag{3.7.8}$$

which is nothing but the dispersion equation used for the determination of the wave number $\mathbf{K}$ in the crystal. Putting $\Omega = (\hbar/2m)k_0^2 - V_0/\hbar$ in equation 3.6.12 the dispersion equation 3.7.8 is the same one derived by using the pole of the Green function. It should be noted that the dispersion equation, which represents

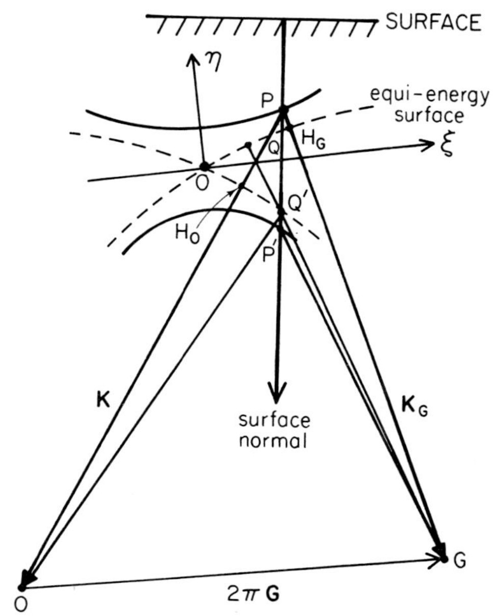

Figure 3.8. Dispersion curves for diffraction in **K** space. The points P and P′ indicate the points satisfied by the dispersion equation.

the energy surface in **K** space of solid state physics, is in principle equation 3.7.8; but in our case of high energy electrons we need a different analysis.

For high energy cases where $K \sim k_0$ equation 3.7.8 becomes

$$4K_0^2(K - K_0)(K_{\mathbf{G}} - K_0) = (2m/\hbar)^2 \, V_{\mathbf{G}} V_{-\mathbf{G}} \tag{3.7.9}$$

where we defined $K_0^2 = k_0^2 - (2m/\hbar^2)V_0$. We can now consider the Cartesian co-ordinates (ξ, η) in **K** space, where the ξ axis is parallel to the reciprocal lattice vector $2\pi\mathbf{G}$ and the origin of the co-ordinates is the Laue–Bragg point $|\mathbf{K}_0| = |\mathbf{K} + 2\pi\mathbf{G}|$ (see figure 3.8). There are two values of K which satisfy the equation 3.7.9 which we call P and P′ in **K** space. So, for example, $|\mathbf{K} - \mathbf{K}_0|$ is $\overline{\mathrm{PH}}_0$ and $|\mathbf{K}_G - \mathbf{K}_0|$ is $\overline{\mathrm{PH}}_G$ in figure 3.8. Using this geometrical consideration we obtain $\overline{\mathrm{PQ}} = \eta - \xi \tan\theta_B$, where Q is the crossing point of the equi-energy surface and the surface normal line (see figure 3.8) and

$$\overline{\mathrm{PH}}_G = \overline{\mathrm{PQ}} \cos\theta_B = (\eta - \xi \tan\theta_B)\cos\theta_B$$

$$\overline{\mathrm{PH}}_0 = (\eta + \xi \tan\theta_B)\cos\theta_B$$

Thus equation 3.7.9 turns out to be

$$\eta^2 - \xi^2 \tan^2\theta_{\mathrm{B}} = \frac{U_{\mathbf{G}} U_{-\mathbf{G}}}{4K_0^2 \cos^2\theta_{\mathrm{B}}} \tag{3.7.10}$$

where we define $U_{\mathbf{G}} = (2m/\hbar^2)V_{\mathbf{G}}$. Equation 3.7.10 shows that our dispersion relation 3.7.9 is nothing but a hyperbola with the asymptotic lines $\eta = \pm\xi \tan\theta_B$. We represent the two branches, corresponding to two hyperbolae, by $j=1$ and 2.

The minimum distance between the two hyperbolae is obtained by putting $\xi = 0$ in equation 3.7.10

$$\eta^{(1)} - \eta^{(2)} = \frac{\sqrt{U_{\mathbf{G}}U_{-\mathbf{G}}}}{K_0 \cos\theta_{\mathrm{B}}} \quad \left(\equiv \frac{2\pi}{\xi_{\mathbf{G}}}\right) \tag{3.7.11}$$

where $\xi_{\mathbf{G}} = 2\pi K_0 \cos\theta_B / \sqrt{U_{\mathbf{G}}U_{-\mathbf{G}}}$ is known as the extinction distance. This is equal to one wavelength of the standing wave, as will be explained later.

Now we assume that the current density of the electrons is conserved at the surface, where electrons are not accumulated. Since the current density is represented by ψ and $\frac{\partial\psi}{\partial z}$, ψ and $\frac{\partial\psi}{\partial z}$ are conserved at the surface ($z=0$).

We represent the incident electron by the plane wave $\exp(i\mathbf{k}_0 \cdot \mathbf{r})$. There are two types of reflected waves at the surface; one is the simple mirror (or specular) reflected wave and the other is the Bragg reflected wave. The wave number of the specular reflected wave is

$$\tilde{\mathbf{k}}_0 = (k_{0x}, k_{0y}, -k_{0z})$$
$$|\mathbf{k}_0| = \tilde{\mathbf{k}}_0$$

However, we may neglect the specular reflected wave for high energy cases, since the refractive index n is ~ 1 (for example, for 10 keV electrons and $V_0 \approx 10\,\mathrm{eV}$, $n \approx K/k_0 \approx 1 + V_0/E_{\mathrm{in}} \approx 1 + 10^{-3}$). Thus, outside the crystal we need consider only two waves, the incident wave and the Bragg reflected wave. This is called the Bragg case. There is another case where there is no Bragg reflected wave, but there is another reflected wave outside the crystal, which is on the side opposite to that of the incident electrons. This case is called the Laue case.

In the Laue case we have to combine the incident wave with the wave function ψ in the crystal. If we represent the wave function by

$$\psi = \sum_{j=1,2} \chi^{(j)} [C_0^{(j)} \exp(i\mathbf{K}^{(j)} \cdot \mathbf{r}) + C_G^{(j)} \exp(i\mathbf{K}_G^{(j)} \cdot \mathbf{r})]$$

we obtain

$$\exp(i\mathbf{k}_0'' \cdot \mathbf{r}'') = \sum_{j=1,2} \chi^{(j)} [C_0^{(j)} \exp(i\mathbf{K}^{(j)''} \cdot \mathbf{r}'') + C_G^{(j)} \exp(i\mathbf{K}_G^{(j)''} \cdot \mathbf{r}'')] \tag{3.7.12}$$

where the symbol $''$ refers to the component parallel to the crystal surface.

If we take $\mathbf{K}^{(j)''} = \mathbf{k}_0'', \mathbf{K}^{(1)''} = \mathbf{K}^{(2)''}$ and $\mathbf{K}_G^{(1)''} = \mathbf{K}_G^{(2)''}$, equation 3.7.12 becomes

$$\left[\sum_j \chi^{(j)} C_0^{(j)} - 1\right] \exp(i\mathbf{k}_0'' \cdot \mathbf{r}'') + \left[\sum_j \chi^{(j)} C_{\mathbf{G}}^{(j)}\right] \exp(i\mathbf{K}_{\mathbf{G}}^{(1)''} \cdot \mathbf{r}'') = 0 \tag{3.7.13}$$

Since equation 3.7.13 is satisfied for all values of $\mathbf{r}''$, we obtain†

$$\sum_j \chi^{(j)} C_0^{(j)} = 1$$

$$\sum_j \chi^{(j)} C_G^{(j)} = 0 \qquad (3.7.14)$$

or

$$\sum_j \chi^{(j)} C_{\mathbf{G}}^{(j)} = \delta_{\mathbf{G},0}$$

It is interesting to note that the conservation of $\partial\psi/\partial z$ gives the same result.

Using equation 3.7.6 the ratio $\psi_{\mathbf{G}}^{(j)}/\psi_0^{(j)}$ is determined. Then the ratio $C_{\mathbf{G}}^{(j)}/C_0^{(j)}$ is also obtained

$$C_{\mathbf{G}}^{(j)}/C_0^{(j)} \doteqdot -\frac{2K_0(K^{(j)}-K_0)}{(2mV_{-\mathbf{G}}/\hbar^2)}$$

For crystals with a centre of symmetry $V_{\mathbf{G}} = V_{-\mathbf{G}}^*$ is real, and‡

$$C_{\mathbf{G}}^{(j)}/C_0^{(j)} = +\frac{\xi \tan\theta_{\mathrm{B}} \pm \eta}{\pi}\xi_G \quad (j=1,2)$$

$$= +(W \pm \sqrt{1+W^2}) \quad (j=1,2) \qquad (3.7.15)$$

where we define $W = \xi \tan\theta_{\mathrm{B}}\xi_G/\pi$, known as the 'resonance error'. The non-dimensional quantity W is proportional to ξ, which measures the deviation from the Laue–Bragg condition. Namely, $W=0$ means that the Laue–Bragg condition is satisfied, where $C_{\mathbf{G}}^{(j)}/C_0^{(j)} = \pm 1$. Thus, C_0 and $C_{\mathbf{G}}$ are in phase for $j=2$ and out of phase for $j=1$, as shown in figure 3.9. The solid lines represent the $\mathbf{K}_{\mathrm{G}}$ wave which is in phase with the incident wave, and the dotted lines the $\mathbf{K}_{\mathrm{G}}$ wave which is out of phase.

Now, normalizing $C_0^{(j)}$ and $C_{\mathbf{G}}^{(j)}$ by

$$|C_0^{(j)}|^2 + |C_{\mathbf{G}}^{(j)}|^2 = 1 \qquad (3.7.16)$$

and defining β by

$$W = \cot\beta \qquad (3.7.17)$$

it is easy to show that

$$C_{\mathbf{G}}^{(1)}/C_0^{(1)} = -\tan\frac{\beta}{2}$$

$$C_{\mathbf{G}}^{(2)}/C_0^{(2)} = \cot\frac{\beta}{2} \qquad (3.7.18)$$

†Equation 3.7.14 is also obtained in multi-beam cases where $j=1, 2, \ldots, N$.
‡Since $V_{-\mathbf{G}}$ is negative, $K_0 \cos\theta_B/(2mV_{-\mathbf{G}}/\hbar^2) = -K_0\cos\theta_B/2m(-V_{-\mathbf{G}})/\hbar^2 = -K_0\cos\theta_B/\sqrt{U_{\mathrm{G}}\,U_{-\mathbf{G}}} = -\xi_{\mathrm{G}}/2\pi$.

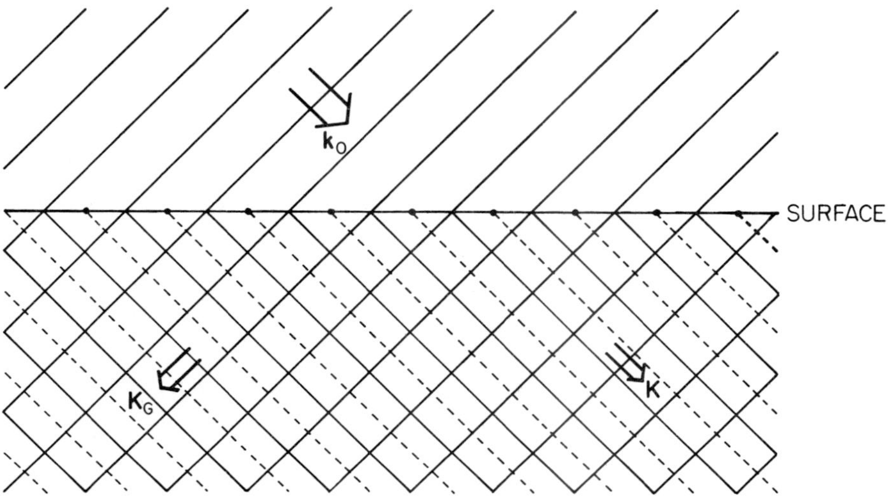

Figure 3.9. Phase relation of two waves corresponding to the branches $j = 1$ and 2 excited in the crystal. The solid and dotted lines in the $\mathbf{K}_G$ direction are out-of-phase.

Inserting equation 3.7.18 into equation 3.7.16 we obtain

$$C_0^{(1)} = C_{\mathbf{G}}^{(2)} = \cos\frac{\beta}{2}$$
$$C_0^{(2)} = -C_{\mathbf{G}}^{(1)} = \sin\frac{\beta}{2} \tag{3.7.19}$$

from which the following ortho-normal relation is derived

$$\sum_{\mathbf{H}} C_{\mathbf{H}}^{(j)} C_{\mathbf{H}}^{(j')} = \delta_{j,j'} \tag{3.7.20}$$

Inserting equation 3.7.19 into equation 3.7.14 the coefficients $\chi^{(j)}$ are determined

$$\chi^{(1)} = \cos\frac{\beta}{2}$$
$$\chi^{(2)} = \sin\frac{\beta}{2} \tag{3.7.21}$$

or

$$\chi^{(j)} = C_0^{(j)} \tag{3.7.22}$$

We have thus obtained the complete state of the electron beam in the crystal with the dispersion relation and appropriate boundary conditions.

It is not so difficult to calculate the intensity of the wave $I_0(\mathbf{r})$ whose wave number is **K** and the intensity $I_{\mathbf{G}}(\mathbf{r})$ corresponding to the wave number $\mathbf{K}_G$

$$I_0(\mathbf{r}) = \left| \sum_i \chi^{(i)} C_0^{(i)} \exp(i\mathbf{K}^{(i)}\cdot\mathbf{r}) \right|^2$$

$$= |\chi^{(1)} C_0^{(1)}|^2 + |\chi^{(2)} C_0^{(2)}|^2 + \chi^{(1)}\chi^{(2)} C_0^{(1)} C_0^{(2)}$$

$$\times \{\exp[i(K^{(1)} - \mathbf{K}^{(2)})\cdot\mathbf{r}] + \exp[-i(\mathbf{K}^{(1)} - \mathbf{K}^{(2)})\cdot\mathbf{r}]\} \qquad (3.7.23)$$

and

$$I_{\mathbf{G}}(\mathbf{r}) = \left| \sum_i \chi^{(i)} C_{\mathbf{G}}^{(i)} \exp(\mathrm{i}\mathbf{K}_{\mathbf{G}}^{(i)}\cdot\mathbf{r}) \right|^2$$

$$= |\chi^{(1)} C_{\mathbf{G}}^{(1)}|^2 + |\chi^{(2)} C_{\mathbf{G}}^{(2)}|^2 + \chi^{(1)}\chi^{(2)} C_{\mathbf{G}}^{(1)} C_{\mathbf{G}}^{(2)}$$

$$\times \{\exp[i(\mathbf{K}^{(1)} - \mathbf{K}^{(2)})\cdot\mathbf{r}] + \exp[-i(\mathbf{K}^{(1)} - \mathbf{K}^{(2)})\cdot\mathbf{r}]\} \qquad (3.7.24)$$

Inserting equations 3.7.19 and 3.7.21 into equations 3.7.23 and 3.7.24, respectively, we obtain

$$I_0(\mathbf{r}) = \left(\cos^4\frac{\beta}{2} + \sin^4\frac{\beta}{2}\right) + \tfrac{1}{2}\cos[(\mathbf{K}^{(1)} - \mathbf{K}^{(2)})\cdot\mathbf{r}]\sin^2\beta \qquad (3.7.25)$$

and

$$I_{\mathbf{G}}(\mathbf{r}) = +\tfrac{1}{2}\{1 - \cos[(\mathbf{K}^{(1)} - \mathbf{K}^{(2)})\cdot\mathbf{r}]\}\sin^2\beta \qquad (3.7.26)$$

It is obvious that

$$I_0(\mathbf{r}) + I_G(\mathbf{r}) = 1$$

It is worth noting that the difference between the two vectors $\mathbf{K}^{(1)}$ and $\mathbf{K}^{(2)}$ is the difference in their z components. According to figure 3.8 the z component of the vector $(\mathbf{K}^{(1)} - \mathbf{K}^{(2)})$ is represented by

$$(\mathbf{K}^{(1)} - \mathbf{K}^{(2)})_z\cdot z = \frac{2|\eta|}{\cos(\Delta\theta)} = \frac{2\pi}{\xi_G}\sqrt{1+W^2}\,\frac{1}{\cos(\Delta\theta)}$$

where the angle $\Delta\theta$ is the angle between the crystal surface and the reciprocal lattice vector $2\pi\mathbf{G}$. Equations 3.7.25 and 3.7.26 become

$$I_0(\mathbf{r}) = \frac{1}{2(1+W^2)}\left\{\cos\left(\frac{2\pi\sqrt{1+W^2}}{\xi_G\cos(\Delta\theta)}z\right) + \frac{(W-\sqrt{1+W^2})^2 + (W-\sqrt{1+W^2})^2}{2}\right\} \qquad (3.7.27)$$

and

$$I_{\mathbf{G}}(\mathbf{r}) = +\frac{1}{2(1+W^2)}\left\{1 - \cos\left(\frac{2\pi\sqrt{1+W^2}}{\xi_G\cos(\Delta\theta)}z\right)\right\} \qquad (3.7.28)$$

If we consider the case where $W=0$ and $\Delta\theta \simeq 0$, it is easy to see that the intensities I_0 and I_{G}, or the energy of the **K** wave and the $\mathbf{K}_{\mathrm{G}}$ wave are exchanged

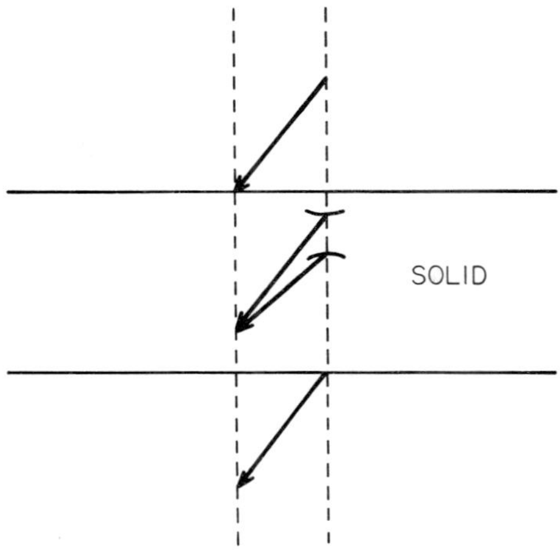

Figure 3.10. Two waves excited in the parallel-shaped crystal become one wave at the exit surface.

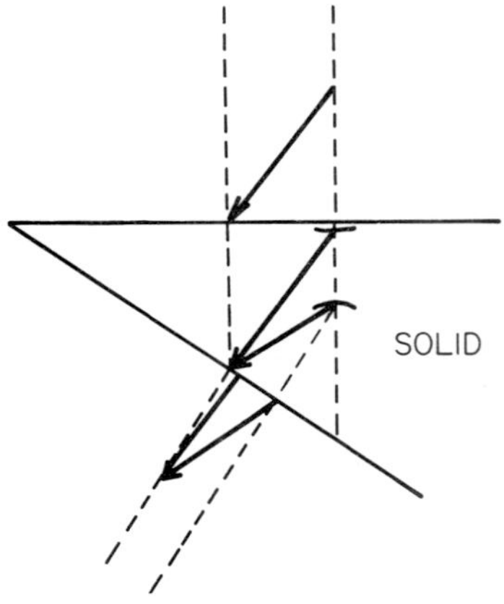

Figure 3.11. Two waves excited in the wedge-shaped crystal remain separate at the exit surface.

alternately along the z axis with period ξ_G. So, we see that electron beams propagate at least a distance of ξ_G in order to form the standing wave $\mathbf{K}_G$. For this reason ξ_G should be the minimum distance of electron penetration in the crystal for the application of our dynamical theory of electron diffraction.

As shown in figure 3.10, the two separated waves, corresponding to branches j in the crystal which have parallel surface planes, become one wave again at the exit surface. However, for the wedge-shaped crystal, as shown in figure 3.11, the two curves† appear outside the exit surface. We can see two curves in a

†These are not the two spots but two curves, since the photograph was taken changing the crystal orientation. The curves therefore represent the dispersion curves.

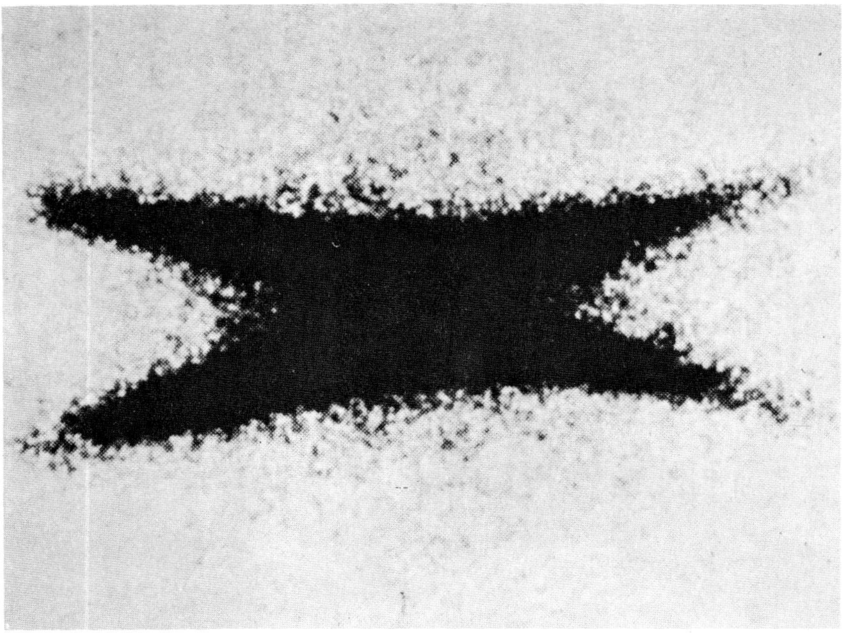

Figure 3.12. The two curves in the photograph correspond to the two branches of the dispersion surface for 90° wedge-shaped MgO crystal, according to Lehmpfuhl and Reissland.[28]

photograph taken by Lehmpfuhl and Reissland,[28] figure 3.12, corresponding to the branches *j*, for a 90° wedge-shaped MgO crystal ($\mathbf{G} = (220)$).

3.8. Correction to the two-wave theory

3.8.1. *Many-beam effects*

There are conditions where a large number N of reciprocal lattice points are close to the Ewald sphere. Under these condition N waves are excited simultaneously, and the problem must be studied using the N-wave approximation.

However, there are cases where only two of the N waves are strongly excited, in and the problem can then be solved by a perturbation method. The other N-2 waves are weakly excited. Equation 3.7.4 is rewritten in general as

$$(k_0^2 - K_{\mathbf{H}}^2)\psi_{\mathbf{H}} = \sum_{\mathbf{H}} U_{\mathbf{H}'}\psi_{\mathbf{H}-\mathbf{H}'} \tag{3.8.1}$$

Now, if we assume the amplitudes ψ_0 and $\psi_{\mathbf{G}}$ are large, then we obtain to a first order approximation

$$\psi_{\mathbf{H}} = \frac{U_{\mathbf{H}}\psi_0 + U_{\mathbf{H}-\mathbf{G}}\psi_{\mathbf{G}}}{k_0^2 - K_{\mathbf{H}}^2} \tag{3.8.2}$$

Further, equation 3.8.1 with $\mathbf{H} = 0$ and $\mathbf{G}$ leads to

$$\begin{aligned} (K^2 - k_0^2 + U_0)\psi_0 + U_{-\mathbf{G}}\psi_{\mathbf{G}} + \sum_{\mathbf{H}'\neq 0,-\mathbf{G}}' U_{\mathbf{H}'}\psi_{-\mathbf{H}'} &= 0 \\ U_{\mathbf{G}}\psi_0 + (K_{\mathbf{G}}^2 - k_0^2 + U_0)\psi_{\mathbf{G}} + \sum_{\mathbf{H}'\neq 0,\mathbf{G}}' U_{\mathbf{H}'}\psi_{\mathbf{G}-\mathbf{H}'} &= 0 \end{aligned} \tag{3.8.3}$$

The third terms in each line of equation 3.8.3 are the ones that can be neglected in the two-wave approximation. Inserting $\psi_{-\mathbf{H}'}$ and $\psi_{\mathbf{G}-\mathbf{H}'}$ in the form of equation 3.8.2 into equation 3.8.3 we obtain

$$\begin{aligned} (K^2 - k_0^2 + V_{11})\psi_0 + V_{12}\psi_{\mathbf{G}} &= 0 \\ V_{21}\psi_0 + (K_{\mathbf{G}}^2 - k_0^2 + V_{22})\psi_{\mathbf{G}} &= 0 \end{aligned} \tag{3.8.4}$$

where

$$V_{11} = U_0 + \sum_{\mathbf{H}'\neq 0,\mathbf{G}} \frac{U_{\mathbf{H}'}U_{-\mathbf{H}'}}{k_0^2 - K_{\mathbf{H}'}^2} \tag{3.8.5}$$

$$V_{22} = U_0 + \sum_{\mathbf{H}'\neq 0,\mathbf{G}} \frac{U_{\mathbf{H}'-\mathbf{G}}U_{\mathbf{G}-\mathbf{H}'}}{k_0^2 - K_{\mathbf{H}'}^2} \tag{3.8.6}$$

$$V_{12} = U_{-\mathbf{G}} + \sum_{\mathbf{H}\neq 0,\mathbf{G}} \frac{U_{\mathbf{H}'-\mathbf{G}}U_{-\mathbf{H}'}}{k_0^2 - K_{\mathbf{H}'}^2} \tag{3.8.7}$$

and

$$V_{21} = U_{\mathbf{G}} + \sum_{\mathbf{H}\neq 0,\mathbf{G}} \frac{U_{\mathbf{G}-\mathbf{H}'}U_{\mathbf{H}'}}{k_0^2 - K_{\mathbf{H}'}^2} \tag{3.8.8}$$

Hence we see that the N-wave problem has been reduced to the two-wave problem (equation 3.8.4) if we define the effective potentials V_{11}, V_{22}, V_{12} and V_{21} instead of U_0, $U_{-\mathbf{G}}$. This is called the 'Bethe correction'.

It should be noted, however, that this correction is only concerned with the scattering potential, and hence the dispersion surface. On the other hand, there are a few other methods of solving the N-wave problem which make use of the 'scattering matrix' to obtain the diffraction intensity without the determination of the dispersion surface. Here we introduce the theory due to Fujimoto.[29,30,31]

We can rewrite equation 3.7.4 in the matrix form

$$\begin{pmatrix} K^2 - K_0^2 & U_{\mathbf{H}} & U_{\mathbf{G}}\ldots\ldots \\ U_{-\mathbf{H}} & K_{\mathbf{H}}^2 - K_0^2 & U_{\mathbf{H}-\mathbf{G}}\ldots\ldots \\ U_{-\mathbf{G}} & U_{\mathbf{G}-\mathbf{H}} & K_{\mathbf{G}}^2 - K_0^2\ldots \\ \ldots\ldots & \ldots & \ldots \end{pmatrix} \begin{pmatrix} \psi_0 \\ \psi_H \\ \psi_G \\ \vdots \end{pmatrix} = 0 \tag{3.8.9}$$

For the Laue case, it is easy to see that

$$K_{\mathbf{G}} \simeq K_0 - \rho_{\mathbf{G}} - \zeta$$

at $W \simeq 0$ in figure 3.8; where ρ_G is the distance from the point Q' and the 'equi-energy sphere' (dotted line in figure 3.8) and ζ is the distance $\overline{P'Q'}$. Taking into account $\zeta \ll K_0$, we obtain

$$\begin{aligned} K_{\mathrm{G}}^2 - K_0^2 &= \rho_{\mathbf{G}}^2 + \zeta^2 - 2K_0\rho_{\mathbf{G}} + 2\rho_{\mathbf{G}}\zeta - 2K_0\zeta \\ &\simeq -2K_0\zeta - 2\rho_{\mathbf{G}}K_0 + \rho_{\mathbf{G}}^2 \\ &= -\lambda + p_{\mathbf{G}} \end{aligned}$$

where $\lambda = 2K_0\zeta$ and $p_{\mathbf{G}} = \rho_{\mathbf{G}}^2 - 2\rho_{\mathbf{G}}K_0$. Roughly speaking, as seen in figure 3.8, $p_{\mathbf{G}} \propto \xi \propto W$. So, $p_{\mathbf{G}}$ represents the deviation from the Laue–Bragg condition.

Now, equation 3.8.9 turns out to be

$$\bar{M}\psi = \lambda\psi \tag{3.8.10}$$

$$\bar{M} = \begin{pmatrix} p_0 & U_{\mathbf{H}} & U_{\mathbf{G}}\cdots\cdots \\ U_{-\mathbf{H}} & p_{\mathbf{H}} & U_{\mathbf{H}-\mathbf{G}}\cdots\cdot \\ U_{-\mathbf{G}} & U_{-\mathbf{G}} & p_{\mathbf{G}}\cdots\cdots \\ \cdots\cdot & \cdots & \cdots \end{pmatrix} \tag{3.8.11}$$

and

$$\psi = \begin{pmatrix} \psi_0 \\ \psi_H \\ \psi_G \\ \vdots \end{pmatrix} \tag{3.8.12}$$

Thus, if we solve equation 3.8.10 we may obtain the eigenvalue $\lambda^{(i)}$ and the eigenvector $\psi^{(i)}$ simultaneously.

The Bloch wave function for the electron beam in the crystal is rewritten, making use of equation 3.7.22 and 3.8.10, as

$$\psi = \sum_i \sum_{\mathbf{G}} C_0^{(i)} C_{\mathbf{G}}^{(i)} \exp\left(i\frac{\lambda^{(i)}}{2K_0} z \right) \exp\left[i(\mathbf{K}_0 + 2\pi\mathbf{G})\cdot\mathbf{r}\right]$$

Making use of

$$\bar{M}^n \psi^{(i)} = (\lambda^{(i)})^n \psi^{(i)}$$

and

$$\begin{aligned} \sum_i (\lambda^{(i)})^n C_0^{(i)} C_{\mathbf{H}}^{(i)} &= \sum_{\mathbf{G}} (\bar{M}^n)_{\mathbf{HG}} \left(\sum_i C_{\mathbf{G}}^{(i)} C_0^{(i)} \right) \\ &= \sum_{\mathbf{G}} (\bar{M}^n)_{\mathbf{HG}} \delta_{\mathbf{G},0} = (\bar{M}^n)_{\mathbf{H}0} \end{aligned}$$

we obtain

$$\psi_{\mathbf{H}} = \sum_i C_0^{(i)} C_{\mathbf{H}}^{(i)} \exp\left(i\frac{\lambda^{(i)}}{2K_0} z \right) \exp\left[i(\mathbf{K}_0 + 2\pi\mathbf{H})\cdot\mathbf{r}\right]$$

$$= \sum_i C_0^{(i)} C_{\mathbf{H}}^{(i)} \sum_n \frac{1}{n!} \left(i \frac{\lambda^{(i)}}{2K_0} z \right)^n \exp[i(\mathbf{K}_0 + 2\pi\mathbf{H})\cdot\mathbf{r}]$$

$$= S_{\mathbf{H}0} \exp[i(\mathbf{K}_0 + 2\pi\mathbf{H})\cdot\mathbf{r}] \tag{3.8.13}$$

where we defined the scattering matrix as

$$\bar{S} = \exp\left(i \frac{z}{2K_0} \bar{M} \right) \tag{3.8.14}$$

Then, the intensity $I_{\mathbf{H}}(\mathbf{r})$ is obtained

$$I_{\mathbf{H}}(\mathbf{r}) = |\psi_{\mathbf{H}}|^2 = |\bar{S}_{\mathbf{H}0}|^2 \tag{3.8.15}$$

It is worth noting that the intensity $I_{\mathbf{H}}(\mathbf{r})$ is obtained directly from the matrix $\bar{M}$ without deriving the dispersion surface.

When we expand the scattering matrix in powers of $(z/2K_0)$, the intensity is calculated as follows

$$I_{\mathbf{H}}(\mathbf{r}) = \left(\frac{z}{2K_0} \right)^2 |M_{\mathbf{H}0}|^2 + \frac{i}{2} \left(\frac{z}{2K_0} \right)^3 [M_{\mathbf{H}0} M_{\mathbf{H}0}^* - M_{\mathbf{H}0} (M^{*2})_{\mathbf{H}0}] + \dots \tag{3.8.16}$$

The first term of equation 3.8.16 is proportional to $|V_{\mathbf{H}}|^2$, corresponding to the kinematic intensity, which we denote as $I_{\mathbf{H}}^K$. We may calculate the intensity $I_{\mathbf{H}}(\mathbf{r})$ as a function of the crystal thickness D for electrons with the wave number $k_0 = 1000\,\mathrm{nm}^{-1}$ into Au (111).[16] This is shown in figure 3.13.

Other methods for calculating the many-beam effects start with the integral equation

$$\psi(\mathbf{r}) = \psi^0(\mathbf{r}) + \int G(\mathbf{r}, \mathbf{r}')\, V(\mathbf{r}')\, \psi(\mathbf{r}')\, \mathrm{d}^3\mathbf{r}'$$

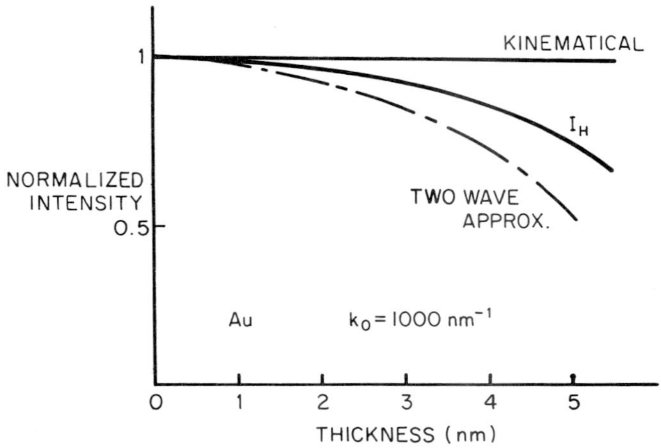

Figure.3.13. Intensity profile as a function of the crystal thickness D for the incident wave number $k_0 = 1000\,\mathrm{nm}^{-1}$ into Au (111), showing the deviation from the kinematical intensity for large D.

as developed by Tournarie,[32,33] Kambe,[34] Fujiwara,[35] and Fengler.[36] Howie and Whelan's technique[37] is based on the multi-slice method, and Cowley and Moodie[38,39,40] developed the phase-lattice diffraction method. However, it is now possible to study the problem using large capacity computers which make it easier to solve the above integral equation directly.

3.8.2. *Contribution of lattice vibration*

The scattering time of electrons is very short compared to the lattice vibration period, and we can assume that the electrons are scattered by static lattice points deviated from the equilibrium position by thermal lattice vibrations. However, since the observation time of electrons is very large compared to the lattice vibration period, the measured electron intensity is calculated by averaging over the statistically distributed positions of the lattice.

For isotropic crystals the equilibrium positions are located at the original lattice points, and with the exception of the scattering factor our diffraction theory is not modified by the lattice vibration. Only the charge distribution around the lattice point is changed by the lattice vibration, and this change then modifies the scattering factor. This has been shown by Parthasarathy[41] for nonabsorption cases and by Kambe[42] for absorption cases.

Now, for simplicity, we shall consider the case of one atom per unit cell. Denoting the displacement of the atom in the lth cell as $\delta\mathbf{R}_l$, the scattering factor is modified by the factor $\Delta(\mathbf{Q})$, defined as

$$\Delta(\mathbf{Q}) = \frac{1}{N}\sum_l \exp\left[i\mathbf{Q}\cdot(\mathbf{R}_l + \delta\mathbf{R}_l)\right] \tag{3.8.17}$$

Here we have to average over the thermal equilibrium state

$$\langle \Delta(\mathbf{Q}) \rangle = \frac{1}{N}\sum_l \exp(i\mathbf{Q}\cdot\mathbf{R}_l)\langle \exp(i\mathbf{Q}\cdot\delta\mathbf{R}_l)\rangle \tag{3.8.18}$$

in which we may employ cumulant expansion, which is well known in the field of statistical mechanics

$$\langle e^x \rangle = \exp\left(\sum_n \frac{1}{n!} M_n\right) \tag{3.8.19}$$

where

$$M_1 = \langle x \rangle$$
$$M_2 = \langle x^2 \rangle - \langle x \rangle^2$$
$$M_3 = \langle x^3 \rangle - 3\langle x^2 \rangle\langle x \rangle + 2\langle x \rangle^3$$
$$M_4 = \langle x^4 \rangle - 4\langle x^3 \rangle\langle x \rangle - 3\langle x^2 \rangle^2 + 12\langle x^2 \rangle\langle x \rangle^2 - 6\langle x \rangle^4$$
$$\vdots$$

The displacement $\delta\mathbf{R}_l$ is represented in a quantized form by the phonon creation and annihilation operators, $a^{+}_{-q\lambda}$ and $a_{q\lambda}$, respectively

$$\delta\mathbf{R}_l = \sum_{q\lambda} \sqrt{\frac{\hbar}{2NM\omega_{q\lambda}}}\,\mathbf{e}_{q\lambda}(a_{q\lambda} + a^{+}_{-q\lambda})\exp(i\mathbf{q}\cdot\mathbf{R}_l) \qquad (3.8.20)$$

where $\mathbf{e}_{q\lambda}$ and $\hbar\omega_{q\lambda}$ are the unit polarization vector and the energy of a phonon with wave number $\mathbf{q}$ and branch λ, respectively. M is the mass of the atom.

It is easy to see that the average of the odd powers, for example $\langle x \rangle$, $\langle x^3 \rangle$, $\langle x^5 \rangle, \ldots$, vanishes exactly, since odd numbers of the creation and/or annihilation operators are included in the above average of odd powers. For example, in $\langle x^3 \rangle$ we take an average like $\langle a_{q\lambda}\, a^{+}_{q'\lambda'}\, a^{+}_{q''\lambda''} \rangle$, which vanishes exactly, because there is no way to get back to the same state by exciting two phonons and de-exciting (annihilating) one phonon. As a result, we obtain

$$\langle \Delta(\mathbf{Q}) \rangle = \frac{1}{N}\left(\sum_l \exp(i\mathbf{Q}\cdot\mathbf{R}_l)\right)\exp\left\{\frac{1}{2}\langle (i\mathbf{Q}\cdot\mathbf{R}_l)^2 \rangle + \frac{1}{4!}[\langle (i\mathbf{Q}\cdot\delta\mathbf{R}_l)^4 - 3\langle (i\mathbf{Q}\cdot\delta\mathbf{R}_l)^2 \rangle^2] + \cdots\right\} \qquad (3.8.21)$$

In our case $\langle (i\mathbf{Q}\cdot\delta\mathbf{R}_l)^2 \rangle$ has a value of the order of 0·1, so we may neglect powers higher than $\langle (i\mathbf{Q}\cdot\delta\mathbf{R}_l)^2 \rangle$.

The diffraction intensity is represented by the scattering factor $V_{\mathbf{G}}$, modified by the correction factor $\langle \Delta(2\pi\mathbf{G}) \rangle$. The temperature dependence of the scattering factor is written by

$$\langle V_{\mathbf{G}} \rangle = V_{\mathbf{G}}\exp[-M(\mathbf{G})] \qquad (3.8.22)$$

where

$$M(\mathbf{G}) = \tfrac{1}{2}\langle (2\pi\mathbf{G}\cdot\delta\mathbf{R}_l)^2 \rangle \qquad (3.8.23)$$

The factor exp $[-M(\mathbf{G})]$ is called the 'Debye–Waller factor'. It should be noted that the thermal average of $(2\pi\mathbf{G}\cdot\delta\mathbf{R}_l)^2$ in equation 3.8.23 does not depend on l, and the summation over l in equation 3.8.21 is performed without the Debye–Waller factor.

The Debye–Waller factor will now be studied in more detail. Inserting equation 3.8.20 into equation 3.8.23 we obtain

$$M(\mathbf{G}) = (2\pi)^2 \sum_{q\lambda} \frac{\hbar}{2NM\omega_{q\lambda}}(\mathbf{G}\cdot\mathbf{e}_{q\lambda})^2(\langle a^{+}_{q\lambda} a_{q\lambda} \rangle + \tfrac{1}{2}) \qquad (3.8.24)$$

In the above, we used the commutation relation

$$[a_{q\lambda}, a^{+}_{q'\lambda'}] = \delta_{\mathbf{q},\mathbf{q}'}\,\delta_{\lambda,\lambda'}$$

and the relation

$$\langle a_{q\lambda}\, a_{q'\lambda'} \rangle = \langle a^{+}_{q\lambda}\, a^{+}_{q'\lambda'} \rangle = 0$$

Since the term $\langle a^{+}_{q\lambda} a_{q\lambda} \rangle$ in equation 3.8.24 is the occupation number of phonons, we may represent it by the expression

$$\langle a_{q\lambda}^{+} a_{q\lambda} \rangle = \frac{1}{\exp(\hbar\omega_{q\lambda}/kT) - 1} \tag{3.8.25}$$

where k is the Boltzmann constant. Inserting equation 3.8.25 into equation 3.8.24 we obtain

$$M(\mathbf{G}) = \pi^2 \sum_{q\lambda} \frac{\hbar(\mathbf{G}\cdot\mathbf{e}_{q\lambda})^2}{NM\omega_{q\lambda}} \coth\left(\frac{\hbar\omega_{q\lambda}}{2kT}\right) \tag{3.8.26}$$

Taking into account the Debye model for acoustic phonons

$$\omega_{q\lambda} = c_\lambda q$$

and replacing the summation over q by the integral over the first Brillouin zone, equation 3.8.26 becomes

$$M(\mathbf{G}) = \frac{V_c}{(2\pi)^3} \sum_{\lambda} \int_0^{q_{\max}} \int \frac{\hbar(2\pi\mathbf{G}\cdot\mathbf{e}_{q\lambda})^2}{2NMc_\lambda} \coth\left(\frac{\hbar c_\lambda q}{2kT}\right) q\,\mathrm{d}q\,\mathrm{d}\Omega \tag{3.8.27}$$

where $q_{\max}$ is the maximum wave number of the phonon, which is given in the Debye model by

$$q_{\max}^3 = \frac{3}{4\pi} \frac{(2\pi)^3}{V_c} \tag{3.8.28}$$

V_c being the volume of the unit cell.

If we define the Debye temperature for the longitudinal and the transverse phonons, denoted as Θ_l and Θ_t, respectively, as

$$k\Theta_l = \hbar c_l q_{\max}$$
$$k\Theta_t = \hbar c_t q_{\max} \tag{3.8.29}$$

we obtain the average Debye temperature Θ_M

$$\frac{1}{\Theta_M^2} = \frac{1}{3}\left(\frac{1}{\Theta_l^2} + \frac{2}{\Theta_t^2}\right) \tag{3.8.30}$$

Making use of this average Debye temperature we may rewrite the coefficient $M(\mathbf{G})$ as

$$M(\mathbf{G}) = \frac{3h^2G^2}{2Mk\Theta_M}\left(\frac{1}{x}\Phi(x) + \frac{1}{4}\right) \tag{3.8.31}$$

where the function $\Phi(x)$ is given by

$$\Phi(x) = \frac{1}{x}\int_0^x \frac{\xi}{\exp(\xi - 1)}\,\mathrm{d}\xi \tag{3.8.32}$$

$$(x = \Theta_M/T)$$

In real crystals the dispersion relation $\omega_{q\lambda} = c_\lambda q$ is not quite correct. Since the phonon spectrum $g(\nu)$ as a function of the phonon frequency is known experimentally (e.g., from the inelastic scattering of neutrons), we can calculate the Debye temperature by the formula

$$\Theta_{\mathrm{M}}^2 = \frac{3hT[\Phi(\beta h\nu) + \frac{1}{4}\beta h\nu]\int g(\nu)\mathrm{d}\nu}{k\int q(\nu)\nu^{-1}\{\frac{1}{2} + [\exp(\beta h\nu) - 1]^{-1}\}\mathrm{d}\nu} \tag{3.8.33}$$

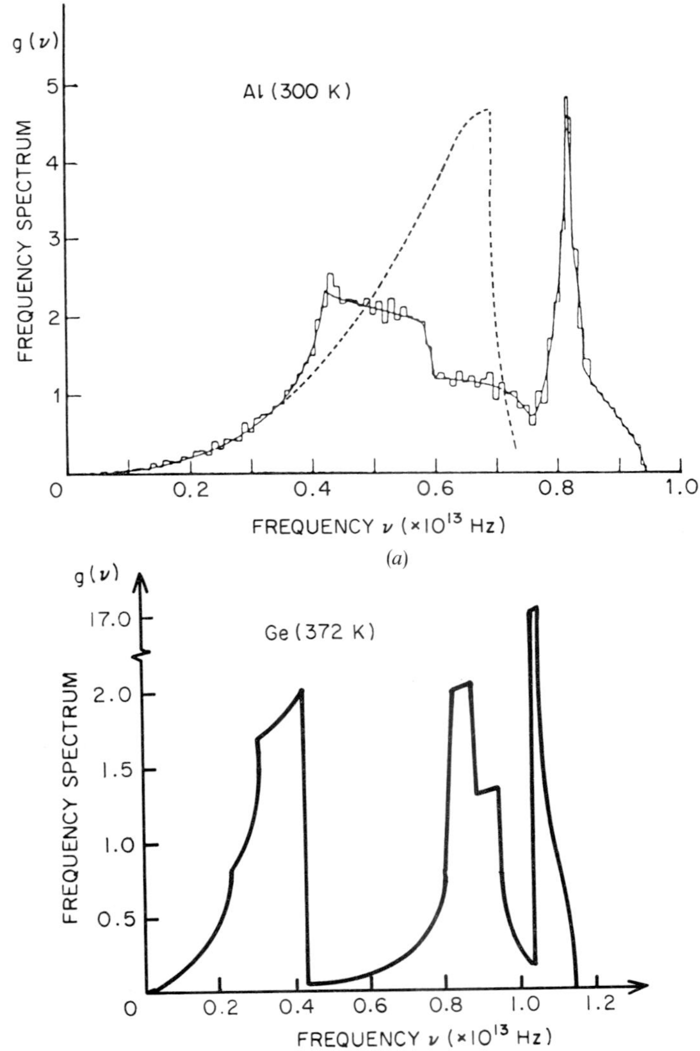

Figure 3.14. Phonon spectra for (*a*) Al and (*b*) Ge. For Al, the spectrum deviates greatly from the Debye model (dotted line), according to Walker[43] and Phillips.[44]

In figure 3.14, the phonon spectra for Al and Ge are shown,[43,44] from which we can see that the phonon spectrum for Ge has deviated greatly from the Debye model. Batterman and Chipman[45] calculated the Debye temperature for Ge according to the formula 3.8.33, and this is shown in figure 3.15 along with the Debye temperature Θ_D of the specific heat. It is worth noting that the temperature dependence of Θ_M and Θ_D is very different in the low temperature region; and even at room temperature Θ_M differs from Θ_D by about 100°C. In table 3.2 the experimental and theoretical values of Θ_M are given. Note that the Debye temperature of a diatomic crystal like NaCl is defined for each lattice deviation, $\delta\mathbf{R}_{Na}$ and $\delta\mathbf{R}_{Cl}$. However, in this case the interaction potentials between the same kind of atoms and those between different kinds of atoms are the same, making the Debye temperatures for Na and Cl atoms equal.

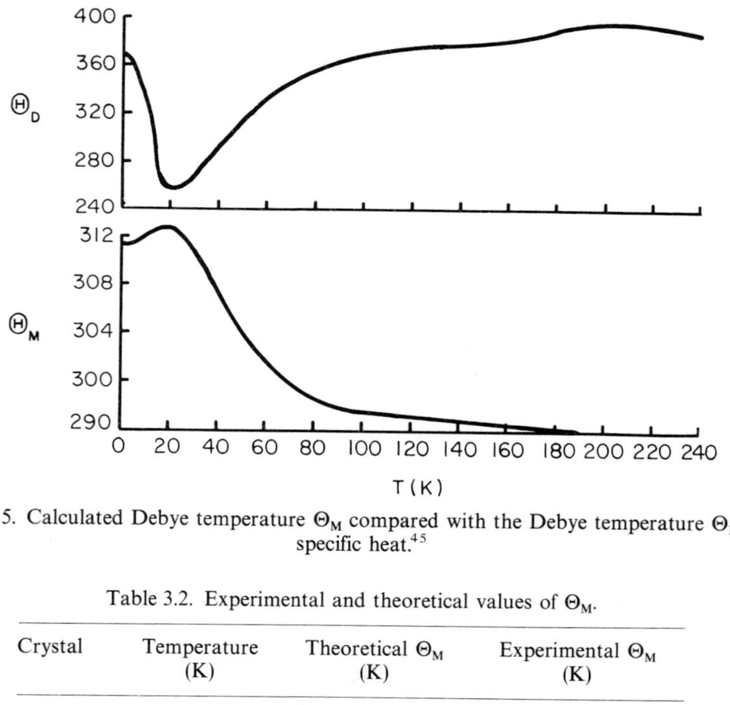

Figure 3.15. Calculated Debye temperature Θ_M compared with the Debye temperature Θ_D of the specific heat.[45]

Table 3.2. Experimental and theoretical values of Θ_M.

Crystal	Temperature (K)	Theoretical Θ_M (K)	Experimental Θ_M (K)
Al	300	386	402[42]
Cu	300	307	298[46]
Ge	300	296[45]	290[47]
NaCl	290	279[48]	292

References

1. Chadderton, L.T., *Phil. Mag.* **18**, 1017 (1968).
2. Gemmell, D.S., *Rev. Mod. Phys.* **46**, No. 1 (1974).
3. Yamamura, Y. and Ohtsuki, Y.H., *Phys. Rev.* **B24**, 3430 (1981).

4. Yamamura, Y. and Ohtsuki, Y.H., *Rad. Eff.* **56**, 1 (1981).
5. Nip, H.C.H. and Kelley, J.C., *Atomic Collision Phenomena* (Amsterdam: North-Holland, 1971).
6. Uggerhøj, E., *Phys. Letters* **22**, 382 (1966).
7. Uggerhøj, E. and Fraddsen, F., *Phys. Rev.* **2B**, 582 (1970).
8. Humphreys, C.J. and Lally, J.S., *J. Appl. Phys.* **41**, 232 (1970).
9. Humphreys, C.J., Thomas, L.E., Lally, J.S. and Fisher, R.M., *Phil. Mag.* **23**, 87 (1970).
10. Fujimoto, F., Takagi, S., Komaki, K., Koike, H. and Uchida, Y., *Rad. Effects* **12**, 153 (1972).
11. Kubota, Y. and Ohtsuki, Y.H., *Phys. Stat. Sol.* **B52**, 271 (1972).
12. Tamura, A. and Ohtsuki, Y.H., *Phys. Stat. Sol.* **B62**, 477 (1974).
13. Yamamura, Y. and Ohtsuki, Y.H., *Phys. Rev.* **B24**, 3430 (1981).
14. Yamamura, Y. and Ohtsuki, Y.H., *Rad. Eff.* **56**, 1 (1981).
15. Ibers, J.A., *Acta Cryst.* **10**, 86 (1957).
16. Forsyth, J.B. and Wells, M., *Acta Cryst.* **12**, 412 (1959).
17. Kittel, C., *Quantum Theory of Solids* (New York: Wiley, 1967).
18. Schultz, T.D., *Quantum Field Theory and the Many Body Problem* (New York: Gordon and Breach).
19. Ohtsuki, Y.H. and Yanagawa, S., *J. Phys. Soc. Japan* **21**, 326 (1966).
20. Ewald, P.P., *Ann. Phys.* **54**, 519 (1917).
21. Ewald, P.P., *Ann. Phys.* **54**, 577 (1917).
22. Ewald, P.P., *Z. Phys.* **2**, 332 (1920).
23. Ewald, P.P., *Z. Phys.* **30**, 1 (1924).
24. Ewald, P.P., *Phys. Z.* **26**, 29 (1925).
25. Ewald, P.P., *Z. Krist.* **97**, 1 (1937).
26. Bethe, H., *Ann. Phys.* **4**, 55 (1928).
27. Bethe, H., *Ann. Phys.* **4**, 87 (1928).
28. Lehmpfuhl, G. and Reissland, A., *Z. Naturforsch.* **23a**, 544 (1968).
29. Fujimoto, F., *J. Phys. Soc. Japan* **14**, 1558 (1959).
30. Fujimoto, F., *J. Phys. Soc. Japan* **15**, 859 (1960).
31. Fujimoto, F., *J. Phys. Soc. Japan* **15**, 1022 (1960).
32. Tournarie, M., *C. R. Acad. Sci.* **252**, 2862 (1961).
33. Tournarie, M., *J. Phys. Soc. Japan* **17**, Supplement BII, 98 (1962).
34. Kambe, K., *Z. Naturforsch.* **22a**, 422 (1967).
35. Fujiwara, K., *J. Phys. Soc. Japan* **14**, 1513 (1959).
36. Fengler, H., *Z. Naturforsch.* **16a**, 1205 (1961).
37. Howie, A. and Whelan, M.J., *Proc. Roy. Soc. London* **A263**, 217 (1961).
38. Cowley, J.M. and Moodie, A.F., *Acta Cryst.* **10**, 609 (1957).
39. Cowley, J.M. and Moodie, A.F., *Acta Cryst.* **12**, 353 (1959).
40. Cowley, J.M. and Moodie, A.F., *Acta Cryst.* **12**, 360 (1959).
41. Parthasarathy, R., *Acta Cryst.* **13**, 802 (1960).
42. Kambe, K., *Z. Naturforsch*, **20a**, 770 (1965).
43. Walker, C.B., *Phys. Rev.* **103**, 547 (1956).
44. Phillips, J.C., *Phys. Rev.* **113**, 147 (1959).
45. Batterman, B.W. and Chipman, D.R., *Phys. Rev.* **127**, 690 (1962).
46. Borie, B., *Acta Cryst.* **9**, 617 (1956).
47. Batterman, B.W., *Phys. Rev.* **127**, 686 (1962).
48. Sen, R.K., *Physica* **30**, 1655 (1964).

CHAPTER 4

inelastic scattering of electrons and ions in crystals

4.1. Introduction

Classical mechanics does not give a complete description of the inelastic scattering (i.e., stopping power, absorption coefficients, and so on) of ion and electron beams in crystals. Even when the channelled trajectories of high energy ions are calculated using classical mechanics we have to use quantum mechanics to describe the inelastic scattering of the channelled ions. To use an analogy, a normally driven car can be understood with classical mechanics, but the collision of cars in an accident and the resulting energy loss requires quantum mechanics for a complete description.

Since an electron's mass is very small, strong inelastic scattering of electrons in crystals is very prominent. So, by studying the inelastic scattering of electrons we may readily understand the inelastic scattering of ions. Therefore, in this chapter we will first study the inelastic scattering of electrons in detail, and then, based on the inelastic scattering factor introduced, we will discuss the inelastic scattering of ion channelling. This will be important for the study of dechannelling in the next chapter. At the end of this chapter we shall discuss the process of energy emission by high-energy channelled electrons and positrons.

4.2. Anomalous absorption

First we shall study the absorption of electron beams due to inelastic scattering in the crystal. As shown in the previous chapter, when electron waves in the crystal penetrate as far as the extinction distance, standing waves appear, which correspond to the branches of the dispersion surfaces. In general, the mean free paths of inelastic scattering are much larger than the extinction distance, from which we could presume that the electron waves are already standing waves with dispersion branches before inelastic scattering occurs.

We calculate the transition probability per unit time $P^{(j)}$ of all inelastic scattering of an electron beam which is in the jth branch by using the Born approximation

$$P^{(j)} = \frac{2\pi}{\hbar} \sum_{f} |\langle \psi_{\rm f} | H' | \psi_{\rm i} \rangle|^2 \delta(E_{\rm f} - E_{\rm i}) \tag{4.2.1}$$

where $\psi_{\rm i}$ and $\psi_{\rm f}$ are the initial state with energy $E_{\rm i}$, and the final state with energy $E_{\rm f}$, respectively.

Now we study the decrease in the electron beam intensity with thickness, where the intensity of the incident electron beam is assumed to be I_0. The electron beam intensity at a depth z from the surface will be represented by

$$I = I_0 \exp(-\mu^{(j)} z)$$

in other words, the electron beam intensity on the jth branch decreases by the absorption coefficient $\mu^{(j)}$.

When we consider inelastic scattering within a small depth dz in which the electron velocity v is not changed during the process, the inelastic scattering probability per unit time z/v, is $P^{(j)}z/v$, and the intensity of the inelastically scattered electron beam is represented by $I_0 P^{(j)} z/v$. This intensity is the difference between the initial beam intensity and the decreased intensity at a depth z in the crystal. Thus

$$I_0 - I_0 \exp(-\mu^{(j)} z) = I_0 P^{(j)} z/v \tag{4.2.2}$$

Assumming $\mu^{(j)} z \ll 1$, which means that we are studying thin samples or weak inelastic scattering, we may expand the exponential on the left side of equation 4.2.2 obtaining[1]

$$\mu^{(j)} = P^{(j)}/v \tag{4.2.3}$$

Equation 4.2.3 shows that the absorption coefficient $\mu^{(j)}$ is calculated directly by equation 4.2.1.

Note that if the electron beam is incident on the surface at an angle θ, the real pass length of the electron beam is $z/\cos\theta$. So, the absorption coefficient along the crystal's normal direction is represented by $\mu^{(j)}/\cos\theta$.

Let us use the standing waves derived from the two-wave approximation in section 3.4 of the previous chapter as the initial state of the electron wave function

$$\varphi_i^{(j)} = C_0^{(j)} \exp(i\mathbf{K}^{(j)} \cdot \mathbf{r}) + C_G^{(j)} \exp(i\mathbf{K}_G^{(j)} \cdot \mathbf{r}) \tag{4.2.4}$$

The final state is taken as $\exp(i\mathbf{K}' \cdot \mathbf{r})$, corresponding to the simple plane wave describing the inelastically scattered wave with wave number $\mathbf{K}'$. This assumption will be justified by actual calculations given in the following sections. However, it is worth noting that even if the final state is given by Bloch waves of the type given by equation 4.2.4, the transition probability, calculated by summing over all directions of the final states, is not significantly different from our present calculation which uses $(\exp(i\mathbf{K}' \cdot \mathbf{r})$ as the final state function.

Finally, the initial and the final states in our system are given by

$$|\Psi_i\rangle = \phi_i^{(j)}|\phi_0\rangle \tag{4.2.5}$$

and

$$|\Psi_f\rangle = \exp(i\mathbf{K}'\cdot\mathbf{r})|\phi_n\rangle \tag{4.2.6}$$

where $|\varphi_0\rangle$ and $|\varphi_n\rangle$ refer to the ground state with energy E_0 and the nth excited state with energy E_n of the crystal, respectively. When equations 4.2.5 and 4.2.6 are inserted into equation 4.2.3 we obtain

$$\begin{aligned}\mu^{(j)} = \frac{2\pi}{\hbar v}\sum_{n\neq 0}\int\frac{\mathrm{d}\mathbf{K}'}{(2\pi)^3}\{&|H'_{n0}(\mathbf{K}^{(j)}-\mathbf{K}')|^2 C_0^{(j)2} + |H'_{n0}(\mathbf{K}_G^{(j)}-\mathbf{K}')|^2 C_G^{(j)2} \\ &+ C_0^{(j)}C_G^{(j)}[H'_{n0}(\mathbf{K}^{(j)}-\mathbf{K}')H'_{0n}(\mathbf{K}-\mathbf{K}_G^{(j)}) \\ &+ H'_{0n}(\mathbf{K}'-\mathbf{K}^{(j)})H'_{n0}(\mathbf{K}_\mathbf{G}^{(j)}-\mathbf{K}')]\}\,\delta\left(\frac{\hbar^2\mathbf{K}'^2}{2m}+E_n-\frac{\hbar^2K^2}{2m}-E_0\right)\end{aligned} \tag{4.2.7}$$

where

$$H'_{n0}(\mathbf{Q}) = \int\langle\psi_n|H'|\psi_0\rangle\exp(i\mathbf{Q}\cdot\mathbf{r})\,\mathrm{d}\mathbf{r} \tag{4.2.8}$$

The first and second terms in the integrand of equation 4.2.7 correspond to the normal absorption coefficients of the 0th and **G**th wave in the jth branch. For isotropic crystals the absorption coefficient does not depend on the wave direction, which means that

$$\sum_{n\neq 0}\int\frac{\mathrm{d}\mathbf{K}'}{(2\pi)^3}|H'_{n0}(\mathbf{K}^{(j)}-\mathbf{K}')|^2\delta(\ldots\ldots) = \sum_{n\neq 0}\int\frac{\mathrm{d}\mathbf{K}}{(2\pi)^3}|H'_{n0}(\mathbf{K}_G^{(j)}-\mathbf{K}')|^2\delta(\ldots\ldots)$$

On the other hand, the coefficients $C_0^{(j)}$ and $C_G^{(j)}$ are normalized so that $C_0^{(j)2}+C_G^{(j)2}=1$ (see section 3.4), and the first and the second terms in equation 4.2.7 represent the normal absorption coefficient in its usual meaning

$$\mu_0 = \frac{2\pi}{\hbar v}\sum_{n=0}\frac{\mathrm{d}\mathbf{K}'}{(2\pi)^3}|H'_{n0}(\mathbf{K}^{(j)}-\mathbf{K}')|^2\delta\left(\frac{\hbar^2K'^2}{2m}+E_n-\frac{\hbar^2K^2}{2m}-E_0\right) \tag{4.2.9}$$

The third term in the integrand of equation 4.2.7 represents the interference effect between the 0th wave and the **G**th wave in the jth branch, which corresponds to the inelastic scattering in the Umklapp process. According to equation 3.7.19 we obtain

$$C_0^{(j)}C_G^{(j)} = \begin{cases} -\frac{1}{2}\sin\beta & (\text{for } j=1) \\ +\frac{1}{2}\sin\beta & (\text{for } j=2)\end{cases}$$

which is written as

$$C_0^{(j)} C_G^{(j)} = \pm \frac{1}{2\sqrt{1+W^2}} \qquad (\text{for } j=1,2) \tag{4.2.10}$$

since $W = \cot \beta$. If we insert equation 4.2.10 into equation 4.2.7 we obtain

$$\mu^{(j)} = \mu_0 \pm \frac{1}{\sqrt{1+W^2}} \Delta\mu_G \tag{4.2.11}$$

where

$$\Delta\mu_G = \frac{\pi}{\hbar v} \sum_{n \neq 0} \int \frac{d\mathbf{K}'}{(2\pi)^3} [H'_{n0}(\mathbf{K}^{(j)} - \mathbf{K}')H'_{0n}(\mathbf{K}' - \mathbf{K}_G^{(j)})$$
$$+ H'_{0n}(\mathbf{K}' - \mathbf{K}^{(j)})H'_{n0}(\mathbf{K}_G^{(j)} - \mathbf{K}')]\delta\left(\frac{\hbar^2 K'^2}{2m} - \frac{\hbar^2 K^2}{2m} + \hbar\omega_{n0}\right) \tag{4.2.12}$$

In the above we define $\hbar\omega_{n0} = E_n - E_0$.

As is shown in equation 4.2.11, in the case of the Laue–Bragg condition $W = 0, \mu^{(1)} = \mu_0 - \Delta\mu_G \sim 0$, if $\Delta\mu_G \sim \mu_0$. Here the electron beams penetrate deeply due to the small absorption coefficient. This corresponds to the case which we studied in figure 3.9, where two waves interfere with each other and the standing waves are formed with the node being at the net planes or the internet planes. $\mu^{(1)} \sim 0$ refers to the case in which the electron current density becomes small at the net planes corresponding to the small transition probability of inelastic scattering, resulting in transmission well into the crystal. We call this 'anomalous transmission', which was first found by Borrmann [2,3,4] in X-ray diffraction, where it is called the 'Borrmann effect'. The expression of equation 4.2.12, called the 'anomalous absorption coefficient', was derived by Yoshioka[5] in a different manner.

We can easily calculate $I_0(z)$ and $I_G(z)$, the intensities of the elastic 0th and **G**th waves, respectively, by taking into account the anomalous absorption. This is done by replacing $K^{(j)}$ by $K^{(j)} + i\mu^{(j)}/2$ in equations 3.7.25 and 3.7.26.†

Then, using equations 3.7.23 to 3.7.28 and equation 4.2.11, we obtain[6]

$$I_0 = \frac{\exp(-\mu_0 z)}{2(1+W^2)} \left[\frac{(W - \sqrt{1+W^2})^2}{2} \exp\left(\frac{\Delta\mu_G z}{\sqrt{1+W^2}}\right) \right.$$
$$+ \frac{(W + \sqrt{1+W^2})^2}{2} \exp\left(-\frac{\Delta\mu_G z}{\sqrt{1+W^2}}\right)$$
$$\left. + \cos\left(\frac{2\pi\sqrt{1+W^2}}{\xi_G} z\right) \right] \tag{4.2.13}$$

---

†We choose $i\mu^{(j)}/2$ in order that the intensity decreases as $|\exp(iK^{(j)}z)|^2 \to |\exp[i(K^{(j)} + i\mu^{(j)}/2)z]|^2 = \exp(-\mu^{(j)}z)$.

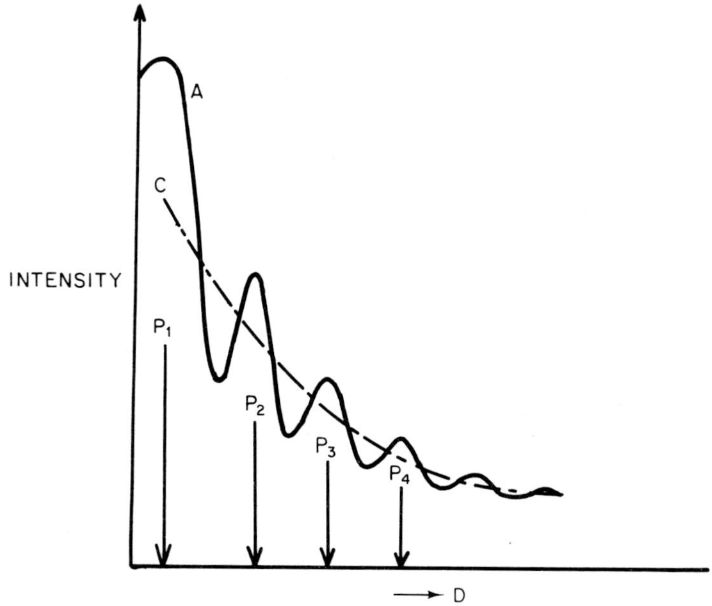

Figure 4.1. Electron beam intensity as a function of the thickness D for a wedge-shaped crystal. The amplitude function (curve A) oscillates about the centre curve C.

and

$$I_G = \frac{\exp(-\mu_0 z)}{2(1+W^2)}\left[\cosh\left(\frac{\Delta\mu_G z}{\sqrt{1+W^2}}\right) - \cos\left(\frac{2\pi\sqrt{1+W^2}}{\xi_G} z\right)\right] \quad (4.2.14)$$

from which we readily note that the relation $I_0 + I_G = 1$, given in Chapter 3, is not the case here.

Thus, the intensities of the 0th and **G**th waves are complex functions of the deviation parameter W from the Laue–Bragg condition and the crystal thickness D.

Figure 4.1 is a plot of intensity as a function of D for a wedge-shaped crystal. The curve A, which corresponds to the oscillates around the centre curve C. So the electron diffraction fringe pattern appears corresponding to peaks P_1, P_2, P_3, and so on, which are called the 'equal thickness fringes'. According to equations 4.2.13 and 4.2.14 the decrease in intensity of the equal thickness fringes for increases in crystal thickness D gives the absorption coefficient.

4.3. Calculation of absorption coefficients

It is not so easy to calculate the normal absorption coefficient μ_0 and the anomalous absorption coefficient $\Delta\mu_G$ for each elementary excitation in the crystal. When scattering centres for inelastic scattering are located randomly (or homogeneously)

$$H'_{n0}(\mathbf{Q}) \propto \int H'(\mathbf{r})_{n0} \exp(i\mathbf{Q}\cdot\mathbf{r})\, d\mathbf{r}$$

and

$$H'_{n0}(\mathbf{Q})H'_{n0}(\mathbf{Q}+2\pi\mathbf{G}) \propto \delta(\mathbf{Q})\delta(\mathbf{Q}+2\pi\mathbf{G})$$

from which we get

$$\Delta\mu_G \sim 0, \text{for } G \neq 0$$

As will be shown in detail later, this result is due to plasmon excitation of the valence electrons (or electron gas). On the other hand, when scattering centres for inelastic scattering are located at crystal lattice sites, say

$$H'(\mathbf{r})_{n0} \propto \sum_l \delta(\mathbf{r}-\mathbf{r}_l)$$

we obtain

$$H'_{n0}(\mathbf{Q}+2\pi\mathbf{G}) \propto \left\{\sum_l \exp[i(\mathbf{Q}+2\pi\mathbf{G})\cdot\mathbf{r}_l]\right\}_{0n} = \left[\sum_l \exp(i\mathbf{Q}\cdot\mathbf{r}_l)\right]_{0n}$$

and

$$H'_{0n}(\mathbf{Q}+2\pi\mathbf{G}) = H'_{n0}(\mathbf{Q})$$

which means that $\Delta\mu_G = \mu_0$. Accordingly, the ratio $\varepsilon_G = \Delta\mu_G/\mu_0$ represents the localization of the scatterer at the lattice points. In what follows an elementary method for the calculation of μ_0 and $\Delta\mu_G$ is described.

4.3.1. *Valence electron excitations*

The interaction Hamiltonian between electron beams and valence electrons in the crystal is the Coulomb interaction of the type

$$H' = \sum_i \frac{e^2}{|\mathbf{r}-\mathbf{R}_i|} \tag{4.3.1}$$

where $\mathbf{R}_i$ corresponds to the co-ordinates of the valence electrons. Inserting equation 4.3.1 into equation 4.2.8 we obtain

$$H'_{n0}(\mathbf{Q}) = \frac{4\pi e^2}{Q^2}(\rho_Q)_{n0} \tag{4.3.2}$$

where

$$(\rho_Q)_{n0} = \left\langle \varphi_n \left| \sum_i \exp(i\mathbf{Q}\cdot\mathbf{R}_i) \right| \varphi_0 \right\rangle \tag{4.3.3}$$

In the above $\rho_Q = \sum_i \exp(i\mathbf{Q}\cdot\mathbf{R}_i) = \int \sum_i \delta(\mathbf{r}-\mathbf{R}_i)\exp(i\mathbf{Q}\cdot\mathbf{r})\,d\mathbf{r}$ which is the Fourier component of the electron density $\sum_i \delta(\mathbf{r}-\mathbf{R}_i)$, and is called the 'density operator'.

By inserting equation 4.3.2 into equation 4.2.9 we obtain[7]

$$\mu_0 = \frac{2}{\hbar v}\int \frac{d\mathbf{K}'}{(2\pi)^3}\frac{4\pi e^2}{(\mathbf{K}-\mathbf{K}')^2} Im \frac{-1}{\varepsilon(\mathbf{K}'-\mathbf{K};\omega)} \tag{4.3.4}$$

where $\varepsilon(\mathbf{K};\omega)$ is the complex dielectric constant for valence electrons. In the above we ignored the branch suffix j in $K^{(j)}$ since the integral is taken over a large region of $\mathbf{K}'$. The value $\hbar\omega$ represents the energy change, and

$$\omega = \frac{\hbar K^2}{2m} - \frac{\hbar K'^2}{2m} \approx \frac{\hbar K}{m}(K-K') = v(K-K') \tag{4.3.5}$$

The anomalous absorption coefficient $\Delta\mu_G$ is also calculated by using the generalized complex dielectric constant $\varepsilon(\mathbf{Q},\mathbf{Q}_G;\omega)$

$$\Delta\mu_G = \frac{2}{\hbar v}\int \frac{d\mathbf{K}'}{(2\pi)^3}\frac{2\pi e^2}{(\mathbf{K}_G-\mathbf{K}')^2} Im\left[\frac{-1}{\varepsilon(\mathbf{K}'_G-\mathbf{K}',\mathbf{K}-\mathbf{K}';\omega)} + \frac{-1}{\varepsilon(\mathbf{K}-\mathbf{K}',\mathbf{K}_G-\mathbf{K}';\omega)}\right] \tag{4.3.6}$$

where

$$Im\frac{-1}{\varepsilon(\mathbf{Q},\mathbf{Q}_G;\omega)} = \frac{4\pi^2 e^2}{\hbar Q^2}\sum_{n\neq 0}(\rho_Q^+)_{0n}(\rho_{Q_G})_{n0}\,\delta(\omega-\omega_{n0}) \tag{4.3.7}$$

$$Im\frac{-1}{\varepsilon(\mathbf{Q}_G,\mathbf{Q};\omega)} = \frac{4\pi^2 e^2}{\hbar Q^2}\sum_{n\neq 0}(\rho_{Q_G}^+)_{0n}(\rho_Q)_{n0}\,\delta(\omega-\omega_{n0}) \tag{4.3.7$'$}$$

is the imaginary part of the generalized dielectric constant. For $\mathbf{G}=0$, $\varepsilon(\mathbf{Q},\mathbf{Q}_G;\omega) = \varepsilon(\mathbf{Q},\mathbf{Q};\omega) = \varepsilon(\mathbf{Q};\omega)$. If we divide $\varepsilon(\mathbf{Q};\omega)$ into the real and the imaginary parts $\alpha_1(\mathbf{Q},\omega)$ and $\alpha_2(\mathbf{Q};\omega)$ (see section 4.8), the dielectric constants are shown[8] in figure 4.2 as a function of ω for $Q \simeq 0$. As is well known,[7] plasmon excitation occurs when $\alpha_1(\mathbf{Q};\omega)=0$ and $\alpha_2(\mathbf{Q},\omega) \ll 1$. In figure 4.2 the point $\omega = \omega_p$ corresponds to the plasmon frequency. In the region of $\omega \simeq \omega_p$, α_1 is represented by $1-\omega_p^2/\omega^2$. We therefore obtain

$$Im\frac{-1}{\varepsilon(\mathbf{Q};\omega)} = \frac{\pi\delta(\omega-\omega_p)}{\left.\dfrac{\partial\alpha_1}{\partial\omega}\right|_{\omega=\omega_p}} = \frac{\pi}{2}\omega[\delta(\omega-\omega_p)+\delta(\omega+\omega_p)] \tag{4.3.8}$$

In these discussions we have neglected the $\mathbf{Q}$ dependence of the dielectric constant since it is very weak near the plasmon frequency.

Taking into account the upper limit[7] of the wave number k_c we may perform the integration of equation 4.3.4 after inserting equation 4.3.8. This results in

$$\mu_0 = \frac{\hbar\omega_p}{2a_B E_0} \log\left(\frac{2E_0}{\hbar\omega_p}\theta_c\right) \tag{4.3.9}$$

where a_B is the first Bohr radius, $E_0 = \hbar^2 K^2/2m$ is the energy of the incident electron beam; and $\theta_c = k_c/K$ is the upper limit of the scattering angle. Equation 4.3.9 is correct for $vk_c \gg \omega_p$. Equation 4.3.9 is the inverse of the mean free path of a single plasmon emission derived first by Ferrell.[9]

We shall now consider the contribution of the valence electron excitations to the complex generalized dielectric constant. Valence electron excitation contributes very little to the anomalous absorption coefficients, as explained in the first part of this section. For this reason there are no calculations which separate the valence electron's contribution from the total electron excitations in determining the anomalous absorption coefficient.

4.3.2. *Core electron excitations*

Gjønness[10] has shown that the matrix element $(\rho_Q^+)_{0n}(\rho_{Q'})_{0n}$ for the core electron excitations is represented by the Compton scattering factor for X-rays by making use of the Hartree–Fock approximation. Since the incident energy of the electron beam is larger than the excitation energy of the core electrons, we may neglect ω_{n0} as compared to ω in equation 4.3.7. Making use of the tight binding model the excited states are represented by the independent excited atomic states given by $|n\rangle$ once again. We then obtain

$$\sum_{n\neq 0} (\rho_Q^+)_{0n}(\rho_{Q_G})_{n0} = \sum_l \sum_{n\neq 0} \langle 0|\rho_Q^+(l)|n\rangle\langle n|\rho_{Q_G}(l)|0\rangle \tag{4.3.10}$$

where l stands for the atomic position, and

$$\rho_Q^+(l) = \sum_i \exp[-i\mathbf{Q}\cdot(\mathbf{r}_i + \mathbf{R}_l)]$$

$\mathbf{r}_i$ being the co-ordinate of the ith electron in the lth atom.

Defining $\bar{\rho}_Q$ by $\sum_i \exp(i\mathbf{Q}\cdot\mathbf{r}_i)$, equation 4.3.10 is rewritten as

$$\sum_l \sum_{n\neq 0} \langle 0|\bar{\rho}_Q^+|n\rangle\langle n|\bar{\rho}_Q|0\rangle \exp(2\pi i\mathbf{G}\cdot\mathbf{R}_l) \tag{4.3.11}$$

Using $\exp(2\pi i\mathbf{G}\cdot\mathbf{R}_l) = 1$ and letting N equal the total number of atoms in the crystal, equation 4.3.11 becomes

$$N\sum_{n\neq 0} \langle 0|\rho_Q^+|n\rangle\langle n|\bar{\rho}_{Q_G}|0\rangle$$
$$= N(\langle 0|\bar{\rho}_Q^+\bar{\rho}_{Q_G}|0\rangle - \langle 0|\bar{\rho}_Q^+|0\rangle\langle 0|\bar{\rho}_{Q_G}|0\rangle) \tag{4.3.12}$$

In the above, the ground state $|0\rangle$ of the crystal is represented by the single electron wave function $\phi_i(\mathbf{r}_i)$

$$|0\rangle = \frac{1}{\sqrt{N_A!}} \sum_P (-1)^P P[\phi_1(\mathbf{r}_1)\phi_2(\mathbf{r}_2)\ldots.\phi_{N_A}(\mathbf{r}_{N_A})] \tag{4.3.13}$$

where P is the permutation operator, and N_A is the total number of electrons in the atom.

Inserting equation 4.3.13 into equation 4.3.12 we obtain

$$\sum_{n\neq 0} (\rho_Q^+)_{0n}(\bar{\rho}_{Q_G})_{n0} = N\mathscr{S}(\mathbf{Q},\mathbf{Q}_G) \tag{4.3.14}$$

where

$$\mathscr{S}(\mathbf{Q},\mathbf{Q}_G) = \sum_i [f_{ii}(2\pi\mathbf{G}) - f_{ii}(\mathbf{Q})f_{ii}(\mathbf{Q}_G)] - \sum_{i\neq j} f_{ij}(\mathbf{Q})f_{ij}(\mathbf{Q}_G) \tag{4.3.15}$$

In the above, $f_{ij}(\mathbf{Q})$ is defined by

$$f_{ij}(\mathbf{Q}) = \int \phi_i^*(\mathbf{r}) \exp(i\mathbf{Q}\cdot\mathbf{r}\phi_j(\mathbf{r})\,\mathrm{d}\mathbf{r} \tag{4.3.16}$$

Inserting equations 4.3.14 and 4.3.15 into equation 4.3.6 the anomalous absorption coefficient $\Delta\mu_G$ is represented in terms of the $\mathscr{S}$ function

$$\Delta\mu_G = \frac{4N}{a_B^2 K^2} \int \frac{\mathscr{S}(\mathbf{K}-\mathbf{K}',\mathbf{K}_G-\mathbf{K}')}{(\mathbf{K}-\mathbf{K}')^2(\mathbf{K}_G-\mathbf{K}')^2}\,\mathrm{d}\Omega_{K'} \tag{4.3.17}$$

where the integration is performed over the sphere $|\mathbf{K}| = |\mathbf{K}'|$.

Thus we may calculate the anomalous absorption coefficient due to the core electron excitations if we have the data for the function $f_{ij}(\mathbf{Q})$, which is the Compton scattering factor for X-rays, since this function also appears in the expression for the Compton scattering intensity.[11]

The calculations of the Compton scattering factor have already been performed. Making use of the values for $f_{ij}(\mathbf{Q})$ calculated by Freeman,[12] Whelan[13] and Radi[14] calculated values for $\Delta\mu_G$ which will be summarized in the following section.

It is worth noting that the factor $\exp(i2\pi\mathbf{G}\cdot\mathbf{R}_l)$ in equation 4.3.11 does not become unity if there are different kinds of atoms in the unit cell. However, there are two terms in the integrand of equation 4.3.6, from which we obtain the factor

$$\tfrac{1}{2}[\exp(2\pi i\mathbf{G}\cdot\mathbf{R}_k) + \exp(-2\pi i\mathbf{G}\cdot\mathbf{R}_k)] = \cos(2\pi\mathbf{G}\cdot\mathbf{R}_k) \tag{4.3.18}$$

where the suffix k stands for the kth type of atom in the unit cell.

Furthermore, we may include the effect of the lattice vibrations in our calculation of the anomalous absorption coefficient. Introducing $\delta\mathbf{R}_l$ as the displacement from the equilibrium position of the lth atom we have a new factor, $\exp(2\pi i\mathbf{G}\cdot\delta\mathbf{R}_l)$, in the expression for $\Delta\mu_G$. Performing the thermal average $\langle\ldots\rangle$, we obtain[15]

$$\begin{aligned}\langle\Delta\mu_G\rangle &= \Delta\mu_G\cdot\langle\exp(2\pi i\mathbf{G}\cdot\delta\mathbf{R}_l)\rangle \\ &= \Delta\mu_G\cdot\exp[-M(\mathbf{G})]\end{aligned} \tag{4.3.19}$$

where $\exp[-M(\mathbf{G})]$ is the usual Debye–Waller factor.

4.3.3. *Phonon excitations* (*TDS*)

Inelastic scattering of electron beams due to phonon excitation is called thermal diffuse scattering (TDS) in diffraction theory. Though the energy loss due to phonon excitation is negligibly small ($\sim 0{\cdot}1\,\mathrm{eV}$) when compared to the electron beam energy, the scattering angle is larger than that due to single electron excitations. In addition, the transition probability of phonon excitations is not small compared to that of the single electron excitations. For this reason the phonon excitations make an important contribution to the calculation of the absorption coefficient.

We assume that the interaction Hamiltonian between electron beams and atoms in the crystal is of the form

$$V = \sum_l V(\mathbf{r} - \mathbf{R}_l)$$

where $\mathbf{r}$ and $\mathbf{R}_l$ are the co-ordinates of electron beams and atoms in the crystal, respectively. Here we introduce once again (as was done in the previous section) the deviation vector $\delta\mathbf{R}_l$ of the atom from its equilibrium lattice position due to lattice vibrations.

Accordingly, the difference in the interaction Hamiltonian between the excited and unexcited lattice vibrations is written as

$$H' = \sum_l [V(\mathbf{r} - \mathbf{R}_l - \delta\mathbf{R}_l) - V(\mathbf{r} - \mathbf{R}_l)] \tag{4.3.20}$$

When the crystal temperature is not too high we may expand the interaction Hamiltonian H' in powers of $\delta\mathbf{R}_l$; we should remember that $\delta\mathbf{R}_l$ is the linear combination of the creation and annihilation operators. The first order term in the expanded Hamiltonian H' now represents the interaction between the electron beam and the single phonon. Similarly, higher terms in H' correspond to the interaction between the electron beam and the double, triple, and so on, phonons. As a result, equation 4.3.20 represents the general interaction between the electron beam and the phonon.

In order to calculate the anomalous absorption coefficient due to phonon excitations we neglect ω_{n0} in equation 4.2.12 and insert equation 4.3.20 into equation 4.2.12 (or equation 4.2.8). By defining $\mathbf{Q} = \mathbf{K}' - \mathbf{K}^{(j)}$ we obtain

$$\begin{aligned}
&\sum_{n \neq 0} H'_{0n}(\mathbf{Q}) H'_{n0}(-\mathbf{Q}_{-G}) \\
&= V(\mathbf{Q}) V(-\mathbf{Q}_{-G}) \Bigg\{ \sum_{n \neq 0} \left\langle \varphi_0 \left| \sum_l \exp[i\mathbf{Q}\cdot(\mathbf{R}_l + \delta\mathbf{R}_l)] \right| \varphi_n \right\rangle \\
&\quad \times \left\langle \varphi_n \left| \sum_l \exp[-i\mathbf{Q}_{-G}\cdot(\mathbf{R}_l + \delta\mathbf{R}_l)] \right| \varphi_0 \right\rangle \Bigg\} \\
&= V(\mathbf{Q}) V(-\mathbf{Q}_{-G}) \Bigg\{ \sum_{n} \left\langle \varphi_0 \left| \sum_l \exp[i\mathbf{Q}\cdot(\mathbf{R}_l + \delta\mathbf{R}_l)] \right| \varphi_n \right\rangle \\
&\quad \times \left\langle \varphi_n \left| \sum_l \exp[-i\mathbf{Q}_{-G}\cdot(\mathbf{R}_l + \delta\mathbf{R}_l)] \right\} \right| \varphi_0 \right\rangle \\
&\quad - \left\langle \varphi_0 \left| \sum_l \exp[i\mathbf{Q}\cdot(\mathbf{R}_l + \delta\mathbf{R}_l)] \right| \varphi_0 \right\rangle \\
&\quad \times \left\langle \varphi_0 \left| \sum_l \exp[-i\mathbf{Q}_{-G}\cdot(\mathbf{R}_l + \delta\mathbf{R}_l)] \right| \varphi_0 \right\rangle \\
&\to V(\mathbf{Q}) V(-\mathbf{Q}_{-G}) \sum_{ll} \exp(i\mathbf{Q}\cdot\mathbf{R}_l - i\mathbf{Q}_{-G}\cdot\mathbf{R}_{l'}) \\
&\quad \times [\langle \exp(i\mathbf{Q}\cdot\delta\mathbf{R}_l) \exp(-i\mathbf{Q}_{-G}\cdot\delta\mathbf{R}_{l'})\rangle - \langle \exp(i\mathbf{Q}\cdot\delta\mathbf{R}_l)\rangle \\
&\quad \times \langle \exp(-i\mathbf{Q}_{-G}\cdot\delta\mathbf{R}_{l'})\rangle]
\end{aligned} \tag{4.3.21}$$

where $V(\mathbf{Q})$ is the Fourier component of the atomic potential, which is proportional to the atomic scattering factor introduced in Chapter 3. In order to derive equation 4.3.21 we also employed the relation

$$\sum_{n \neq 0} A_n = \sum_n A_n - A_0$$

and replaced the expectation value $\langle \varphi_0 | \ldots | \varphi_0 \rangle$ by the thermal average $\langle \ldots \rangle$. Furthermore, we used a formula which is known in statistical mechanics as the 'cumulant expansion'[16]

$$\langle e^x e^y \rangle = \exp(\tfrac{1}{2}\langle x^2 \rangle + \tfrac{1}{2}\langle y^2 \rangle + \langle xy \rangle + \ldots) \tag{4.3.22}$$

and the centre-of-symmetry assumption for the crystal, i.e.

$$H'_{n0}(-\mathbf{Q}) H'_{0n}(\mathbf{Q}_G) = H'_{0n}(\mathbf{Q}) H'_{n0}(-\mathbf{Q}_{-G})$$

Thus equation 4.2.12 becomes

$$\begin{aligned}
\Delta\mu_G = \frac{2\pi}{\hbar v} \int \frac{d\mathbf{K}'}{(2\pi)^3} \tilde{V}(\mathbf{K}' - \mathbf{K}^{(j)}) \tilde{V}(\mathbf{K}_G^{(j)} - \mathbf{K}') \\
\times \sum_{ll'} \exp[i(\mathbf{K}' - \mathbf{K}^{(j)})\cdot\mathbf{R}_l + i(\mathbf{K}_G^{(j)} - \mathbf{K}')\cdot\mathbf{R}_{l'}] \\
\times \exp[F_{ll'}(\mathbf{K}' - \mathbf{K}^{(j)}, \mathbf{K}_G^{(j)} - \mathbf{K}') - 1]\, \delta\left(\frac{\hbar^2 K'^2}{2m} - \frac{\hbar^2 K^{(j)2}}{2m}\right)
\end{aligned} \tag{4.3.23}$$

where $F_{ll'}(\mathbf{Q}, \mathbf{Q}')$ is the correlation function between the lattice deviations of the lth and l'th atoms defined by

$$F_{ll'}(\mathbf{Q}, \mathbf{Q}') = \langle i(\mathbf{Q} \cdot \delta\mathbf{R}_l)(i\mathbf{Q}' \cdot \delta\mathbf{R}_{l'}) \rangle \tag{4.3.24}$$

In equation 4.3.23 $\tilde{V}(\mathbf{Q})$ is $V(\mathbf{Q})$ multiplied by the usual Debye–Waller factor $\exp[-M(\mathbf{Q})]$.

It is not easy to calculate equation 4.3.23 in general. For homogeneous crystals the correlation function $F_{ll'}(\ldots)$ depends only on the difference $l - l'$. Further, if we assume the Einstein model for lattice vibrations, the summation over l and l' will be simplified. In this case, using the relation

$$F_{ll'}(\ldots) = F_{ll}(\ldots)\,\delta_{ll'} \quad (\equiv F(\ldots)\delta_{ii'}) \tag{4.3.25}$$

we may perform the double summation in equation 4.3.23 and obtain

$$\begin{aligned}
&\sum_{ll'} \exp\,[i(\mathbf{K}' - \mathbf{K}^{(j)}) \cdot \mathbf{R}_l + i(\mathbf{K}_G^{(j)} - \mathbf{K}') \cdot \mathbf{R}_{l'}] \\
&\qquad \times \{\exp\,[F_{ll}(\mathbf{K}' - \mathbf{K}^{(j)}, \mathbf{K}_G^{(j)} - \mathbf{K}')\delta_{ll'}] - 1\} \\
&\quad = \sum_l \exp(2\pi i \mathbf{G} \cdot \mathbf{R}_l)[\exp F(\mathbf{K}' - \mathbf{K}^{(j)}, \mathbf{K}_G^{(j)} - \mathbf{K}') - 1] \\
&\quad = N[\exp F(\mathbf{K}' - \mathbf{K}^{(j)}, \mathbf{K}_G^{(j)} - \mathbf{K}') - 1]
\end{aligned} \tag{4.3.26}$$

Inserting equation 4.3.26 into equation 4.3.23, and taking account of the relation

$$M_{(\mathbf{K}' - \mathbf{K}^{(j)})/2\pi} + M_{(\mathbf{K}_G^{(j)} - \mathbf{K}')/2\pi} - F(\mathbf{K}' - \mathbf{K}^{(j)}, \mathbf{K}_G^{(j)} - \mathbf{K}') = M(G)$$

we obtain the anomalous absorption coefficient $\Delta\mu_G$

$$\begin{aligned}
\Delta\mu_G = &\frac{2\pi N}{\hbar v} \int \frac{d\mathbf{K}'}{(2\pi)^3} V(\mathbf{K}' - \mathbf{K}^{(j)})\, V(\mathbf{K}_G^{(j)} - \mathbf{K}') \\
&\times \{\exp[-M(\mathbf{G})] - \exp(-M_{(\mathbf{K}' - \mathbf{K}^{(j)})/2\pi} - M_{(\mathbf{K}_G^{(j)} - \mathbf{K}')/2\pi})\} \\
&\times \delta\left(\frac{\hbar^2 K'^2}{2m} - \frac{\hbar^2 \mathbf{K}^{(j)^2}}{2m}\right)
\end{aligned} \tag{4.3.27}$$

Since the integrand of equation 4.3.27 includes only the scattering factor and the Debye–Waller factor, the numerical calculation is not so difficult; it was first done by Hall and Hirsch.[17]

It should be noted that the expansion of the electron interaction (equation 4.3.20) in powers of $\delta\mathbf{R}_l$, or single phonon, double phonon, and so on, expansion, corresponds to the expansion of $\exp[F_{ll'}(\ldots)]$ in powers of $F_{ll'}(\ldots)$ in equation 4.3.23. Accordingly, the anomalous absorption coefficient given by equation 4.3.27 includes the many phonon excitations, though the basic method was based on the Einstein model.

The calculations of the anomalous absorption coefficient have been performed by Hall[18] and Yoshioka and Kainuma[19] making use of the Debye model. We shall now consider single phonon excitation using the Debye model, with which we obtain the correlation function

$$\langle(\mathbf{Q}\cdot\delta\mathbf{R}_l)(\mathbf{Q}'\cdot\delta\mathbf{R}_{l'})\rangle$$

which is rewritten as

$$\frac{1}{\omega_{q\lambda}}\langle a_{q\lambda}^{+}a_{q\lambda}\rangle \approx \frac{1}{\beta\hbar\omega_{q\lambda}^2}$$

where $\mathbf{q}=\mathbf{K}^{(j)}-\mathbf{K}'$, referring to the method shown in equation 3.8.26. If we consider the acoustic phonon,† where $\omega_{q\lambda}\sim c_\lambda q$, ($c_\lambda$ is the acoustic velocity) the integral for the anomalous absorption coefficient becomes

$$\int(\ldots)\frac{q}{q^2}\mathrm{d}\Omega_q$$

where the integration is performed over the sphere $|\mathbf{K}'|=|\mathbf{K}|=|\mathbf{K}^{(j)}|$. In general, $\tilde{V}(q)$ does not diverge for $q\to 0$.

Since the above integral diverges at $q=0$, we have to take a minimum wave number q_{min} as the lower limit of the integration over q. In the calculation for core electron excitation the lower limit of the integration q_{min} was attributed to the threshold energy of the excitation energy derived from energy conservation and the upper limit q_{max} was $2K$. For plasmon excitation the q_{max} to be used is the plasmon cut-off k_{c}. In the present case of phonon excitation q_{max} should be taken as the Debye cut-off k_{D}. The problem is in the determination of q_{min}. It is nonsense to take q_{min} as the threshold due to energy conservation because it is zero at $q\sim 0$. This is also true for the absorption coefficient due to bremsstrahlung, shown in following section.

We may choose q_{min} as follows.

(*a*) Only those electron beams emerging from the apparatus with an aperture angle θ_{a} will be detected for the absorption measurement. Therefore in this case we can have

$$q_{\mathrm{min}}\sim K\theta_{\mathrm{a}}$$

(*b*) Even if the aperture is very small, the time $\Delta t=(D+L)/v$ in which the electron beam whose wave packet has length L stays in a crystal of thickness D is finite, since L and D are finite. So, the excitation frequency $2\pi\omega^{-1}=2\pi(c_\lambda q)^{-1}$ has to be satisfied by

$$\Delta t\geq 2\pi(c_\lambda q)^{-1}$$

or

$$(D+L)/v\geq 2\pi(c_\lambda q)^{-1}$$

---

†Since the frequency $\omega_{q\lambda}$ of the optical phonon is much larger than that of the acoustic phonon near $q\sim 0$, we may neglect the contribution of the optical phonon to the correlation function.

This leads to

$$q_{\min} = \frac{v}{c_\lambda}\frac{2\pi}{D+L}$$

Unfortunately, the length of the electron wave packet L is not known for any apparatus used so far.

(*c*) Yoshioka and Kainuma[19] chose

$$q_{\min} \sim 2\pi/(\text{crystal size})$$

taking into account the phonons with wavelengths smaller than the crystal size.

(*d*) Hall[18] chose

$$q_{\min} \sim 2\pi/\xi_G$$

ξ_G being the extinction distance, the inverse distance between the two dispersion surfaces in the dynamical theory. This means that the electron beam interacts only with those phonons that are within the extinction distance (see section 4.6).

At the moment we have no exact idea as to which is the correct value of $q_{\min}$. However, in ordinary calculations we take $q_{\min} \sim 2\pi/\xi_G$.

4.4. Complex potential (Yoshioka term)

Electron beams in the crystal could be scattered elastically by the potential $V_{\mathbf{G}}$ and also by the field of polarization of the crystal system. The polarization process is divided into two parts: one is the real process in which the definite energy is transferred and the other is the virtual process which occurs before the energy transfer. The real process corresponds to the imaginary scattering potential. The complex potential is called the 'optical potential' in the field of atomic collisions or nuclear physics for homogeneous distributions of scatterers (i.e. without lattice periodicity in which the reciprocal lattice vector $\mathbf{G}$ is zero). Yoshioka[5] first introduced the complex potential in crystals where the reciprocal lattice vector is not zero.

We may expand the total wave function ϕ by the excited states of the crystal $|\varphi_n\rangle$

$$\phi = \sum_n \psi_n(\mathbf{r})|\varphi_n\rangle \tag{4.4.1}$$

where $\mathbf{r}$ is the co-ordinate of the electron beam. Inserting equation 4.4.1 into the Schrödinger equation

$$\left(-\frac{\hbar^2}{2m}\nabla^2 + V(\mathbf{r}) + H'\right)\phi = E\phi \tag{4.4.2}$$

and pre-multiplying by $\langle \varphi_0|$ and $\langle \varphi_n|$ we obtain

$$\left[\nabla^2 + k_0^2 - \frac{2m}{\hbar^2}V(\mathbf{r})\right]\psi_0 = \frac{2m}{\hbar^2}\sum_{n\neq 0} H'_{0m}\psi_m \tag{4.4.3}$$

and

$$\left[\nabla^2 + k_n^2 - \frac{2m}{\hbar^2}V(\mathbf{r})\right]\psi_n = \frac{2m}{\hbar^2}\sum_{m\neq n} H'_{nm}\psi_m \tag{4.4.4}$$

where $k_n^2 = (2m/\hbar^2)(E - E_n)$. In the above we assumed $V(\mathbf{r}) = V(\mathbf{r})_{00} = V(\mathbf{r})_{nn}$.

Equations 4.4.3 and 4.4.4 are the Schrödinger equations for electrons scattered elastically and inelastically, respectively. The inhomogeneous terms on the right sides of equations 4.4.3 and 4.4.4 connect elastic scattering with inelastic scattering.

Here we assume that the inelastic beams are not diffracted by the potential $V(\mathbf{r})$, to be clarified later, and that $H'_{nm}\psi_m \ll H'_{n0}\psi_0$. The solution to equation 4.4.4 then becomes[20]

$$\psi_n = \frac{-m}{2m\hbar^2}\int H'_{n0}(\mathbf{r}')\psi_0(\mathbf{r}')\frac{\exp(ik_n|\mathbf{r}-\mathbf{r}'|)}{|\mathbf{r}-\mathbf{r}'|}\mathrm{d}^3\mathbf{r}' \tag{4.4.5}$$

Inserting equation 4.4.5 into equation 4.4.3, we obtain

$$\left[\nabla^2 + k_0^2 - \frac{2m}{\hbar^2}V(\mathbf{r})\right]\psi_0 = \frac{2m}{\hbar^2}\int A(\mathbf{r},\mathbf{r}')\psi_0(\mathbf{r}')\mathrm{d}^3\mathbf{r}' \tag{4.4.6}$$

where

$$A(\mathbf{r},\mathbf{r}') = \frac{-m}{2\pi\hbar^2}\sum_{n\neq 0} H'_{0n}(\mathbf{r})H'_{n0}(\mathbf{r}')\frac{\exp(ik_n|\mathbf{r}-\mathbf{r}'|)}{|\mathbf{r}-\mathbf{r}'|} \tag{4.4.7}$$

Equation 4.4.6 shows that the non-local potential[20,21] $A(\mathbf{r},\mathbf{r}')$ is added to the normal potential $V(\mathbf{r})$.

Even if $A(\mathbf{r},\mathbf{r}')$ is not known in detail, the solution ψ_0 can be expanded as in equation 3.7.3 if we assume $V(\mathbf{r}) \gg A(\mathbf{r},\mathbf{r}')$, i.e.

$$\psi_0(\mathbf{r}) = \sum_{\mathbf{G}}\psi_{\mathbf{G}}\exp(i\mathbf{K}_{\mathbf{G}}\cdot\mathbf{r})$$

Inserting this into equation 4.4.6, and pre-multiplying by $\exp(i\mathbf{K}_G\cdot\mathbf{r})$ we obtain

$$(K_{\mathbf{G}}^2 - k_0^2)\psi_{\mathbf{G}} + \sum_{\mathbf{H}}\frac{2m}{\hbar^2}(V_{\mathbf{G}-\mathbf{H}} + C_{\mathbf{G},\mathbf{H}})\psi_{\mathbf{H}} = 0 \tag{4.4.8}$$

where

$$C_{\mathbf{G},\mathbf{H}} = \int A(\mathbf{r},\mathbf{r}')\exp(i\mathbf{K}_{\mathbf{H}}\cdot\mathbf{r}' - i\mathbf{K}_{\mathbf{G}}\cdot\mathbf{r})\,\mathrm{d}^3\mathbf{r}\,\mathrm{d}^3\mathbf{r}'$$

$$= -\frac{m}{4\pi^3\hbar^2}\sum_{n\neq 0}\int\frac{H'_{0n}(\mathbf{K}'-\mathbf{K}_\mathbf{G})H'_{n0}(\mathbf{K}_\mathbf{H}-\mathbf{K}')}{K'^2-k_n^2+i\delta}\mathrm{d}^3\mathbf{K}' \tag{4.4.9}$$

In the above we used the Fourier transform of the form

$$\frac{1}{4\pi}\frac{\exp(ik_n|\mathbf{r}-\mathbf{r}'|)}{|\mathbf{r}-\mathbf{r}'|} = \frac{1}{(2\pi)^3}\lim_{\delta\to 0}\int\frac{\exp[i\mathbf{K}'.(\mathbf{r}-\mathbf{r}')]}{K'^2-k_n^2+i\delta}\mathrm{d}^3\mathbf{K}' \tag{4.4.10}$$

Making use of the identity relation

$$\frac{1}{x+i\delta} = \mathrm{P}\frac{1}{x} + i\pi\delta(x) \tag{4.4.11}$$

equation 4.4.9 is rewritten as

$$C_{\mathbf{G},\mathbf{H}} = C^r_{\mathbf{G},\mathbf{H}} + iC^i_{\mathbf{G},\mathbf{H}} \tag{4.4.12}$$

$$C^r_{\mathbf{G},\mathbf{H}} = -\frac{m}{4\pi^3\hbar^2}\sum_{n\neq 0}\int\mathrm{P}\frac{H'_{0n}(\mathbf{K}'-\mathbf{K}_\mathbf{G})H'_{n0}(\mathbf{K}_\mathbf{H}-\mathbf{K}')}{K'^2-k_n^2}\mathrm{d}^3\mathbf{K}' \tag{4.4.13}$$

$$C^i_{\mathbf{G},\mathbf{H}} = \frac{-m}{4\pi^2\hbar^2}\sum_{n\neq 0}\int H'_{0n}(\mathbf{K}'-\mathbf{K}_\mathbf{G})H'_{n0}(\mathbf{K}_\mathbf{H}-\mathbf{K}')\delta(K'^2-k_n^2)\mathrm{d}^3\mathbf{K}'$$

$$= -\frac{m}{8\pi^2\hbar^2}\sum_{n\neq 0}k_n\int[H'_{0n}(\mathbf{K}'-\mathbf{K}_\mathbf{G})H'_{n0}(\mathbf{K}_\mathbf{H}-\mathbf{K}')]_{K'=k_n}\mathrm{d}\Omega_{K'} \tag{4.4.14}$$

The angular integration in equation 4.4.14 is performed over the sphere $|\mathbf{K}'| = k_n$ in wave number space. The potentials $C^r_{\mathbf{G},\mathbf{H}}$ and $C^i_{\mathbf{G},\mathbf{H}}$ have real values in crystals with a centre of symmetry. The polarization of the real process corresponds to $C^i_{\mathbf{G},\mathbf{H}}$, which is calculated using energy conservation of the type $\delta(K'^2 - k_n^2)$. On the other hand, the principal value $\mathrm{P}(1/(K'^2 - k_n^2)$ in $C^r_{\mathbf{G},\mathbf{H}}$ corresponds to the polarization of the virtual process. We call $C^r_{\mathbf{G},\mathbf{H}}$ and $C^i_{\mathbf{G},\mathbf{H}}$ the 'Yoshioka terms'.

Now, due to the imaginary potential $C^i_{G,H}$ the wave number of the electron beam becomes a complex number. As a result the beams are absorbed in the crystal. Taking into account the complex potential $C^r_{G,H} + iC^i_{G,H}$ in the dispersion relation of the two-wave approximation (cf., equation 3.7.8) and assuming $V_{G-H} \gg C^i_{G,H}$, the imaginary part of the wave vector in the *j*th branch is

$$Im K^{(j)} = -\frac{m}{\hbar^2 k_0}\left(C^i_{0,0} - \frac{m}{\hbar^2 k_0}\frac{V_\mathbf{G}C^i_{\mathbf{G},0} + V_{-\mathbf{G}}C^i_{0,\mathbf{G}}}{2\eta^{(j)}\cos\theta_B}\right) \tag{4.4.15}$$

From equation 3.7.10 we have

$$\eta^2 = \xi^2\tan^2\theta_B + \frac{U_G U_{-G}}{4K_0^2\cos^2\theta_B} = \frac{\pi^2}{\xi_G^2}(W^2+1) \tag{4.4.16}$$

We therefore proceed to calculate the absorption coefficient $\mu^{(j)} = 2\,\mathrm{Im}\,K^{(j)}$

$$\mu^{(j)} = \mu_0 \pm \frac{1}{\sqrt{1+W^2}}\Delta\mu_G \tag{4.4.17}$$

where

$$\mu_0 = -\frac{2m}{\hbar^2 k_0} C^i_{0,0} \tag{4.4.18}$$

and

$$\Delta\mu_G = \frac{-m}{\hbar^2 k_0}(C^i_{\mathbf{G},0} + C^i_{0,\mathbf{G}}) \tag{4.4.19}$$

These results coincide with equations 4.2.9, 4.2.11 and 4.2.12 derived already using the transition probability.

4.5. Comparison with experiments

4.5.1. *Measurements of complex potentials*

We shall now study a few typical experiments devised to measure the complex potentials and the anomalous absorption coefficients.

The method of Kohra and Watanabe

This method is based on the decrease in intensity of the equal-thickness fringe along a line perpendicular to the edge of the wedge-shaped crystal which we have studied in section 4.2. Kohra and Watanabe[22] first used this method to measure the anomalous absorption coefficient for the MgO crystal. Many authors[23–27] followed this method to obtain normal and anomalous absorption coefficients for different crystals. The experimental results obtained are smaller (by ~ 50 per cent) than the values expected from theory and the values obtained by other experimental methods. The reason for this is as follows. The inelastically scattered electron beams also contribute to the equal-thickness fringes, thus adding to the fringe intensity; this results in good transparency of the electron beams in the crystal.[28]

The method of Lehmpfuhl and Molière

As is shown in figure 3.11 (p.43), two waves excited in the wedge-shaped crystal, corresponding to two branches, are connected to the two spots outside the crystal. Since the separation of the two spots is proportional to the separation of the dispersion surface, we may measure the extinction distance or $C^r_{\mathbf{G},0}$ from the separation of the two spots. This method was developed by Lehmpfuhl and Molière,[29,30] and is not disturbed by the inelastically scattered electrons.

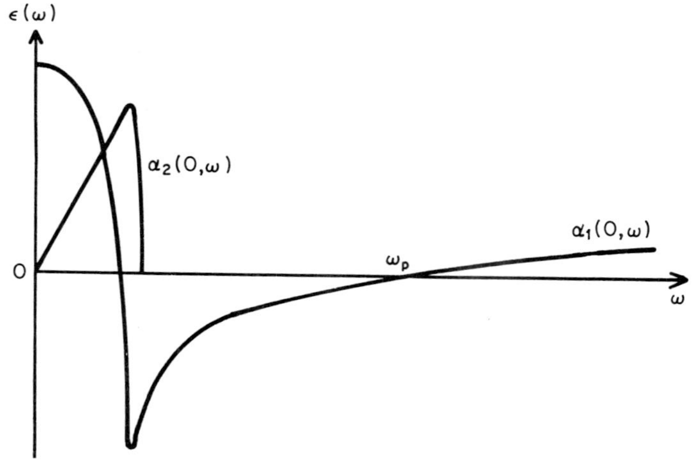

Figure 4.2. The real and imaginary parts of the dielectric constants, $\alpha_1(0, \omega)$ and $\alpha_2(0, \omega)$. The plasmon excitation appears when $\alpha_1 = 0$ and $\alpha_2 \ll 1$.

Other methods

Other techniques have been proposed by Goodman and Lehmpfuhl,[31] Hashimoto *et al.*,[32,33] and Mayer.[34,35] Many authors[36-40] have shown that the polycrystal method is also useful for the measurement of $C^r_{\mathbf{G},0}$.

4.5.2. *Measurements of absorption coefficients*

Kohra and Watanabe,[22] Mayer,[34,35] Hashimoto,[33] and many other authors have tried to determine the values of the normal and anomalous coefficients for various crystals. Table 4.1 shows a comparison of the calculated values of Radi[14] and Whelan[13] with their experimental results.

4.6. Diffuse patterns

In the preceding section we studied the secondary effects of inelastic scattering on the elastically scattered waves. In this and in the next section we shall discuss inelastically scattered waves directly.

Inelastically scattered waves are modified in general by the periodic crystal fields (diffraction). However, the intensity pattern of waves diffracted after inelastic scattering differs greatly from the intensity pattern of waves scattered inelastically after diffraction, where the original diffraction spot is modified by the small-angle inelastic scattering. This modification of the diffraction spot results from a blueing of the fine diffraction spot on the film. We call the modified spot a 'diffuse pattern'.

Table 4.1. A comparison between the calculated values and experimental results of Radi[14] and Whelan (W).[13]

Crystal	G	V_G	Core ($\mu_0, \Delta\mu_h$)	Plasmon	TDS	Total	Exp
Al	000	14·91	0.221	0·38	0·086	0·69	0·50
			0.58(W)		0·20(H.H)		
	111	5·93	0·034	0	0·071	0·11	0·23
			0·032(W)		0·153		
	200	4·98	0·027	0	0·067	0·09	
			0·027(W)		0·297		
	220	3·17	0·016(W)	0	0·054	0·07	
Cu	000	20·05	~2·15	0·2	0·414	~2·76	1·35
	111	11·26	0·121	0	0·356	0·48	0·81
	200	10·15	0·092	0	0·340	0·43	0·92
	220	7·54	0·042	0	0·285	0·33	
Si	000	12·21	0·133	0·40	0·058	0·60	0·72
	111	5·25	0·037	0	0·038	0·07	
	220	4·36	0·019	0	0·046	0·07	0·14
	311	2·49	0·010	0	0·031	0·04	0·08
Ge	000	14·00	0·59	0·38	0·256	1·23	1·25
	111	6·77	0·086	0	0·170	0·26	
	220	6·50	0·055	0	0·218	0·27	0·52
	311	3·94	0·029	0	0·145	0·17	
MgO	000	13·10	1·70				1·5
	111	1·26	0·001				0·13
	200	6·79	0·057				0·55
	220	4·99	0·023				
Li	000		0·21(W)				
	110		0·002(W)				
	200		0·001(W)				

The intensity pattern of waves diffracted after inelastic scattering, occurring at the net planes, is known as a 'Kikuchi pattern', which was first observed by Kikuchi in 1928.

The diffuse pattern intensity will be obtained by the calculation of the probability of inelastic scattering of the waves onto the diffraction dispersion branches. Making use of the transition probability $P^{(j)}$ of waves on the jth branch, which was discussed in the previous section, we obtain the intensity of the diffuse pattern. In the two-wave approximation the electron waves on the branch for $j=1$ or 2 are scattered inelastically into waves with wave number $\mathbf{K}'$, which once more form standing waves within the extinction distance with the other branches as shown in figure 4.3.

Therefore, we picture this process as inelastic scattering from branches of elastic waves to branches of inelastic waves. As is shown in figure 4.3, there are two steps in the transition from branch 1: $1 \rightarrow 1$ and $1 \rightarrow 2$. These are called

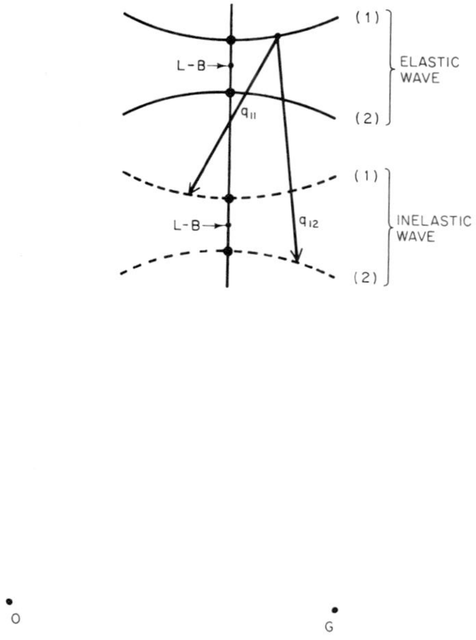

Figure 4.3. Intraband and interband transitions between two branches of the elastic wave and two branches of the inelastic wave.

intraband transition with momentum transfer of $\hbar\mathbf{q}_{11}$ and interband transition with momentum transfer $\hbar\mathbf{q}_{12}$, respectively.

We shall now calculate in more detail the inter- and intraband transition probability. For simplicity we assume that the exact Bragg condition is satisfied for the elastic wave (i.e., $W = 0$). Denoting the transition probability from the branch i to the branch j by $P^{(ij)}$, we obtain the total transition probability P

$$P = \sum_{i,j} P^{(ij)} \tag{4.6.1}$$

The calculation of $P^{(ij)}$ was performed in section 4.3 except for the consideration of the boundary condition at the exit surface of the crystal

$$P^{(ij)} = \frac{2\pi}{\hbar} \sum_n |M_{n0}^{(ij)}|^2 \delta\left(\frac{\hbar^2 K^{(i)2}}{2m} - \frac{\hbar^2 K'^{(j)2}}{2m}\right) \tag{4.6.2}$$

where $M_{n0}^{(ij)}$ is the matrix element for the transition $(i \rightarrow j)$

$$M_{n0}^{(ij)} \propto \sum_{\mathbf{G},\mathbf{H}} C_{\mathbf{G}}^{\prime(j)*} C_{\mathbf{H}}^{(i)} H'_{n0}(\mathbf{K}'^{(j)} - \mathbf{K}^{(i)} + 2\pi\mathbf{G} - 2\pi\mathbf{H}) \tag{4.6.3}$$

When the scattering centre is distributed throughout the crystal and bound weakly like plasmon excitations, we may neglect all the terms except $\mathbf{G} = \mathbf{H}$

in equation 4.6.3 since

$$H'_{n0}[\mathbf{q}+2\pi(\mathbf{G}-\mathbf{H})] \ll H'_{n0}(\mathbf{q})$$

for $\mathbf{G} \neq \mathbf{H}$. In this case we obtain

$$M_{n0}^{(ij)} \propto H'_{n0}(\mathbf{K}'^{(j)}-\mathbf{K}^{(i)}) \sum_{\mathbf{G}} C'^{(j)}_{\mathbf{G}} C^{(i)}_{\mathbf{G}}$$

Now we assume that the inelastic wave number included in $C'^{(j)}_{\mathbf{G}}$ is very close to the elastic wave number in $C^{(j)}_{\mathbf{G}}$, resulting in $C'^{(j)}_{\mathbf{G}} = C^{(j)}_{\mathbf{G}}$. According to equation 3.7.20, we obtain

$$\sum_{\mathbf{G}} C'^{(j)}_{\mathbf{G}} C^{(i)}_{\mathbf{G}} \simeq \sum_{\mathbf{G}} C^{(j)}_{\mathbf{G}} C^{(i)}_{\mathbf{G}} = \delta_{ji}$$

and equation 4.6.3 turns out to be

$$M_{n0}^{(ij)} \propto \delta_{ij} H'_{n0}(\mathbf{K}'^{(j)}-\mathbf{K}^{(i)})$$

where only the intraband transitions are effective.

Next we study the case where the scattering centre is tightly bound to the crystal atoms. Here we may assume that

$$H'_{n0}(\mathbf{q}+2\pi\mathbf{G}) \gg H'_{n0}(\mathbf{q})$$

for $\mathbf{q} \ll 2\pi\mathbf{G}$.

Taking into account relation 3.7.20, the transition matrix element for $i=j$ is given by

$$\begin{aligned} M_{n0}^{(ii)} &\propto \sum_{\mathbf{G},\mathbf{H}} C'^{(i)}_{\mathbf{G}} C^{(i)}_{\mathbf{H}} H'_{n0}(2\pi\mathbf{G}-2\pi\mathbf{H}) \\ &\simeq \sum_{\mathbf{G},\mathbf{H}} C^{(i)}_{\mathbf{G}} C^{(i)}_{\mathbf{H}} H'_{n0}(2\pi\mathbf{G}-2\pi\mathbf{H}) \\ &= \sum_{\mathbf{H}'}\left(\sum_{\mathbf{H}} C^{(i)}_{\mathbf{H}} C^{(i)}_{\mathbf{H}} C^{(i)}_{\mathbf{H}+\mathbf{H}'}\right) H'_{n0}(2\pi\mathbf{H}') \\ &= \sum_{\mathbf{H}'} \delta_{\mathbf{H}'0} H'_{n0}(2\pi\mathbf{H}') \\ &= H'_{n0}(0) \ll H'_{n0}(2\pi\mathbf{G}) \end{aligned}$$

which indicates that there are no intraband transitions for the scattering centres localized in the lattice atoms.[42] One example of these scattering centres is phonon excitation. The calculation of the intraband transition probability for one phonon excitation shows that this probability is 100 times less than the interband transition probability. Thus the minimum wave number of emitted phonons will be of the same order of magnitude as the interband distance, $\sim 2\pi/\xi_G$, which prevents the divergence of the integration of the phonon emission probability for $q \to 0$.[43]

Core electron excitations are another example of interband transitions.

However, valence electron excitation corresponds both to the inter- and intraband transition.

As is found by direct calculation the intensities of the diffuse pattern due to plasmon and phonon excitation are given by the following expressions[44,45] in terms of the angle θ, the deviation from the Laue–Bragg direction:
(*a*) plasmon excitation

$$P \propto \frac{\theta_E}{\theta^2 + \theta_E^2} \cos^2 \tfrac{1}{2}(\beta - \beta') \cos^2 \tfrac{1}{2}\beta \sin^2 \tfrac{1}{2}\beta' \tag{4.6.4}$$

where $\theta_E^2 = 2m\omega_p/\hbar K^2$;
(*b*) phonon excitation

$$P \propto |V(2\pi G)|^2 (2\pi G)^2 kT(\xi_G/c)^2 [1 - (\xi_G K\theta/2\pi)^2] \tag{4.6.5}$$

There are a few papers[46,47,48] which give detailed calculations of the diffuse pattern for the N-wave approximation.

4.7. Kikuchi patterns

Kikuchi patterns appear in the interaction of electrons with thick crystals. Examples of Kikuchi patterns are shown in the photograph for different crystal thicknesses (figure 4.4). As is shown in the photograph, the ordinary Laue–Bragg spots are obtained for thin crystals, but disappear for thick ones, to be replaced by black and white lines (Kikuchi lines) and bands (Kikuchi bands). Almost all Kikuchi lines split where they cross Kikuchi bands. It is also known that a white Kikuchi band turns into a black one as the crystal thickness increases.

Many authors[59,60,65] have tried to interpret the Kikuchi patterns since their discovery. The first explanation of the Kikuchi lines, which are relatively easier to explain than the bands, was presented by Kikuchi himself. The electron beams whose dispersive waves are diffracted at the net planes are scattered inelastically, as shown in figure 4.5. The inelastic waves which originate from point 0 (dotted lines) are diffracted at the net plane M and propagate in the direction **K**, where the electron intensity is larger than the background intensity. This results in the white line (on positive film). On the other hand, the intensity in the **K**′ direction is smaller than the background intensity, resulting in the black line, since electrons moving in the **K**′ direction in the background are transferred to the **K** direction by diffraction. In three dimensions, the pair of black and white lines appear perpendicular to the **K** and **K**′ directions. It should be noted that the Kikuchi line close to the incident spot I is a black line, as is shown in figures 4.4 and 4.5. The electrons scattered inelastically through large angles (with very small probability compared to small angle scattering) are also diffracted by the opposite side of the net plane (**M**′**K**′ in figure 4.5) and propagate in the **K**′ direction (also with a very small probability).

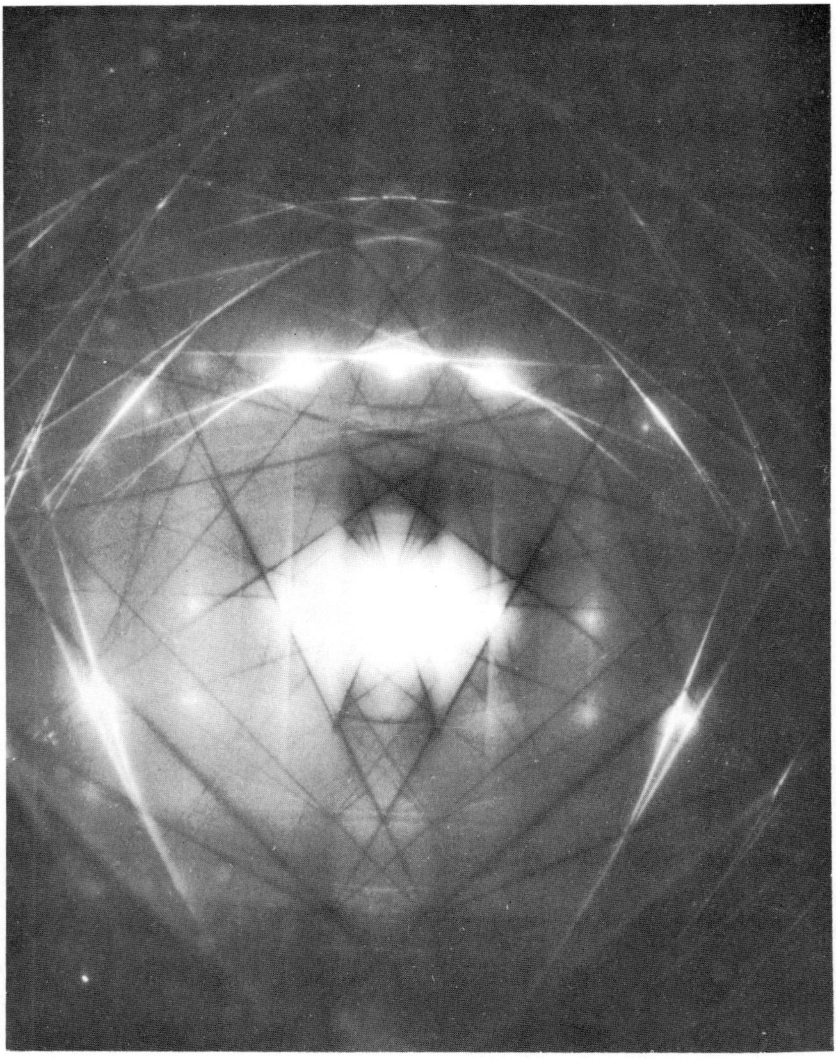

Figure 4.4. The Kikuchi pattern.

Changing the direction of the incident electron beam so as to be parallel to the net plane, the point I in figure 4.5 moves to the midpoint of **K** and **K**′ in figure 4.6, when the Kikuchi line disappear and the Kikuchi bands, which were not seen before, appear. The Kikuchi band is formed by the interference of diffracted inelastic waves (ONM) and non-diffracted (OM) inelastic waves.

The quantitative theory of Kikuchi patterns has been developed by Kainuma[58] and Takagi,[59] who studied it taking into account phonon excitations.

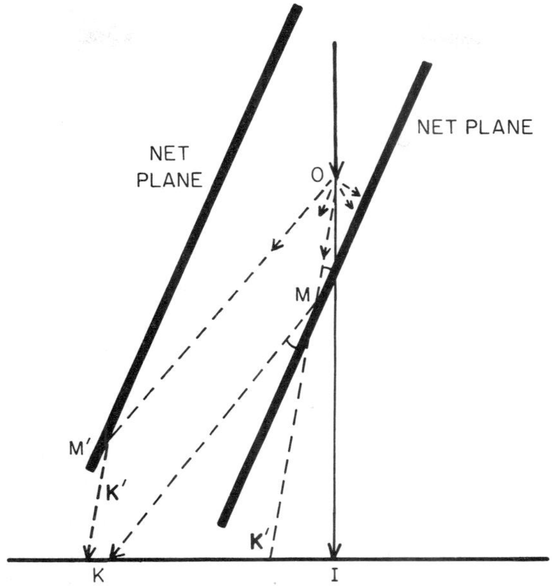

Figure 4.5. Formation of the Kikuchi lines. Inelastically scattered waves (dotted lines) start from the position 0, and are diffracted at the net plane M.

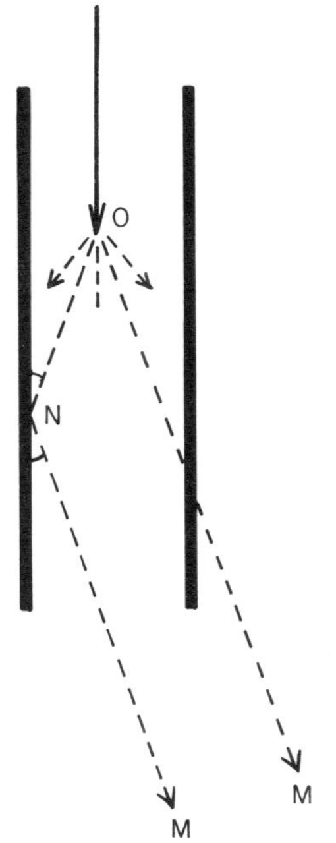

Figure 4.6. Formation of the Kikuchi band, visualized by changing the direction of the incident electron in the previous figure.

For the calculation of the transition probability of Kikuchi bands the diffraction effect is not included in the initial state $|\bar{\psi}_i\rangle$, unlike anomalous absorption, where

$$|\bar{\psi}_i\rangle = \exp(i\mathbf{K}_0\cdot\mathbf{r})|\varphi_0\rangle \tag{4.7.1}$$

The diffraction effect is, however, included in the final state $|\bar{\psi}_f\rangle$

$$|\bar{\psi}_f\rangle = \psi_{K'}|\varphi_n\rangle \tag{4.7.2}$$

where $\psi_{K'}$ is the Bloch function with the inelastic wave number $\mathbf{K}'$

$$\psi_{K'} = \sum_{jG} \chi_j c_G^{(j)} \exp[i(\mathbf{K}'^{(j)} + 2\pi\mathbf{G})\cdot\mathbf{r}] \tag{4.7.3}$$

In the above, χ_j and $c_G^{(j)}$ are the coefficients determined by the dispersion relation and the boundary condition for each of the branches, distinguished from the coefficients for elastic waves by the lower case letters.

For simplicity, let us consider the two-wave approximation ($\mathbf{G}=0$ and $\mathbf{G}=\mathbf{G}$ for $j=1$ or 2). The wave function should be connected to the wave function of 0 and $\mathbf{G}$ waves outside the exit surface. Taking the exit surface at $z=D$, as in Chapter 3, we obtain for the 0 wave

$$\chi_j c_0^{(j)} = \tfrac{1}{2}\left[1 + \frac{(-1)^j w}{\sqrt{1+w^2}}\right] \exp[i(\mathbf{K}' - \mathbf{K}^{(j)})_z D] \tag{4.7.4}$$

$$\chi_j c_{\mathbf{G}}^{(j)} = \frac{(-1)^{j-1}}{2\sqrt{1+w^2}} \exp[i(\mathbf{K}' - \mathbf{K}'^{(j)})_z D] \tag{4.7.5}$$

and for the $\mathbf{G}$ wave

$$\chi_j c_0^{(j)} = \frac{(-1)^{j-1}}{2\sqrt{1+w^2}} \exp[i(\mathbf{K}'_G - \mathbf{K}'^{(j)}_{\mathbf{G}})_z D] \tag{4.7.6}$$

$$\chi_j c_G^{(j)} = \tfrac{1}{2}\left[1 + \frac{(-1)^j w}{\sqrt{1+w^2}}\right] \exp[i(\mathbf{K}'_G - \mathbf{K}'^{(j)}_{\mathbf{G}})_z D] \tag{4.7.7}$$

where w is the deviation parameter W from the exact Laue–Bragg condition (the small letter is again used for the inelastic waves).

Inserting equations 4.7.1 to 4.7.7 into the formula 4.2.1 we obtain matrix elements like equation 4.2.8 for inelastic scattering. We neglect the jth branch's dependence on the wave number $\mathbf{K}^{(j)}$ in the matrix element of the high energy electrons as well as the excitation energy for the high energy electrons. Thus, replacing the value $\mathbf{K}' - \mathbf{K}_0$ by $\mathbf{q}$, we obtain the intensity I_0 for the $\mathbf{K}'$ direction

$$I_0 \propto S(\mathbf{q},\mathbf{q}) + \frac{1}{2(1+w^2)}[S(\mathbf{q}_G,\mathbf{q}_G) - S(\mathbf{q},\mathbf{q})] - \frac{w}{1+w^2} S(\mathbf{q}_G,\mathbf{q}) \tag{4.7.8}$$

and for the $\mathbf{K}'_G$ direction[†]

$$I_G \propto S(\mathbf{q}_G, \mathbf{q}_G) + \frac{1}{2(1+w^2)}[S(\mathbf{q},\mathbf{q}) - S(\mathbf{q}_G,\mathbf{q}_G)] + \frac{w}{1+w^2} S(\mathbf{q}_G,\mathbf{q}) \quad (4.7.9)$$

where the inelastic scattering factor $S(\mathbf{Q},\mathbf{Q}')$ is represented (for crystals with a centre of symmetry) by

$$S(\mathbf{Q},\mathbf{Q}') = \frac{1}{E}\sum_{n\neq 0} H'_{n0}(\mathbf{Q})^* \mathrm{H}_{n0}(\mathbf{Q}')\,\delta\left(\frac{\hbar^2 K_0'^2}{2m} - \frac{\hbar^2 K_0^2}{2m}\right) \quad (4.7.10)$$

which is the same scattering factor that appeared in the detailed calculation of the anomalous absorption coefficient 4.3.17.

The first term of equation 4.7.8 or 4.7.9 is the normal inelastic scattering probability, independent of diffraction, which gives the background of the Kikuchi pattern. The second term reveals the symmetry with respect to w, as shown in figure 4.7(*a*). The black or white of the Kikuchi lines is determined by the sign of the factor

$$S(\mathbf{q}_G, \mathbf{q}_G) - S(\mathbf{q},\mathbf{q})$$

The inelastic scattering factor $S(\mathbf{Q},\mathbf{Q}')$ is proportional to Q^{-2} and $(Q')^{-2}$ for core electron excitations, as shown in section 4.3, and becomes larger with decreasing Q^2 and Q'^2 for phonon excitations. Thus the Kikuchi line closer (for $\mathbf{q} < \mathbf{q}_G$) to the incident spot (the point I in figure 4.5) corresponds to

$$S(\mathbf{q}_G, \mathbf{q}_G) < S(\mathbf{q},\mathbf{q})$$

or

$$S(\mathbf{q}_G, \mathbf{q}_G) - S(\mathbf{q},\mathbf{q}) < 0$$

resulting in a black line, whereas the Kikuchi line further from the spot I results in a white one. Rotating the crystal so that the incident direction becomes parallel to the net plane, as discussed before, we obtain the conditions $\mathbf{q} = \mathbf{q}_G$ and $S(\mathbf{q}_G,\mathbf{q}_G) - S(\mathbf{q},\mathbf{q}) = 0$, and the Kikuchi line term vanishes.

Now, the third term of equations 4.7.8 and 4.7.9 is not symmetrical with respect to w, as is shown in figure 4.7(*b*). Taking into account the intensity profiles for both the $\mathbf{K}'$ and $\mathbf{K}'_G$ directions, we obtain a band-like pattern, a Kikuchi band, with the band edge at $w = 0$. The black or white of the band is characterized by the sign of the function $S(\mathbf{q}_G, \mathbf{q})$. This is explained in more detail below.

The inelastic scattering function $S(\mathbf{q}_G, \mathbf{q})$ is given for core electron excitation

[†]In what follows we neglect the D dependence. The contrast changes of the Kikuchi patterns with various crystal thicknesses shown experimentally by Uyeda *et al.* (*Acta Cryst.* 7, 217 (1954)) are interpreted by the D dependence.

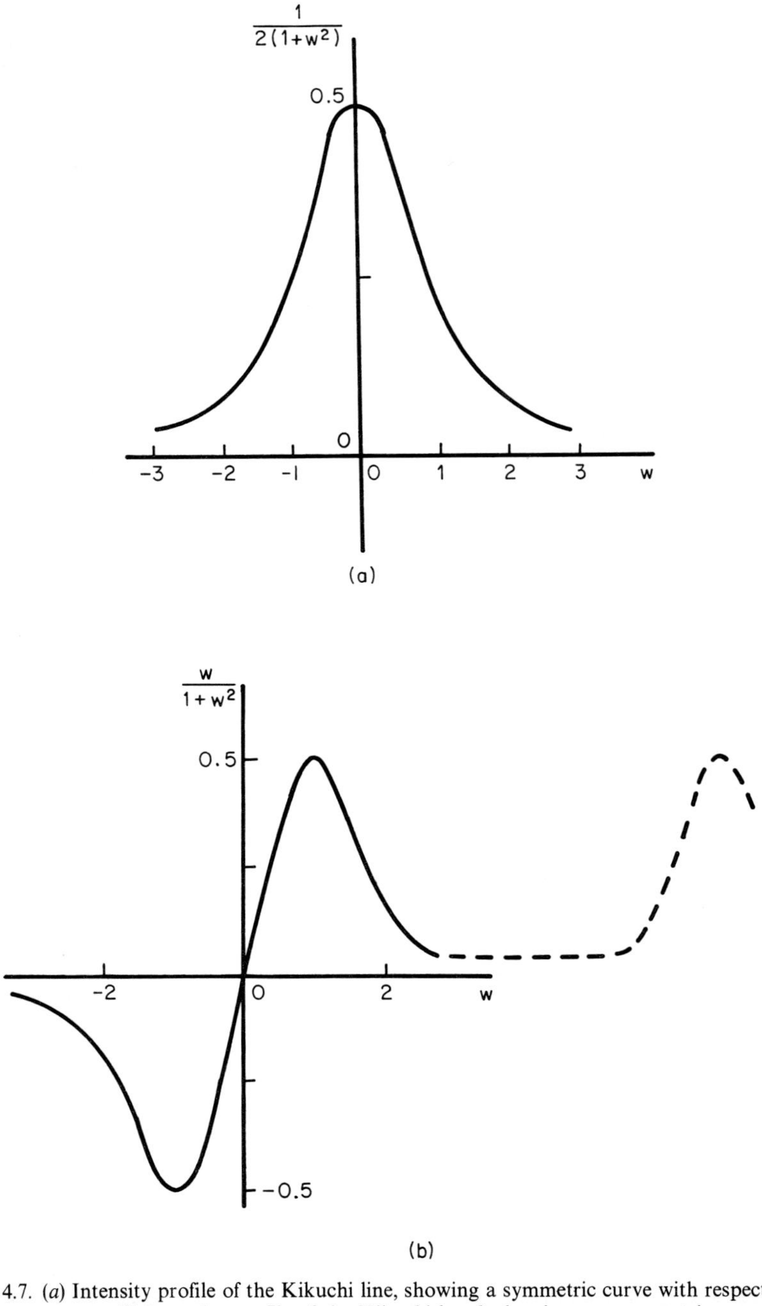

Figure 4.7. (*a*) Intensity profile of the Kikuchi line, showing a symmetric curve with respect to the parameter *w*. (*b*) Intensity profile of the Kikuchi band, showing an asymmetric curve.

by equation 4.3.15

$$S(\mathbf{q}_{\mathbf{G}}, \mathbf{q}) = \frac{(4\pi e^2)^2 N}{q_{\mathbf{G}}^2 q^2 v} \exp[-M(G)] \mathscr{S}(\mathbf{q}_{\mathbf{G}}, \mathbf{q}) \delta\left(q_z - \frac{1}{\hbar v}\right) \qquad (4.7.11)$$

and for phonon excitation (in the Einstein model) by

$$S(\mathbf{q}_{\mathbf{G}}, \mathbf{q}) = V(\mathbf{q}_{\mathbf{G}}) V(\mathbf{q}) N \{\exp(-M(\mathbf{G})) - [\exp(-M(\mathbf{G}))\exp(-M(\mathbf{q}_{\mathbf{G}}))]\} \delta(q_z)/v \qquad (4.7.12)$$

It is not difficult to show by direct calculation of equations 4.7.11 and 4.7.12 that $S(\mathbf{q}_{\mathbf{G}}, \mathbf{q})$ becomes negative within the circle[61,62]

$$q^2 + q_G^2 < |2\pi \mathbf{G}|^2 \qquad (4.7.13)$$

For example, we can show this by using equation 4.7.12. The second term of equation 4.7.12 becomes large compared to the first term for the negative region of $S(\mathbf{q}_{\mathbf{G}}, \mathbf{q})$. This means†

$$M(\mathbf{G}) > M(\mathbf{q}/2\pi) + M(\mathbf{q}_{\mathbf{G}})$$

or

$$(2\pi G)^2 > q^2 + q_G^2$$

since $M(\mathbf{G})$ is proportional to G^2. Within the two circles shown in figure 4.8

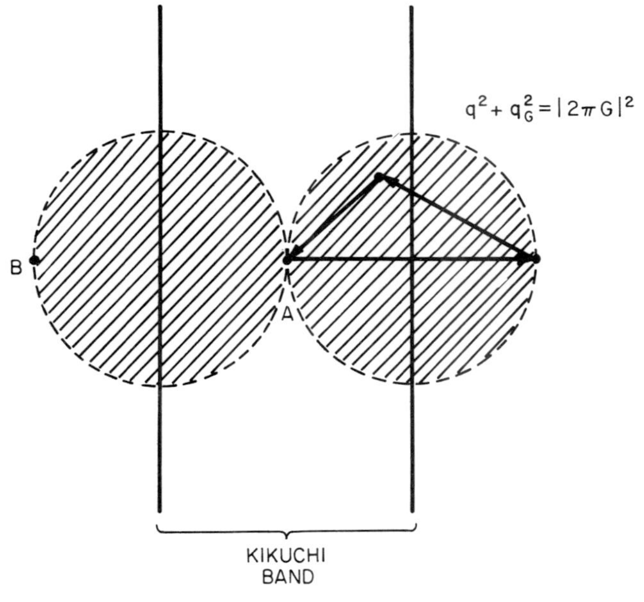

Figure 4.8. Region of negative inelastic scattering function, represented by the two circles in the Kikuchi band.

†We can obtain the same circle of the negative region by using the Debye model for phonon excitations.

$S(\mathbf{q}_G, \mathbf{q})$ becomes negative and the Kikuchi band turns out to be black, as seen in the photograph.

The detailed calculation of the Kikuchi pattern intensity has been performed by Okamoto *et al.*[61,62] and is shown in figures 4.9 and 4.10. Plasmon excitation, valence electron excitation, multi-phonon excitation and core electron excitation were taken into account in these calculations. These calculations agree well with the energy analysis of the Kikuchi pattern.[63]

Next we shall discuss the multi-beam effect on the Kikuchi patterns, from which we may interpret various phenomena of the Kikuchi patterns: (*a*) the

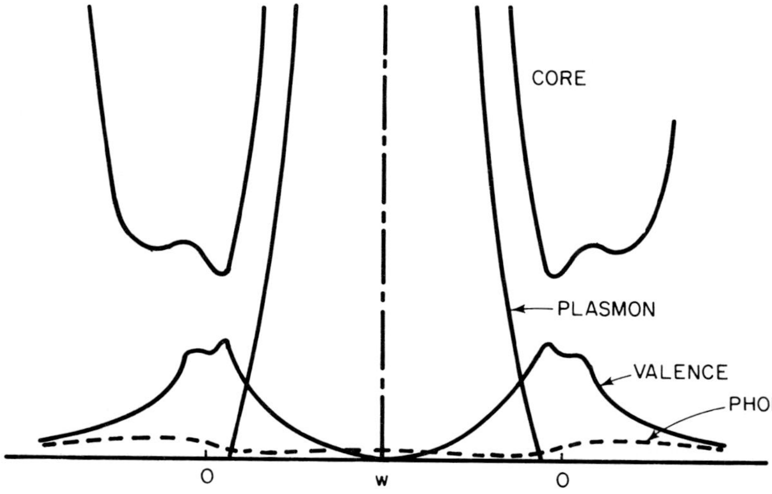

Figure 4.9. Detailed calculations of the Kikuchi pattern due to plasmon excitation, core electron excitation and phonon excitation.

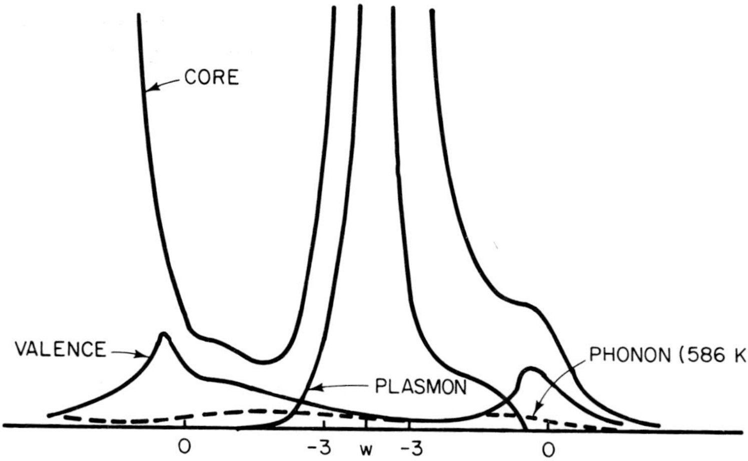

Figure 4.10. The same profile under different conditions.

contrast reversal of the Kikuchi lines in the region where the Kikuchi lines and the Kikuchi bands cross; and (*b*) the appearance of a central line in the Kikuchi band. First of all, we shall start with the general formula. The transition probability for the Kikuchi pattern is derived from equations 4.7.1 and 4.7.2

$$P \propto \sum_{n \neq 0} \left| \sum_{jG} \chi_j c_G^{(j)} U_{\mathbf{G},n}^{(j)} \right|^2 \delta\left(\frac{\hbar^2 K'^2}{2m} - \frac{\hbar^2 K_0^2}{2m} + E_n - E_0\right) \tag{4.7.14}$$

where $U_{G,n}^{(j)}$ is the matrix element of the interaction Hamiltonian

$$U_{G,n}^{(j)} = H'_{n0}(\mathbf{K}_0 - \mathbf{K}^{(j)} - 2\pi\mathbf{G})$$

In equation 4.7.14 the coefficient $x_j c_G^{(j)}$ is determined by the boundary condition at $z = D$, equation 4.7.4, where the phase factor $\exp[i(\mathbf{K}' - \mathbf{K}'^{(j)})_z D]$ is included. Therefore, in actual cases, we obtain the intensity $\langle P \rangle_D$ averaged over the fluctuation in thickness D. Neglecting the branch dependence of the matrix element $U_{G,n}^{(j)}$, we have

$$\langle P \rangle_D \propto \sum_n \sum_{\mathbf{G},\mathbf{G}'} \sum_{jj'} \langle \chi_j \chi_{j'} c_G^{(j)} c_{G'}^{(j)} \rangle_D U_{\mathbf{G},n}^* U_{\mathbf{G}',n} \delta\left(\frac{\hbar^2 K'^2}{2m} - \frac{\hbar^2 K_0^2}{2m} + E_n - E_0\right)$$

$$= \sum_{\mathbf{G},\mathbf{G}'} I_{\mathbf{G},\mathbf{G}'} \sum_n U_{\mathbf{G},n}^* U_{\mathbf{G}',n} \delta\left(\frac{\hbar^2 K'^2}{2m} - \frac{\hbar^2 K_0^2}{2m} + E_n - E_0\right) \tag{4.7.15}$$

where $I_{\mathbf{G},\mathbf{G}'} = \sum_j \chi_j^2 c_{\mathbf{G}}^{(j)} c_{\mathbf{G}'}^{(j)}$. If we consider N waves for the incident electron beam the following factor

$$\sum_{\mathbf{H},\mathbf{H}'} I_{\mathbf{H},\mathbf{H}'}^0 = \sum_{\mathbf{H},\mathbf{H}'} \sum_j \chi_j^2 c_{\mathbf{G}}^{(j)} c_{\mathbf{G}}^{(j)}$$

should be multiplied to equation 4.7.14. In N-wave theory we can calculate $I_{\mathbf{G},\mathbf{G}'}$ directly from the scattering matrix without detailed knowledge of the dispersion branches.[64]

N-Wave theories for the Kikuchi pattern have been developed by Kainuma and Kogiso[65] and many other workers.[66,67,68] The Kikuchi patterns for low energy electron diffraction have been studied theoretically and experimentally by Mosser *et al.*[69]

4.8. Stopping power, the stopping power function and the inelastic scattering function

4.8.1. *Stopping power—the Bethe–Bloch formula*

Under channelling conditions, the inelastic scattering probability turns out to be small. For phonon excitation, the scattering probability of the channelled particles becomes especially small, since the scatterer is localized in the channelling plane or axis. The valence electrons as scatterers, on the other hand, are not so localized in the channelling plane or axis. We shall now introduce the inelastic scattering function which depends on the distance from the channelling plane or the channelling axis. Using the inelastic scattering function we may calculate the channelling stopping power (stopping power function), and also the dechannelling coefficient.

First of all, we will briefly review the ordinary theory of stopping power for a random (electron or ion) beam. Since Bohr's theory in 1913,[70,71,72] ordinary stopping power has been studied by many workers, concentrating on its velocity and atomic number dependency.

The stopping power is considered to originate from electron excitation and nuclear recoil motion in the solid, which may be related to phonon excitation for small energy transfer. The stopping power due to a nuclear recoil becomes dominant for very low energy ions, when the ion cannot readily excite the solid's electrons.

Figure 4.11 shows schematically the velocity dependence of the stopping power $\left(\frac{-\mathrm{d}E}{\mathrm{d}z}\right)$ for ions of atomic number Z_1 in a solid with atomic number

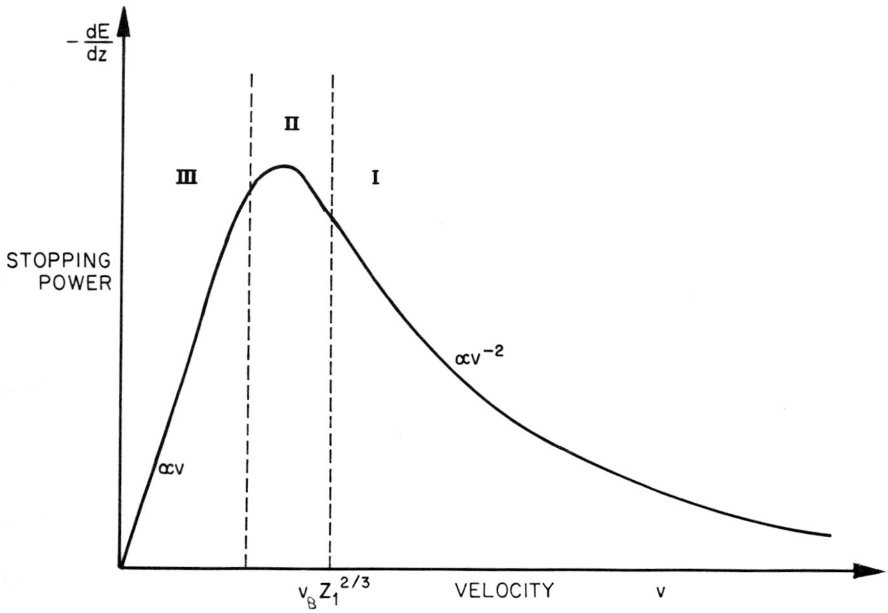

Figure 4.11. The velocity dependence of the stopping power $\left(-\frac{\mathrm{d}E}{\mathrm{d}z}\right)$ for ions of atomic number Z_1 in a solid (atomic number Z_2).

Z_2. For the high energy region labelled I, the electron excitation energy of the solid is replaced by an average value $\overline{\Delta E}$, which is independent of the ion velocity v, and the stopping power can then be represented by the inelastic scattering probability multiplied by $\overline{\Delta E}$. Roughly speaking, the stopping power in this energy region is proportional to the absorption coefficient studied in section 4.3, and to v^{-2}. The detailed calculation of the stopping power in this region was made by Bethe[73] and Bloch.[74] It should be noted that the stopping power is also proportional to $(Z_1 e)^2$, as in the case of the absorption coefficient.

For the very low energy region labelled III $v < v_B Z_2^{2/3}$, v_B being the Bohr velocity $e^2/\hbar$; the stopping power may be considered as the frictional force on a spherical ball in a liquid or gas, which is proportional to v. Region III has not been studied extensively.

We shall start our study of the stopping power with the derivation of the Bethe–Bloch formula. As stated in equation 4.3.4, the transition probability as a function of the excitation energy $\hbar\omega$ takes the form

$$Im \int_0^\infty \frac{dq}{q}\left(\frac{1}{\varepsilon(q,\omega)} - 1\right)$$

The energy loss rate (stopping power) may then be represented by the above transition probability multiplied with $\hbar\omega$

$$-\frac{dE}{dz} = \frac{4\pi(Z_1 e)^2 e^2}{mv^2} nL$$

$$L = \frac{1}{\pi\omega_0^2}\int_0^\infty \frac{dq}{q}\int_{-\infty}^{\infty} \omega \, d\omega \, Im\left(\frac{-1}{\varepsilon(q,\omega)}\right) \tag{4.8.1}$$

where $\omega_0^2 = 4\pi ne^2/m$, n being the electron density in the solid. The energy loss function $Im(-1/\varepsilon)$ was calculated in detail by Lindhard for valence electron excitation, as explained in section 4.3. Therefore the present discussion is concentrated mainly on core electron excitation.

Taking the interaction Hamiltonian H', the energy loss function $Im(-1/\varepsilon)$ takes the form, as defined in section 4.3, of

$$Im(-1/\varepsilon) = \frac{4\pi^2 e^2}{\hbar q^2}\sum_{n\neq 0} |H'_{n0}(\mathbf{q})|^2[\delta(\omega - \omega_{n0}) + \delta(\omega + \omega_{n0})] \tag{4.8.2}$$

Inserting equation 4.8.2 into equation 4.8.1, we obtain

$$L = \frac{1}{\pi\omega_0^2}\int_0^\infty \frac{dq}{q}\int_{-\infty}^{\infty}\sum_{n\neq 0} \omega_{n0}|H'_{n0}(\mathbf{q})|^2\,[\delta(\omega - \omega_{n0}) + \delta(\omega + \omega_{n0})] \tag{4.8.3}$$

In order to calculate equation 4.8.3 we should note the following fact. Taking the physical variable χ, the kinetic equation

$$\frac{\partial \chi}{\partial t} = \frac{i}{\hbar}(H_0, \chi)$$

turns out to be the equation for the matrix element χ_{n0}

$$\dot{\chi}_{n0} = \frac{i}{\hbar}(E_n - E_0)\,\chi_{n0}$$

where the excitation energy is $E_n - E_0 = \hbar\omega_{n0}$. If we replace $\hbar\omega_{n0}$ in $\delta(\omega - \omega_{n0})$ of equation 4.8.3 by the mean excitation energy I, the summation over n may be performed easily

$$\begin{aligned}\sum_{n\neq 0}(E_n - E_0)|H'_{n0}|^2 &= \sum_{n\neq 0}(E_n - E_0)(H'_{n0}H'^{*}_{0n} + H'^{*}_{0n}H'_{n0})/2 \\ &= -i\hbar\sum_{n\neq 0}(\dot{H}'_{n0}H'^{*}_{0n} + H'^{*}_{0n}\dot{H}'_{n0})/2 \\ &= -i\hbar(-\dot{H}'H'^{*} + H'^{*}\dot{H}')_{00}/2 \end{aligned} \tag{4.8.4}$$

In our case, since $H_0 = \sum_i \hbar^2\nabla_j^2/2m$, we obtain

$$\dot{H}' = \frac{i}{\hbar}[H_0, H'] = \frac{i\hbar}{2m}\left[\sum_j \nabla_j^2, H'\right] \tag{4.8.5}$$

From equations 4.8.4 and 4.8.5, the summation in the stopping power equation turns out to be

$$\sum_{n\neq 0}(E_n - E_0)|H'_{n0}(\mathbf{q})|^2\delta(\omega - \omega_{n0}) = \frac{\hbar^2}{m}\left[\sum_j(\nabla_j H'_j)(\nabla_j \mathbf{H}'^{*}_j)\right]_{00} \tag{4.8.6}$$

where we assumed that the interaction Hamiltonian H' is expanded for each electron index j: $H' = \sum_j H'_j$.

Taking H' to be the Coulomb potential, and performing the integral over q, we get

$$\begin{aligned} L &= \int_\eta^\infty [K_0^2(x) + K_1^2(x)]x\,\mathrm{d}x \\ &= \eta K_0(\eta)K_1(\eta) \end{aligned} \tag{4.8.7}$$

where $\eta = b_{\min}I/\hbar v$, $b_{\min}$ being the cut-off of the impact parameter, which has a value of the order of the de Broglie wave length

$$b_{\min} = \frac{\hbar}{mv} \tag{4.8.8}$$

Since the velocity of the incident ion beam is very large, $\eta \ll 1$, and we obtain

$$L = \log(2/\eta) - \gamma \tag{4.8.9}$$

where γ is Euler's constant and will be neglected for high energies. Writing equation 4.8.9 more explicitly, we obtain

$$L = \log\left(\frac{2mv^2}{I}\right)$$

hence

$$-\frac{\mathrm{d}E}{\mathrm{d}z} = \frac{4\pi(Z_1 e)^2 e^2}{mv^2} nL \tag{4.8.10}$$

which is the Bethe–Bloch formula.

When we consider only single electron excitation we can get the well-known form of the stopping power. The stopping power is represented by energy transfer T, with the differential cross-section $\mathrm{d}\sigma$ by the single electron excitation

$$-\frac{\mathrm{d}E}{\mathrm{d}z} = N\int T\,\mathrm{d}\sigma$$

Considering Rutherford scatterings of electrons by an ion, we obtain†

$$\mathrm{d}\sigma = \frac{2\pi Z_1^2 e^4}{mv^2}\frac{\mathrm{d}T}{T^2}$$

and

$$-\frac{\mathrm{d}E}{\mathrm{d}z} = \frac{2\pi Z_1^2 e^4}{mv^2}(NZ_2)\log\frac{T_{\mathrm{max}}}{T_{\mathrm{min}}}$$

$$= \frac{2\pi Z_1^2 e^4}{mv^2}(NZ_2)\log\left(\frac{2mv^2}{I}\right) \tag{4.8.11}$$

where we put $T_{\mathrm{max}} = 2mv^2$ and $T_{\mathrm{min}} = I$. Since the factor NZ_2 is the electron number density, equation 4.8.11 is just one half of the Bethe–Bloch formula (equation 4.8.10). The other half contributing to the Bethe–Bloch stopping power is due to the collective excitation, which is attributed to the long-range Coulomb interaction. Single electron excitation corresponds to close (short-range) Coulomb interaction. So, we may conclude that the close interaction and the long-range interaction contribute equally to the stopping power, which is called the 'equi-partition rule for the stopping power'. This rule will be proved again later.

We shall now return to the Bethe–Bloch formula from the point of view of the complex dielectric function $\varepsilon(q_1, \omega)$. For plasmon excitation of energy $\hbar\omega_{\mathrm{p}}$, we get

$$Im\frac{-1}{\varepsilon(q,\omega)} = \frac{\pi\omega}{2}[\delta(\omega - \omega_{\mathrm{p}}) + \delta(\omega + \omega_{\mathrm{p}})]$$

†See equation 1.3.8.

(see equation 4.3.8), and equation 4.8.1 becomes

$$L = \frac{1}{\pi\omega_p^2}\int_0^\infty \frac{dq}{q}\int_{-\infty}^{\infty} \pi\frac{\omega^2}{2}[\delta(\omega-\omega_p)+\delta(\omega+\omega_p)]d\omega$$
$$= \int_0^\infty \frac{dq}{q} = \log\left(\frac{q_{max}}{q_{min}}\right)$$

For plasmon excitation of plasmon energy $\hbar\omega_p$, q_{min} is $\hbar\omega_p/\hbar v$, and q_{max} is taken to be the plasmon cut-off k_c. For single electron excitation, we may take q_{min} to be $I/\hbar v$, and $\hbar q_{max}$ to be equal to $2mv$, the maximum momentum transfer during the collision. Thus $L = \log\left(\frac{q_{max}}{q_{min}}\right)$ takes the form

$$L = \log\left(\frac{2mv/\hbar}{I/\hbar v}\right) = \log\left(\frac{2mv^2}{I}\right) = \log\left(\frac{k_c \hbar v}{\hbar\omega_p}\right) \tag{4.8.12}$$

which corresponds to equation 4.8.10.

It is noteworthy that the stopping power in the form of the Bethe–Bloch formula is proportional to the electron density n, which came from the calculation of the expectation values by summing quantities for single electrons such as $\left(\sum_i f(\mathbf{r}_j)\right)_{00}$ However, for the calculation of the absorption coefficient the expectation value was represented in the form $\left[\sum_{jj'} f(\mathbf{r}_j, \mathbf{r}_j')\right]_{00}$, which is not proportional to n in general.

A systematic discussion of single electron and plasmon excitation is also made by making use of the complex dielectric function calculated by Lindhard[75] using the electron gas model. The dielectric function $\varepsilon(\mathbf{q},\omega)$ is expressed in the form

$$\varepsilon(\mathbf{q},\omega) = 1 + \xi[\alpha_1(\mathbf{q},\omega) + i\alpha_2(\mathbf{q},\omega)]$$
$$\xi = 4me^2 k_F/\pi\hbar^2 q^2 \tag{4.8.13}$$

from which we obtain

$$L = \frac{1}{\pi\omega_0^2}\int_0^\infty \frac{dq}{q}\int_{-\infty}^{\infty} \omega\, d\omega \frac{\xi\alpha_2(\mathbf{q},\omega)}{[1+\xi\alpha_1(\mathbf{q},\omega)]^2 + \xi^2\alpha_2(\mathbf{q},\omega)^2} \tag{4.8.14}$$

The functions $\alpha_1(\mathbf{q},\omega)$ and $\alpha_2(\mathbf{q},\omega)$ were calculated by Lindhard[75] as follows. In general the complex dielectric function $\varepsilon(\mathbf{q},\omega)$ is expressed by

$$\varepsilon(\mathbf{q},\omega) = 1 - \frac{4\pi e^2}{\hbar q^2}\sum_n f_n\left(\frac{1}{\omega - \omega_{mn} + i\gamma} - \frac{1}{\omega + \omega_{mn} + i\gamma}\right)$$

for the mth excitation with excitation energy $\hbar\omega_{mn}$ and the damping term γ. If we consider the degenerate electron gas, we obtain

$$\varepsilon(\mathbf{q}, \omega) = 1 + \frac{4\pi m e^2}{\hbar^2 q^2} \sum_{k<k_F} \left\{ \frac{1}{q^2 + 2\mathbf{q}\cdot\mathbf{k} - \frac{2m}{\hbar}(\omega + i\gamma)} + \frac{1}{q^2 - 2\mathbf{q}\cdot\mathbf{k} + \frac{2m}{\hbar}(\omega + i\gamma)} \right\} \tag{4.8.15}$$

from which we get

$$\alpha_1(\mathbf{q}, \omega) = \frac{1}{2} + \frac{k_F}{4q}\left[1 - \left(\frac{q}{2k_F} - \frac{\omega}{qv_F}\right)^2\right] \log \left| \frac{\frac{q}{2k_F} - \frac{\omega}{qv_F} + 1}{\frac{q}{2k_F} - \frac{\omega}{qv_F} - 1} \right| + \frac{k_F}{4q}\left[1 - \left(\frac{q}{2k_F} + \frac{\omega}{qv_F}\right)^2\right] \log \left| \frac{\frac{q}{2k_F} + \frac{\omega}{qv_F} + 1}{\frac{q}{2k_F} + \frac{\omega}{qv_F} - 1} \right| \tag{4.8.16}$$

$$\begin{aligned} \alpha_2(\mathbf{q}, \omega) &= \frac{\pi}{2}\frac{\omega}{qv_F} && \text{for } \frac{q}{2k_F} + \frac{\omega}{qv_F} < 1 \\ &= \frac{\pi k_F}{4q}\left[1 - \left(\frac{q}{2k_F} - \frac{\omega}{qv_F}\right)^2\right] && \text{for } \left|\frac{q}{2k_F} - \frac{\omega}{qv_F}\right| < 1 < \frac{q}{2k_F} + \frac{\omega}{qv_F} \\ &= 0 && \text{for } \left|\frac{q}{2k_F} - \frac{\omega}{qv_F}\right| > 1 \end{aligned} \tag{4.8.17}$$

Making use of $\alpha_1(\mathbf{q}, \omega)$ and $\alpha_2(\mathbf{q}, \omega)$, Lindhard and Winter[76] made a numerical integration of equation 4.8.14.

However, it should be noted that there is a sum rule, called Bethe's sum rule, which is derived below

$$\int_{-\infty}^{\infty} \omega \, d\omega \, Im \frac{-1}{\varepsilon(q, \omega)} = \pi \omega_0^2 \tag{4.8.18}$$

and equation 4.8.1 becomes

$$\begin{aligned} L &= \frac{1}{\pi\omega_0^2} \pi\omega_0^2 \int_0^\infty \frac{dq}{q} \\ &= \log\left(\frac{q_{max}}{q_{min}}\right) \end{aligned}$$

corresponding to equation 4.8.12, where q_{max} and q_{min} were given phenomenologically. In Lindhard and Winter's direct integral over ω and q, such phenomenological parameters were avoided. In other words, they determined the mean ionization energy I from fundamentals.

Now we shall derive the sum rule 4.8.18. We take $t=0$ as the time origin when the interaction with the medium begins. For $t<0$ there are no interactions and therefore no responses. This corresponds to the causality relation. If we consider the Fourier transform

$$Im\,\varepsilon^{-1}(q,t)=\int Im\,\varepsilon^{-1}(q,\omega)\exp(i\omega t)\,\mathrm{d}\omega$$

the causality relation $Im\varepsilon^{-1}(q,-\infty)=0$ means that ω should have a negative imaginary part $-\delta(\delta>0)$ where $\exp(i\omega t)$ may be represented by $\exp(i\omega t)\exp(\delta t)\underset{t=-\infty}{\longrightarrow}0$. So, the poles of $Im1/\varepsilon(q,\omega)$ are in the lower half-plane. For the large semicircle in the upper half-plane, the integral vanishes

$$\int_{\mathrm{C}}^{\text{upper half}}\omega\,\mathrm{d}\omega\,Im\frac{-1}{\varepsilon(q,\omega)}=0$$

In more detail

$$\int_{-\infty}^{\infty}\omega\mathrm{d}\omega Im\frac{-1}{\varepsilon(q,\omega)}+\int_{0}^{\pi}\mathrm{d}\theta R^2\exp(i\theta)i\exp(i\theta)Im\frac{-1}{\varepsilon(q,R\,\exp(i\theta))}=0 \quad (4.8.19)$$

For large ω the real part of $\varepsilon(q,\omega)$ may be written in the form $1-\omega_0^2/\omega^2$ and then the second term of 4.8.19 becomes

$$R^2 i\int_0^{\pi}\mathrm{d}\theta\exp(2i\theta)Im\frac{-1}{1-(\omega_0^2/R^2)\exp(-2i\theta)}=-\int_0^{\pi}\omega_0^2\mathrm{d}\theta=-\pi\omega_0^2$$

which leads to the sum rule

$$\int_{-\infty}^{\infty}\omega\,\mathrm{d}\omega\,Im\frac{-1}{\varepsilon(q,\omega)}=\pi\,\omega_0^2$$

We shall now prove the equi-partition rule by making use of the dielectric function. Using the expression for $\alpha_1(\mathbf{q},\omega)$ and $\alpha_2(\mathbf{q},\omega)$ to map $\varepsilon(\mathbf{q},\omega)$ in the complex q plane we find that $\varepsilon(\mathbf{q},\omega)$ has zeros on the positive real axis, corresponding to plasmon excitation and single electron excitation, marked by o and × × × × respectively in figure 4.12. Replacing $\dfrac{q}{2k_{\mathrm{F}}}$ by q' and $\dfrac{\omega}{qv_{\mathrm{F}}}$ by u, we get

$$L\propto\int_0^{\infty}q'\,\mathrm{d}q'\int_0^{v/v_{\mathrm{F}}}u\,\mathrm{d}u\,Im\left(\frac{-1}{\varepsilon(q',u)}\right) \quad (4.8.20)$$

In order to perform the integration over q', we consider the following integral

$$F=\frac{i}{2}\int_C\mathrm{d}x\left(\frac{-1}{\varepsilon(x^{1/2},u)}-1\right) \quad (4.8.21)$$

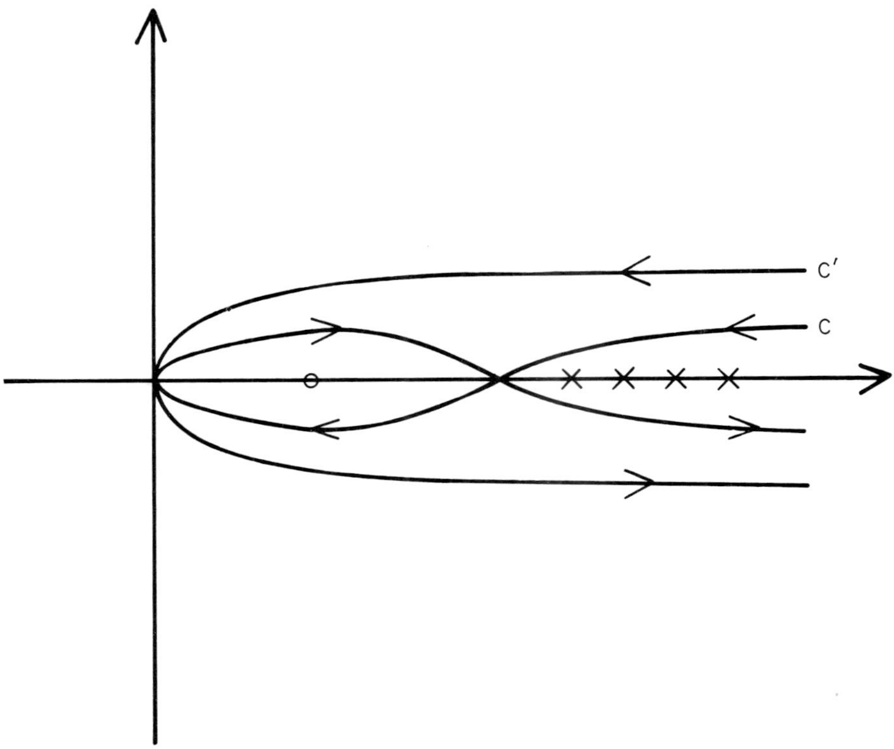

Figure 4.12. The contour integration for the calculation of the stopping power. The points × and o represent poles due to the single electron excitation and plasmon excitation respectively.

where the contour C is taken as shown in figure 4.12. The parts of $\alpha_2(q,\omega)=0$ and $\alpha_2(q,\omega)\neq 0$ in equation 4.8.17 correspond to plasmon excitation and single-electron excitation, respectively. When we consider the complex variable $\omega+i\delta$ $(\delta<0)$ for ω in equation 4.8.17 we can prove that the sign of the imaginary parts of q for plasmon excitation and single excitation become opposite. So, from the causality relation, we can find that the pole of $\dfrac{1}{\varepsilon(q,\omega)}$ due to plasmon excitation is in the lower half-plane, and the poles due to single-electron excitation are in upper half-plane.

The function F can be divided into two parts corresponding to plasmon and single-electron excitations.

$$F = F(\text{plasmon}) + F(\text{single electron})$$

for contour C. If we replace the contour C by the modified contour C′ shown in figure 4.12, we get.

$$F = -F(\text{plasmon}) + F(\text{single electron})$$

for contour C′. Contour C′ is closed at infinity. Within the semicircle there are no poles and, since the integral over the large semicircle vanishes, the integral F vanishes

$$F = -F(\text{plasmon}) + F(\text{single electron}) = 0$$

Thus we obtain

$$F(\text{plasmon}) = F(\text{single electron})$$

Therefore, the equi-partition rule is proved.

4.8.2. *Stopping power for very low energy ions*

For very low energy ions in region III in figure 4.11 we cannot use the Born approximation. Theories for low energy ions in this energy region are very complicated and many authors have tried to develop a systematic one, making use of phenomenological considerations, the complex dielectric constant and Green function methods.

First we shall describe the phenomenological theory of Firsov.[77] In Firsov's model the energy transfer from the ion to the atom in the solid occurs by the transfer of electrons from the ion to the atom. The electrons captured by the ion carry momentum, and the amount of this momentum transferred to the ion is proportional to the ion's velocity v, from which the friction force (stopping power) on the ion originates. Even though the captured electrons are transferred back to the atom in the solid when the ion departs from the atom there is no recoil during this second electron transfer, since the electron is trapped in a higher energy level.

The above model is similar to the simple mechanics of the hopper, drawn in figure 4.13. The sand is dropped on the belt with velocity v at a rate $\dfrac{\Delta m}{\Delta t}$.

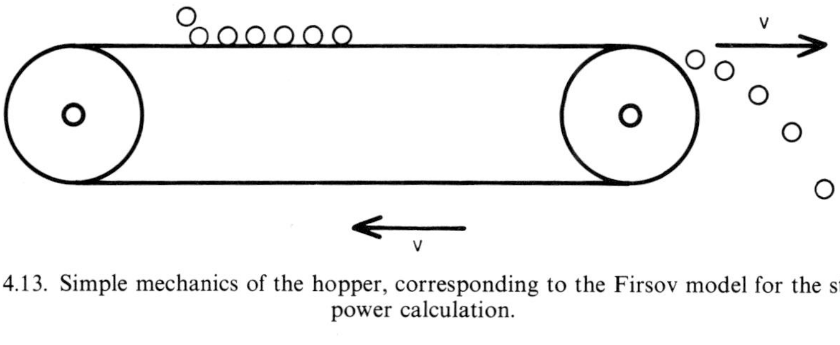

Figure 4.13. Simple mechanics of the hopper, corresponding to the Firsov model for the stopping power calculation.

The friction force F due to the momentum transfer is $v\left(\dfrac{\Delta m}{\Delta t}\right)\Delta t$. However, it should be noted that there are no recoil forces when the sand departs from the belt with the velocity v. The 'frictional' force F (stopping power) on the belt is equal to

$$F = v\Delta m$$

which is proportional to v.

Now we consider the electron transfer from the atom to the ion. The average velocity of the electron in the atom takes the form

$$v(r) = \frac{3}{4m}(3\pi^2)^{1/3}\,\hbar\, n(r)^{1/3} \tag{4.8.22}$$

according to the Thomas–Fermi model, where $n(r)$ is the electron density at the distance r from the nucleus. Consider the plane between the ion and the atom as shown in figure 4.14. The perpendicular component $v_\perp$ of the electron velocity is averaged over all directions

$$\langle v_\perp \rangle = \frac{\displaystyle\int_0^{2\pi}\int_0^{\pi/2} v\cos\theta\sin\theta\,\mathrm{d}\theta\,\mathrm{d}\varphi + \int_0^{2\pi}\int_{\pi/2}^{\pi} v\cos\theta\sin\theta\,\mathrm{d}\theta\,\mathrm{d}\varphi}{\displaystyle\int_0^{2\pi}\int_0^{\pi}\sin\theta\,\mathrm{d}\theta\,\mathrm{d}\varphi}$$

$$= \frac{(v\frac{1}{2}2\pi) - (v\frac{1}{2}2\pi)}{2\times 2\pi} = \frac{1}{4}v - \frac{1}{4}v \tag{4.8.23}$$

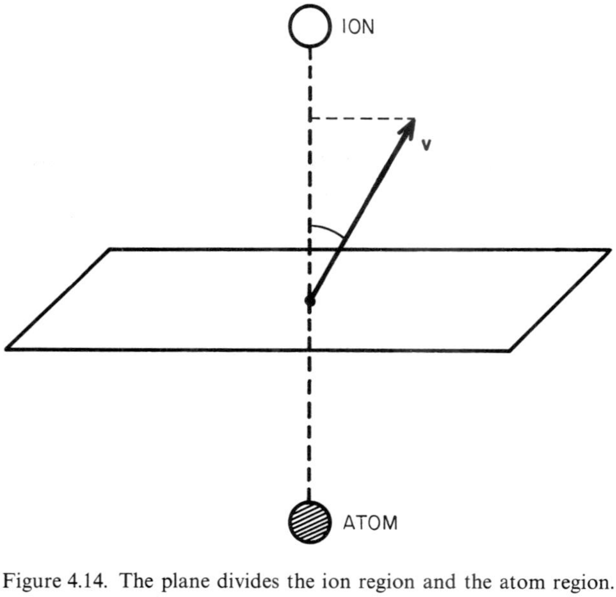

Figure 4.14. The plane divides the ion region and the atom region.

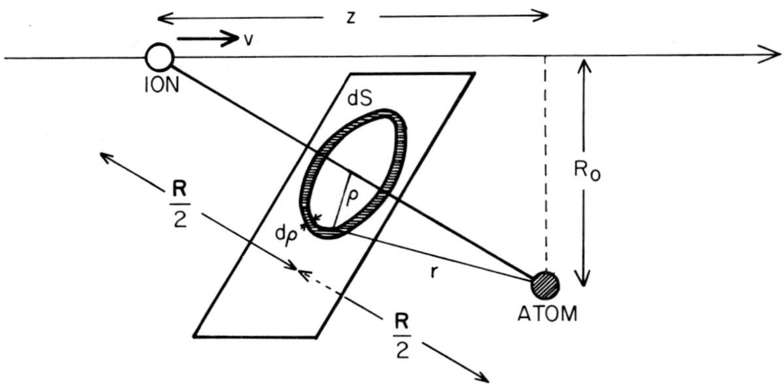

Figure 4.15. The electron flux from the atom to the ion through the small area dS at the position r from the centre of the atom.

The first term corresponds to the electron transfer from the atom to the ion.

Consider the ion to be distant from the atom and the plane at the position $\mathbf{R}/2$ (see figure 4.15). The electron flux from the atom to the ion through the small area dS at the position r from the centre of the atom on the plane is $n(r)[v(r)/4]\,\mathrm{d}S$, which is transferred to the ion with velocity $\dot{\mathbf{R}}$. The momentum transfer per unit time thus becomes the force

$$\mathrm{d}\mathbf{F} = m\dot{\mathbf{R}}\frac{n(r)\,v(r)}{4}\mathrm{d}S$$

or

$$\mathbf{F} = m\dot{\mathbf{R}}\int\frac{n(r)\,v(r)}{4}\mathrm{d}S$$

As before it should be noted that the electron transfer back from the ion to the atom does not give any recoil to the ion.

Thus, the energy loss ΔE takes the form

$$\Delta E = m\int\left(\int\frac{n(r)\,v(r)}{4}\mathrm{d}S\right)(\dot{\mathbf{R}}\cdot\mathrm{d}\mathbf{R}_{\mathrm{in}}) \tag{4.8.24}$$

where $\mathbf{R}_{\mathrm{in}}$ is the vector parallel to the velocity $\mathbf{v}$

$$\begin{aligned}(\dot{\mathbf{R}}\cdot\mathrm{d}\mathbf{R}_{\mathrm{in}}) &= v\,\mathrm{d}z\\ &= v^2\,\mathrm{d}t\end{aligned} \tag{4.8.25}$$

Inserting equations 4.8.22 and 4.8.25 into equation 4.8.24, we obtain

$$\Delta E = \tfrac{3}{16}(3\pi^2)^{1/3}\,\hbar v\int_{-\infty}^{\infty}\left(\int n(r)^{4/3}\,\mathrm{d}S\right)\mathrm{d}z \tag{4.8.26}$$

From figure 4.15 we get the following relations

$$\left.\begin{aligned} R^2 &= R_0^2 + z^2 \\ r^2 &= \frac{R^2}{4} + \rho^2 \\ &= \frac{R_0^2}{4} + \frac{z^2}{4} + \rho^2 \end{aligned}\right\} \qquad (4.8.27)$$

Finally we obtain

$$-\Delta E = \tfrac{3}{16}(3\pi^2)^{1/3}\,\hbar v \iint n\left(\sqrt{\frac{R_0^2}{4} + \frac{z^2}{4} + \rho^2}\right)^{4/3} 2\pi\rho\, \mathrm{d}\rho\, \mathrm{d}z$$

Firsov made the calculation of the integral by making use of the Thomas–Fermi model for electron density

$$-\Delta E = \frac{me^4}{2\hbar^2} \frac{0{\cdot}7(Z_1 + Z_2)^{5/3}}{[1 + 0{\cdot}16(Z_1 + Z_2)^{1/3}(R_0/a_B)]^5} \frac{v}{v_B} \qquad (4.8.28)$$

It should again be noted that the stopping power is proportional to the velocity v.

Experimentally, the proportionality relation has been confirmed by many workers, including in part the dependence on the ion's atomic number Z_1. In experiments in which Z_1 was changed a few workers[78,79,80] detected an oscillation with respect to Z_1, which was explained by Cheshire and Poate,[81,82] who made use of the extended Firsov model. In Figure 4.16 a comparison of Cheshire and Poate's result with experimental results for Au⟨110⟩ is shown.

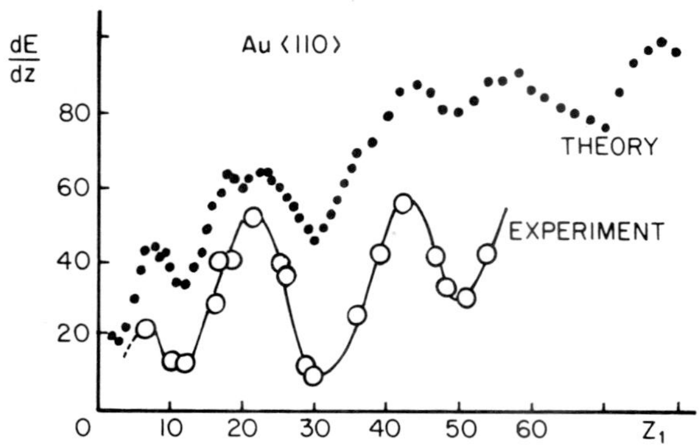

Figure 4.16. The Z_1 dependence of the stopping power according to Cheshire and Poate.[81]

Fundamental theories of the low energy stopping power have been developed from the dielectric constant method, the Green function method and the phase shift method. In the dielectric constant method equation 4.8.14 for the factor L takes the form

$$L = \frac{6}{\pi}\int_0^{v/vF} u\mathrm{d}u \int_0^\infty \mathrm{d}q' \frac{q'^3\alpha_2(u,q')}{[q'^2+\xi\alpha_1(u,q')]^2+\xi^2\alpha_2(u,q')^2}$$

If we consider the case where $(v/v_F) \ll 1$, the imaginary part $\alpha_2(u,q')$ may be expressed in the form

$$\alpha_2(u,q') \sim \pi\, u/2$$

and L becomes

$$L = \frac{6}{\pi}\int_0^{v/v_F} u\,\mathrm{d}u \frac{\pi u}{2}\int_0^\infty \frac{q'^3\,\mathrm{d}q'}{[q'^2+\xi\alpha_1(0,q')]^2}$$

$$= C(\xi)(v/v_F)^3 \qquad (4.8.29)$$

where

$$C(\xi) = \int_0^\infty \frac{q'^3\,\mathrm{d}q'}{[q'^2+\xi\alpha_1(0,q')]^2}$$

From equation 4.8.29 we can easily show that the stopping power for the low energy region is proportional to the velocity v. Extending the above discussion with a phenomenological consideration Lindhard *et al.*[84] obtained the following relation†

$$-\frac{\mathrm{d}E}{\mathrm{d}z} = \zeta_e\, 8\pi n e^2 a_B \frac{Z_1Z_2}{(Z_1^{2/3}+Z_2^{2/3})^{3/2}}\frac{v}{v_B} \qquad (4.8.30)$$

where the parameter ζ_e has a value of $1 \sim 2$.

On the other hand, the Green function method has also been applied in order to calculate the stopping power for the low energy region. Kitagawa and Ohtsuki[85] derived the following formula making use of the retarded Green function $G^R(\mathbf{q},\omega)$

$$-\frac{\mathrm{d}E}{\mathrm{d}z} = 0{\cdot}32\pi^{7/3} Z_1^2 \hbar v\, n^{2/3} \qquad (4.8.31)$$

which is also proportional to the velocity v and the electron density $n^{2/3}$. Though equation 4.8.31 is simple, since the electron gas model was employed in the calculation, it should be noted that it was derived from first principles.

Briggs and Pathak[86] developed the phase shift method, in which scattering of target electrons by the impact of ions is treated by the phase shift η_e (see Chapter 1). They derived the linear dependence with respect to v.

†Lindhard *et al.*[84] did not show the detailed derivation process of equation 4.8.30 in their paper.

4.8.3. *Nuclear stopping power*

Consider a collision of an ion of mass M_1 and energy E with a nucleus with an impact parameter b. It is easy to understand the cases with large impact parameters where the nucleus remains bound to the lattice during the collision and the energy loss is very small ($\sim 0{\cdot}05$ eV). In the case of a small impact parameter collision, however, the nucleus leaves the lattice point, and the energy of the ion may be transferred to the motion of the nucleus, which contributes greatly to the stopping power. A special case of this occurs when the nuclear mass M_2 is equal to the ion mass M_1; the nuclear stopping power then becomes a maximum.

Let us calculate the momentum transfer of the ion moving with impact parameter b and velocity v at a distance $\mathbf{r}$ from the nucleus, as shown in figure 4.17. The force $\mathbf{F}(\mathbf{r})$ is derived from the potential $V(r)$

$$\mathbf{F}(\mathbf{r}) = -\frac{\mathrm{d}V(\mathbf{r})}{\mathrm{d}\mathbf{r}}$$

and the transverse component of the force with respect to the straight trajectory is $F(r)b/r$. The transverse momentum transfer $\Delta p_\perp$ during the path length dz is obtained by

$$\Delta p_\perp = \left(-\frac{\mathrm{d}V(r)}{\mathrm{d}r}\right)\frac{b}{r}\frac{dz}{v}$$

$$= \frac{b}{(r^2-b^2)^{1/2}}\left(-\frac{\mathrm{d}V(r)}{\mathrm{d}r}\right)\frac{\mathrm{d}r}{v}$$

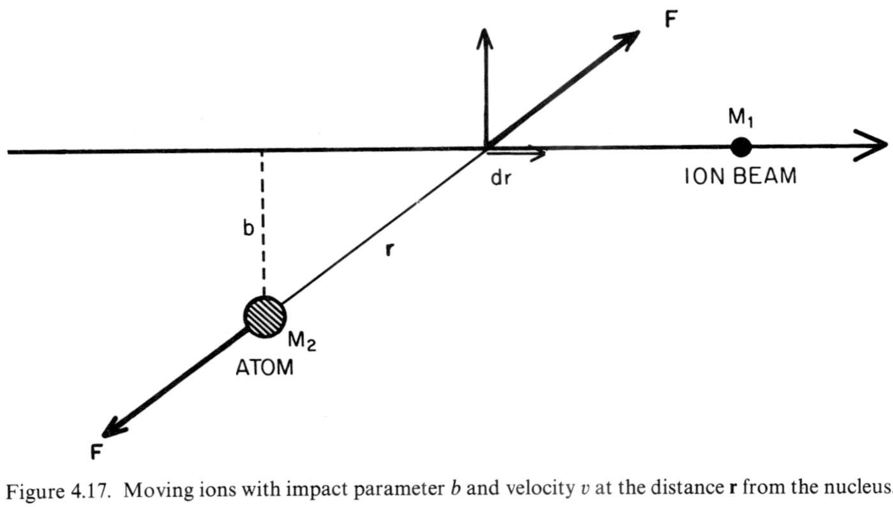

Figure 4.17. Moving ions with impact parameter b and velocity v at the distance $\mathbf{r}$ from the nucleus.

Thus we get the total energy transfer

$$\Delta E = \frac{\Delta p_\perp^2}{2M_2} = \frac{1}{2M_2 v^2}\left[2\int_b^\infty \mathrm{d}r \frac{b}{(r^2 - b^2)^{1/2}}\left(-\frac{\mathrm{d}V(r)}{\mathrm{d}r}\right)\right]^2$$
$$\equiv \frac{M_1}{M_2 E}\kappa(b)^2$$

where $E = \frac{1}{2}M_1 v^2$, and

$$\kappa(b) = \int_b^\infty \frac{b}{(r^2 - b^2)^{1/2}}\left(-\frac{\mathrm{d}V(r)}{\mathrm{d}r}\right)\mathrm{d}r$$

In a path length z there are n atoms with different impact parameters b_j. Therefore the total stopping power becomes

$$-\frac{\mathrm{d}E}{\mathrm{d}z} = \frac{M_1}{M_2 E d}\sum_{j=1} \kappa^2(b_j) \tag{4.8.32}$$

Bohr[70,71,72] calculated the above equation assuming Rutherford scattering for a random atom distribution, obtaining the result

$$-\frac{\mathrm{d}E}{\mathrm{d}z} = \tfrac{1}{4}\pi N_0 \, \mathrm{d}_{\mathrm{clos}}^2 \, T_{\mathrm{max}} \ln \frac{T_{\mathrm{max}}}{T_{\mathrm{min}}} \tag{4.8.33}$$

where $T_{\mathrm{max}} = 4M_1 M_2 E/(M_1 + M_2)^2$ (i.e., maximum energy transfer for head-on collisions), $T_{\mathrm{min}} = M_1 \kappa(a_B)^2/M_2 E$ (i.e., energy loss for grazing collisions at the distance of the atom size a_B), N_0 is the number of atoms per unit volume, and d_{clos} is the 'impact distance' in a head-on collision, defined as

$$d_{\mathrm{clos}} = \frac{Z_1 Z_2 e^2 (M_1 + M_2)}{M_2 E}$$

Nelson[87] also calculated the nuclear stopping power assuming a Coulomb potential, and Lindhard and Scharff[83] assumed a Thomas–Fermi potential in their work, which is

$$-\frac{\mathrm{d}E}{\mathrm{d}z} = \frac{4M_1}{M_1 + M_2}(\pi a_{\mathrm{TF}} N)(Z_1 Z_2 e^2)\frac{1}{\varepsilon}\int_0^\varepsilon f(\zeta)\,\mathrm{d}\zeta$$
$$\left(\varepsilon = \frac{a_{\mathrm{TF}}^2}{Z_1 Z_2}\frac{M_2}{M_1 + M_2}E\right)$$

where $f(\zeta)$ is the function calculated numerically from the Thomas–Fermi function.

According to equation 4.8.33 it is known that the nuclear stopping power becomes larger for decreasing M_2 and E. If

$$\frac{a_{\mathrm{TF}}}{Z_1 Z_2 e^2}\frac{M_2}{M_1 + M_2} < 10$$
$$(a_{\mathrm{TF}} = a_0 0{\cdot}8853(Z_1^{2/3} + Z_2^{2/3})^{-1/2})$$

the nuclear stopping power calculated by making use of the Thomas–Fermi potential is larger than the electronic stopping power. The above condition corresponds to the very low velocity case in which the velocity v is much smaller than the atomic orbit velocity $Z_2^{2/3} e^2/\hbar$.

4.8.4. *The stopping power function for a channelled beam*

As pointed out in the previous section, the stopping power for channelled ions is smaller than that for a random beam. Specifically, the stopping power may be a function of the 'frequency' ω of the channelling. Datz *et al.*[88] measured the stopping power by changing the initial conditions and the frequency ω. They called the ω-dependent stopping power the 'initial stopping power'. Figure 4.18 shows the experimental results for 60 MeV I ions in Au (111) and Au (100) channels. According to the figure the initial stopping power is represented by a linear function of ω

$$-\frac{\mathrm{d}E}{\mathrm{d}z} = A + B\omega \tag{4.8.34}$$

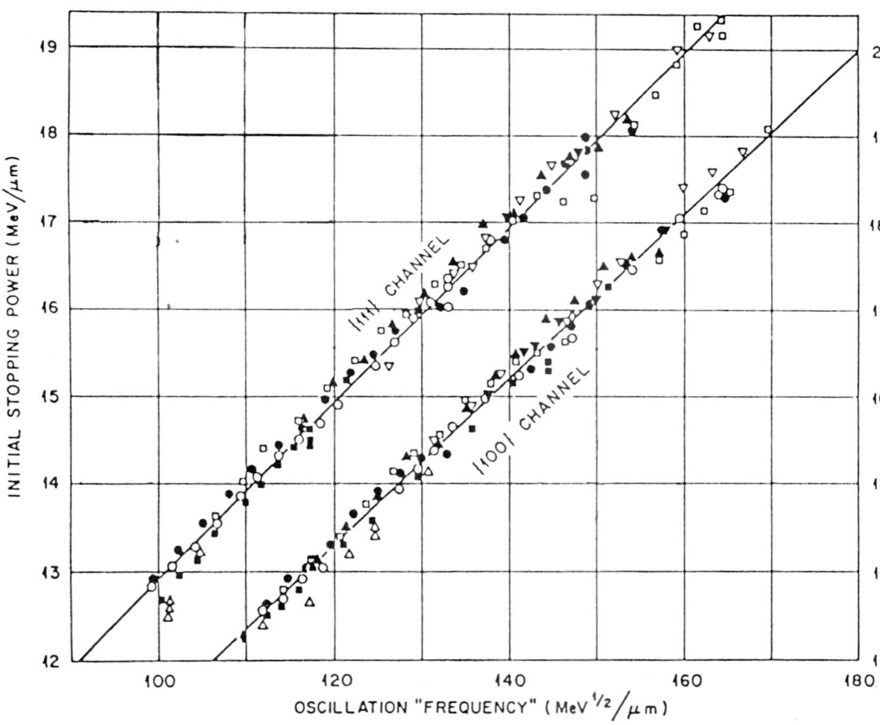

Figure 4.18. The 'initial' stopping power by Datz *et al.*[88] for a 60 MeV I ion into Au (111) and Au (100).

when A and B are constants. We will now try to interpret equation 4.8.34 theoretically.

The stopping power is a function of the position of the ion considered, which we call the 'stopping power function'. For the high energy region, the electron stopping power is proportional to the electron density n. The first question is: could the stopping power function be obtained by replacing the average electron density n by a position dependent density $n(r)$? The stopping power for the channelled ion is obtained by averaging the stopping power function over the channelling trajectories, as was done in the derivation of the Lindhard continuum potential.

First of all, we derive the stopping power function from first principles. As in section 4.4, the Schrödinger equation of the ion including the non-local inelastic scattering potential $A(\mathbf{R}, \mathbf{R}')$ takes the following form

$$\nabla^2\varphi_0 + \left[k_0^2 - \frac{2M_1}{\hbar^2}H'_{00}(\mathbf{R})\right]\varphi_0 - \frac{2M_1}{\hbar^2}\int A(\mathbf{R}, \mathbf{R}')\,\varphi_0(\mathbf{R}')\,\mathrm{d}\mathbf{R}' = 0 \tag{4.8.35}$$

For channelling, φ_0 is given as

$$\varphi_0 = \frac{1}{\sqrt{L}}\varphi(x, y)\exp(iKz) \tag{4.8.36}$$

where $\varphi(x, y)$ is a Bloch wave perpendicular (transverse component) to the channelling z axis with length L. We can now apply the separation of variables technique. Inserting equation 4.8.36 into equation 4.8.35, multiplying both sides by $[\exp(-iKz)]/\sqrt{L}$ and integrating over z we obtain

$$\nabla^2\varphi(x, y) + \left[k_\perp^2 - \frac{2M_1}{\hbar^2}V(x, y)\right]\varphi(x, y) - \frac{2M_1}{\hbar^2 L}\int \exp(-iKz)$$
$$\times\, A(\mathbf{R}, \mathbf{R}')\,\varphi(x', y')\exp(iKz')\,\mathrm{d}R'\,\mathrm{d}z = 0 \tag{4.8.37}$$

where

$$V(x, y) = \frac{1}{L}\int_{-\infty}^{\infty} H'_{00}(R)\,\mathrm{d}z$$

is the Lindhard continuum potential. $\hbar^2k_\perp^2/2M_1 = \hbar^2(k_0^2 - k^2)/2M_1$ is the transverse energy. The last term on the left side in equation 4.8.37 is the average inelastic non-local potential. $A(\mathbf{R}, \mathbf{R}')$ varies more rapidly at $x = x'$ and $y = y'$ than does $\varphi(x, y')$, as will be shown in detail later. Therefore, we may take $\varphi(x', y')$ out of the integral at $x' = x$, $y' = y$; i.e., we may localize the inelastic scattering in the transverse plane

$$\nabla^2\varphi(x, y) + \left\{k_\perp^2 - \frac{2M_1}{\hbar^2}[V(x, y) + C(x, y)]\right\}\varphi(x, y) = 0 \tag{4.8.38}$$

where $C(x, y)$ is the inelastic continuum potential defined by

$$C(x, y) = \frac{1}{L} \iint \exp(-iKz)\, A(\mathbf{R}, \mathbf{R}') \exp(iKz')\, \mathrm{d}R'\, \mathrm{d}z \tag{4.8.39}$$

If we take the interaction Hamiltonian $H'(\mathbf{R})$ between the ion and the crystal as in section 4.4, we obtain the actual form of $C(x, y)$

$$C(x, y) = -\frac{M_1}{4\pi^3\hbar^2 L} \sum_{n \neq 0} \int \mathrm{d}\mathbf{K}' \frac{\exp(-iK'_x x - iK'_y y)}{K'^2 - k_n^2} H'_{0n}(K'_x, K'_y, K'_z - K) \times H'_{n0}(x, y; K - K'_z) \tag{4.8.40}$$

where

$$H'_{n0}(x, y; K - K'_z) = \int_L H'_{n0}(\mathbf{R}) \exp[-i(K - K'_z)z]\, \mathrm{d}z \tag{4.8.41}$$

Since we may obtain the absorption coefficient $\mu(x, y)$ from the imaginary part $C^i(x, y)$ of the potential $C(x, y)$, the transition probability $P_n(\mathbf{K}', \mathbf{K}; x, y)$ at the position (x, y), corresponding to the nth excited state of the crystal, is expressed (assuming $L = 1$) by

$$P_n = (\mathbf{K}', \mathbf{K}; x, y) = \frac{1}{4\pi^2\hbar^2 v} \exp(-iK'_x x - iK'_y y) H'_{0n}(K'_x, K'_y, K'_z - K) \times H_{n0}(x, y; K - K'_z)\, \delta(K' - k_n) \tag{4.8.42}$$

which was introduced first by Ohtsuki[89] and is called the 'inelastic scattering probability function'. The absorption coefficient $\mu(x, y)$ takes the form

$$\mu(x, y) = \frac{1}{v} \int \sum_{n \neq 0} P_n(\mathbf{K}', \mathbf{K}; x, y)\, \mathrm{d}\mathbf{K}'$$

For the planar case, we obtain

$$P_n(\mathbf{K}', \mathbf{K}; y) = \frac{1}{4\pi^2\hbar^2 v} \exp(-iK'_y y) H'_{0n}(K'_x, K'_y, K'_z - K) \times H'_{n0}(-K'_x; y; K - K'_z)\, \delta(K' - k_n) \tag{4.8.43}$$

Making use of the inelastic scattering probability functions 4.8.42 and 4.8.43, we may calculate the stopping power function and the diffusion coefficient function, which play an important role in dechannelling theory. We now derive the stopping power function

$$W(x, y) = \frac{1}{v} \sum_{n \neq 0} \int (E_n - E_0)\, P_n(\mathbf{K}', \mathbf{K}; x, y)\, \mathrm{d}\mathbf{K}' \quad \text{(for axial channelling)}$$

$$W(y) = \frac{1}{v} \sum_{n \neq 0} \int (E_n - E_0)\, P_n(\mathbf{K}', \mathbf{K}; y)\, \mathrm{d}\mathbf{K}' \quad \text{(for planar channelling)} \tag{4.8.44}$$

In the above, $(E_n - E_0)$ is the excitation energy of the crystal. Inserting equation 4.8.43 into equation 4.8.44, we obtain

$$W(y) = \frac{1}{4\pi^2\hbar^2 v^2} \int \mathrm{d}\mathbf{K}' \exp(-iK'_y y) \sum_{n\neq 0} (E_n - E_0) H'_{0n}(K'_x, K'_y, K'_z - K)$$
$$\times H'_{n0}(-K'_x, y; K - K'_z)\,\delta(K' - k_n) \qquad (4.8.45)$$

If we make a calculation similar to the one in the derivation of the Bethe–Bloch formula, equation 4.8.45 becomes

$$W(y) = -\frac{1}{\sigma}\int_{-\infty}^{\infty} \Delta E(y^2 + u^2)^{1/2}\,\mathrm{d}u \qquad (4.8.46)$$

where $\Delta E(x_0)$ is the stopping power function for one atom

$$\Delta E(x_0) = -\frac{2Z_1^2 e^4}{mv^2}\int_{-\infty}^{\infty}\mathrm{d}z\int_0^{2\pi}\mathrm{d}\theta\int_{b_{\min}}^{\infty} p\,\mathrm{d}p\, q^2[K_0^2(qp) + K_1(qp)]$$
$$\times n[(z^2 + p^2 + x_0^2 - 2x_0 p\cos\theta)^{1/2}] \qquad (4.8.47)$$

which was derived by Kitagawa and Ohtsuki[90] using the wave packet method. In the above, K_0 and K_1 are the modified Bessel functions of the second kind, and $n(\mathbf{R})$ is the electron density. The quantities q and $b_{\min}$ are defined as

$$q = I/\hbar v$$
$$b_{\min} = \hbar/mv$$

We may rewrite $W(y)$ defined by equation 4.8.45 for close collisions† in the form

$$W(y) = \frac{1}{4\pi^2\hbar^2 v^2}\int \mathrm{d}\mathbf{K}' \exp(-iK'_y y) \sum_{n\neq 0} \frac{\hbar^2}{2m}(K^2 - k_n^2) H'_{0n}(K'_x, K'_y, K'_z - K)$$
$$\times H'_{n0}(-K'_x, y; K - K'_z)\delta(K' - K_n)$$
$$= \frac{1}{8\pi^2 v^2 m}\int \mathrm{d}\mathbf{K}' (K^2 - K'^2) \sum_{n\neq 0} H'_{0n}(K'_x, K'_y, K'_z - K)$$
$$\times H'_{n0}(-K'_x, y; K - K'_z)\,\delta(K' - k_n)$$

If we expand $W(y)$ with respect to the reciprocal lattice vector G, we obtain

$$W(y) = \sum_G W(G)\exp(iGy)$$
$$W(G) = \frac{1}{8\pi^2 v^2 m}\int_{|\mathbf{K}'| = \bar{k}_n} \mathrm{d}\mathbf{K}'\, K'^2_\perp\, S(\mathbf{q}, \mathbf{q}_G)$$

† We should consider P_n in equation 4.8.45 (see equation 4.8.43) to be the scattering probability of the atomic electrons with velocity $-v$ and mass m.

where $\mathbf{q} = (K'_x, K'_y, K' - K_z)$, $\bar{k}_n$ is the average value and $\mathbf{K}'_\perp$ is the transverse (x, y) component. Substituting equation 4.7.11 for $S(\mathbf{q}, \mathbf{q}_G)$, and neglecting the terms $\sum_i f_{ii}(q) f_{ii}(q_G)$ and $\sum_{ij} f_{ij}(q) f_{ij}(q_G)$, we get

$$W(G) = \frac{1}{8\pi^2 v^2 m}(4\pi e^2)^2 N \int_{|\mathbf{K}'| = k_n} d\mathbf{K}'\, q_\perp^2 \frac{1}{q^2 q_G^2} \sum_i f_{ii}(G)$$

When $G = 0$, $W(G = 0)$ becomes

$$W(G = 0) = \frac{(4\pi e^4)Nn}{v^2 m} \ln\left(\frac{q_{\max}}{q_{\min}}\right)$$

since $\sum_i f_{ii}(0)$ means the electron density. This is ordinary stopping power. If we take the assumption that $\mathbf{q}_G \approx \mathbf{q}$ and $q_{\min} = \hbar G$, we obtain

$$W(G \neq 0) = \frac{4\pi e^4 N}{v^2 m} n(G) \ln\left(\frac{q_{\max}}{\hbar G}\right)$$

Then

$$W(y) = \frac{4\pi Z_1 e^4}{mv^2} N\left[n \ln\left(\frac{2mv^2}{I}\right) + \sum_{\mathbf{G} \neq 0} n(G) \exp(i\mathbf{G}\cdot\mathbf{r}) \ln\left(\frac{2mv}{\hbar G}\right)\right]$$

The above expression for $W(y)$ was derived by Esbensen and Golovchenko and Burenkov *et al.*[92] Ohtsuki and Kitagawa,[93] and Dettmann and Robinson[94] also derived the energy loss function in the high energy region, and studied the channelling stopping power.

It should be noted, however, that the derivation of the above expression for $W(y)$ includes the rough assumption that $\mathbf{q}_G \approx \mathbf{q}$. Here we can point out that $W(y)$ is represented by the electronic diffusion coefficient $D_\perp(y)$ defined in equation 5.2.7

$$W(y) = \frac{N}{8\pi^2 mv^2} 4\pi^2 v^2 D_\perp(y) = \frac{1}{2m} D_\perp(y)$$

Detailed calculation of $D_\perp(y)$ will be given in section 5.2.1 (p.129). The above relation between the stopping power and the diffusion coefficient differs by a factor of $\frac{1}{2}$ from that of Lindhard (see equations 5.1.9 and 5.1.10).

If we want to obtain the simple expression for $W(y)$, we obtain

$$W(y) = \frac{4\pi Z_1^2 e^4}{mv^2}\left\{ L_e n(y) + \sum_{G \neq 0}\left[\tfrac{1}{2}\ln\left|\frac{q_{\max}^2 - G^2 + \beta^2}{2q_{\max}^2}\right| + \tfrac{1}{8}\left(\frac{1}{R(G)} - R(G)\right)\ln\left|\frac{q_{\max}^2 + G^2 R(G)^2 - \beta^2}{q_{\max}^2 - G^2 R(G)^2 - \beta^2}\left(\frac{R(G)+1}{R(G)-1}\right)\right|\right] \times n(G)\exp(iGy)\right\}$$

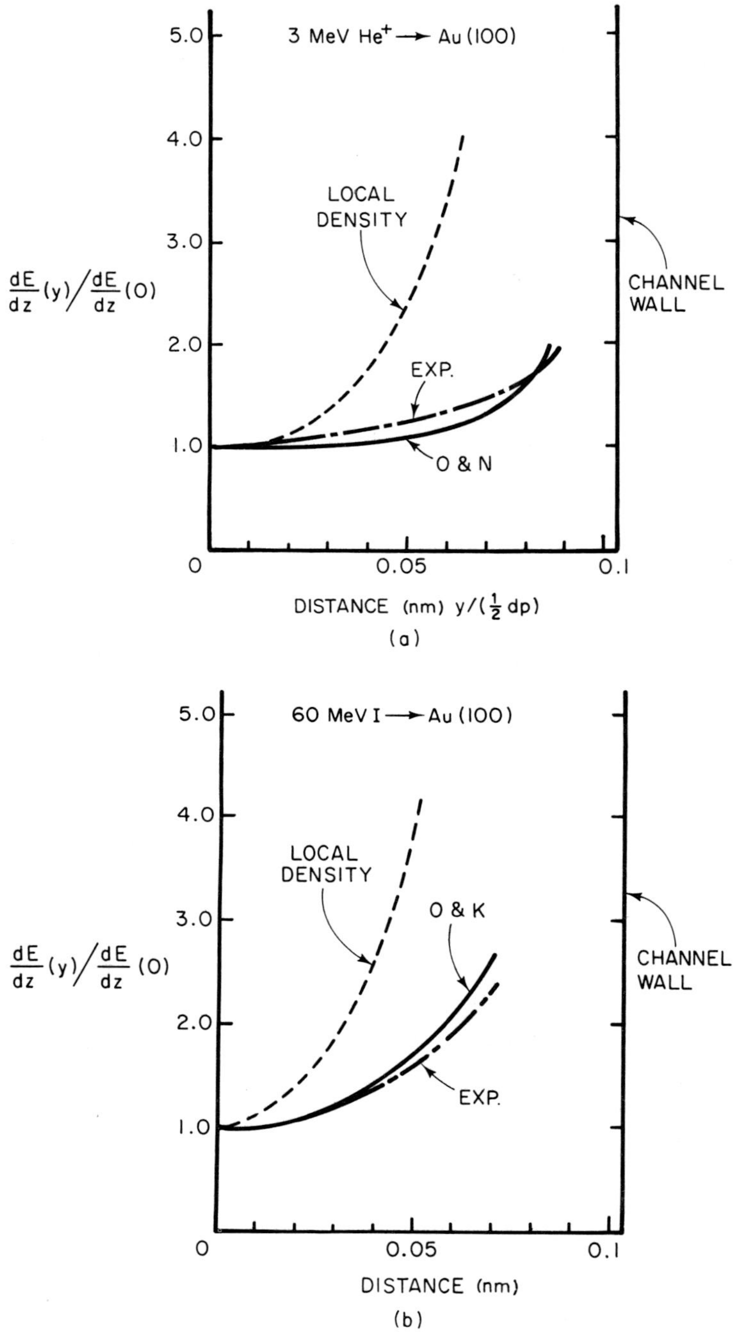

Figure 4.19. Comparisons between experimental and theoretical results of the stopping power functions (*a*) for 3 MeV He ions into Au (100) and (*b*) for 60 MeV I ions into Au (110).

$$R(G) = \sqrt{1 + 4I/(\hbar v G)^2}$$
$$\beta^2 = [(q_{max}^2 - G^2)^2 + 4q_{min}^2 G^2]^{1/2} \tag{4.8.48}$$

A detailed discussion and another simplified formula will be given in section 5.2.1.

In figure 4.19(*a*) numerical results of equation 4.8.48 compared with the experimental curve of Robinson[95] for 3 MeV He ions in Au(100) planar channelling are given.

It should be noted that equation 4.8.43 is correct for the high energy region. It is difficult to obtain the inelastic scattering probability function $P_n(\mathbf{K}', \mathbf{K}; y)$ for the low energy region. In the Firsov theory the stopping power was derived first as a function of position (see equation 4.8.28). If we assume that the electron density $n(r)$ in equation 4.8.28 is calculated by the Moliere potential, we obtain

$$n(r) = \frac{(Z_1 + Z_2)b^2}{4\pi r}[0{\cdot}35 \exp(-br) + 8{\cdot}8 \exp(-4br) + 40 \exp(-20br)]$$

The stopping power for planar channelling of low energy ions then takes the following form[93]

$$W(y) = 2C \exp[-\tfrac{2}{3}by + \tfrac{2}{3}\ln(\tfrac{2}{3}by)]$$
$$C = \frac{81}{512\sigma}\hbar v[0{\cdot}35\pi(Z_1 + Z_2)]^{4/3} \tag{4.8.49}$$

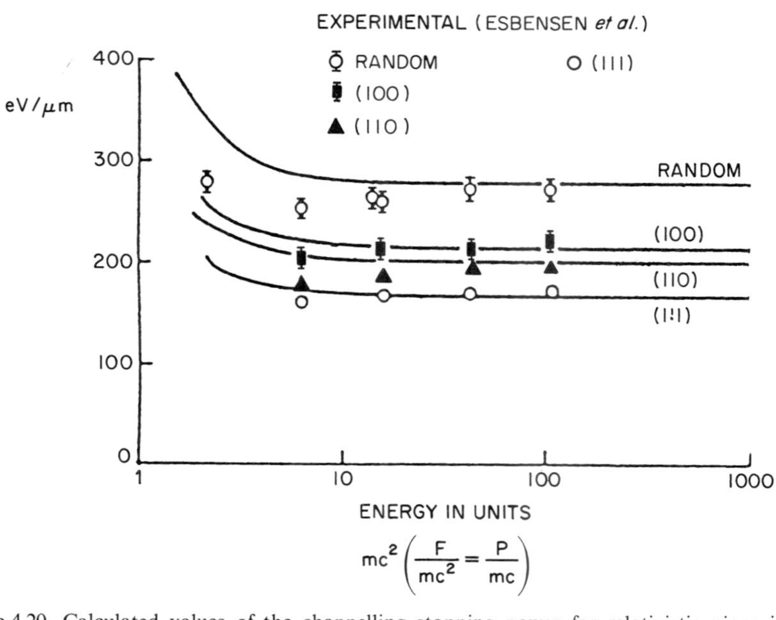

Figure 4.20. Calculated values of the channelling stopping power for relativistic pions into Si according to Burenkov *et al.*[92]

Figure 4.19(*b*) shows the stopping power function in the form of equations 4.8.47 and 4.8.49 for 60 MeV I ions into Au (100) compared with the experimental results by Datz *et al.*[88]

It is easy to obtain the initial stopping power or the channelling stopping power $-\left(\frac{\mathrm{d}E}{\mathrm{d}z}\right)_{\mathrm{ch}}$ from the stopping power function

$$\left(-\frac{\mathrm{d}E}{\mathrm{d}z}\right)_{\mathrm{ch}} = \frac{\int_0^{y_{\max}} W(y)\frac{\mathrm{d}y}{y}}{\int_0^{y_{\max}} \frac{\mathrm{d}y}{y}}$$

If we take the channelling trajectory to be in the form $y = y_{\max}\sin\omega t$, we get

$$\left(-\frac{\mathrm{d}E}{\mathrm{d}z}\right)_{\mathrm{ch}} = \frac{2}{\pi}\int_0^{y_{\max}} W(y)\frac{\mathrm{d}y}{\sqrt{y_{\max}^2 - y^2}}$$

Figure 4.20 shows the calculated values of $\left(-\frac{\mathrm{d}E}{\mathrm{d}z}\right)_{\mathrm{ch}}$ for relativistic pions in Si by Burenkov *et al.*[92] with the experimental data of Esbensen and Golovchenko.[91]

4.8.5. *Small corrections for stopping power*

Relativistic correction

Fano[96] derived the stopping power corrected by the relativistic effect for ultra high energy ions

$$-\frac{\mathrm{d}E}{\mathrm{d}z} = \frac{4\pi Z_1^2 e^4}{mv^2} N Z_2 \left(\ln\frac{2mv^2}{I} + \ln\frac{1}{1-\beta^2} - \beta^2\right) \tag{4.8.50}$$

where $\beta = v/c$. For relativistic electron beams we get

$$-\frac{\mathrm{d}E}{\mathrm{d}z} = \frac{4\pi e^4}{mv^2} N Z_2 \left\{\ln\frac{mv^2 E}{2I^2} + \ln\frac{1}{1-\beta^2} - [2(1-\beta^2)^{1/2} - 1 + \beta^2]\ln 2 + (1-\beta^2) + \tfrac{1}{8}[1-(1-\beta^2)^{1/2}]^2\right\} \tag{4.8.51}$$

Shell correction

There are no exact theories for the stopping power in the energy region labelled II in figure 4.11. However, in this region the Bethe–Bloch formula is also applicable if we exclude contributions due to deeper shell electrons, where the

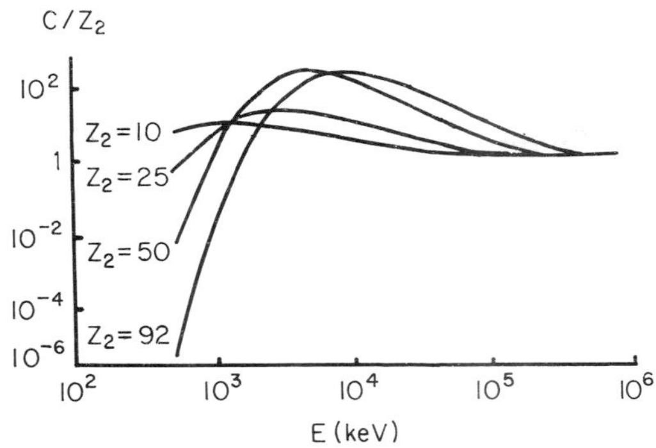

Figure 4.21. Calculated values of the correction term C/Z_2 for H^+ by Andersen and Ziegler.[97]

factor $\ln(2mv^2/I)$ in the stopping power is replaced by

$$\ln\left(\frac{2mv^2}{I}\right) - \frac{C}{Z_2} \tag{4.8.52}$$

The correction term C/Z_2 has been calculated in detail by Andersen and Ziegler.[97] Figure 4.21 shows their results for H^+ ions.

The Bragg rule

The stopping power for the target of chemical compounds is a summation over the stopping powers for component elements. If the target is composed of a chemical compound A_aB_b, the stopping power takes the form

$$-\frac{dE}{dz} = a\left(-\frac{dE}{dz}\right)_A + b\left(-\frac{dE}{dz}\right)_B \tag{4.8.53}$$

where $\left(-\frac{dE}{dz}\right)_A$ is the stopping power for element A. This formula is called the 'Bragg rule for the stopping power'. The Bragg rule is correct when the electron system is not strongly modified by chemical combination.

Z_1^3 correction (Barkas effect)

Barkas studied emulsion tracks of the decay $K \rightarrow 3\pi$, and found indications that the track length of π^- were slightly longer than that of π^+. About ten years later, Barkas and his collaborators[98] measured ranges in more detail. This difference in ranges is attributed to a Z_1^3 correction to the parentheses in equation 4.8.50, where Z_1 is ± 1 for $\pi^\pm$.

Theoretical calculations have been proposed by Ashley *et al.*,[99,100] Ritchie and Brandt,[101] and Lindhard.[102] They took into account the screening or the

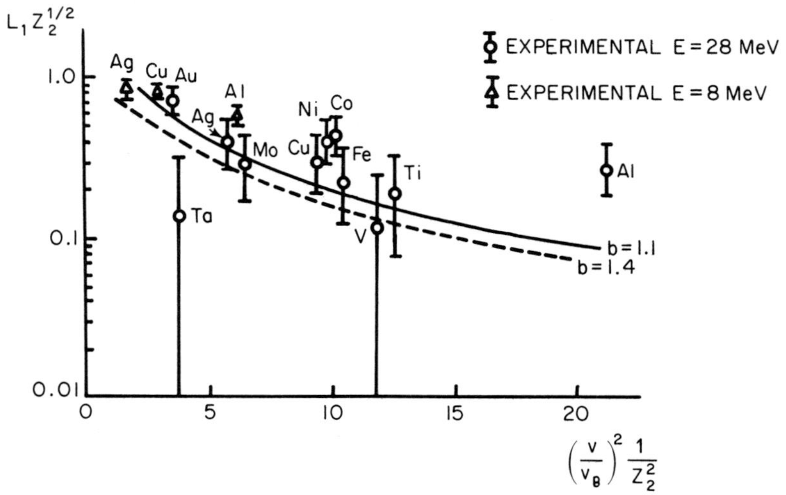

Figure 4.22. Comparison of experimental results of the correction term L_1 by Ishiwari *et al.*[103] and calculated values by Ashley *et al.*[99] The unknown parameter b is taken to be $b = 1{\cdot}1$ and $b = 1{\cdot}4$.

many body effects in the dense medium using the dielectric constant method or classical mechanics.

The Z_1^3 correction is described by $L_1 Z_1$ in the parentheses in equation 4.8.50

$$-\frac{\mathrm{d}E}{\mathrm{d}z} = \frac{4\pi Z_1^2 e^4}{mv^2}\left(\ln\frac{2mv^2}{I} + \ln\frac{1}{1-\beta^2} - \beta^2 + L_1 Z_1\right) \tag{4.8.54}$$

The value of L_1 (as a function of Z_2) calculated by Ashley *et al.* is shown in figure 4.22 with the experimental results of Ishiwari *et al.*[103] In the calculation an unknown parameter b is included, which is taken as $b = 1{\cdot}1$ and $b = 1{\cdot}4$ in the figure.

4.9. Energy loss due to photon emission

We have already studied the energy loss due to single electron excitation, which decreases rapidly with increasing beam energy. However, in the high energy region the energy loss due to photon emission (for example, due to bremsstrahlung) becomes more important. The wavelength of photons emitted in this energy region is in the X-ray or γ-ray region. When the charged particles, especially electrons and positrons, propagate along a crystal axis some kinds of coherent phenomenon can be seen.

4.9.1. *Bremsstrahlung*

It is well known in classical electrodynamics that electromagnetic radiation is emitted from electrons or positrons which are accelerated due to forces originating from the crystal atoms. According to classical electrodynamic theory the radiation energy flow per unit time emitted by the high energy electrons represented by the trajectory $\mathbf{r}(t)$ is given as

$$S=\frac{2}{3c^3}e^{2'}\ddot{\mathbf{r}}^2$$

If we know the radiation energy flow per unit time S for all electrons in the beam, we can determine the stopping power due to bremsstrahlung.

Bethe and Heitler[104] calculated the differential cross-section of the bremsstrahlung taking account of relativity and quantum effects. They considered the interaction Hamiltonian $H_{\text{int}}(\propto \mathbf{p}\cdot\mathbf{A})$ between the electron (with momentum operator $\mathbf{p}$) and the radiation (with vector potential $\mathbf{A}$), and the Coulomb potential V_C between the electron and the atom. The bremsstrahlung is represented by the Feynman diagram drawn in figure 4.23. The left side corresponds to the virtual process in which the Coulomb interaction acts before radiation emission, and the right side corresponds to the process occurring after emission.

Racah[105] calculated the stopping power $-\dfrac{\mathrm{d}E}{\mathrm{d}z}$ due to bremsstrahlung in detail, and obtained

$$-\frac{\mathrm{d}E}{\mathrm{d}z}=\tfrac{16}{3}NE\bar{\phi} \tag{4.9.1}$$

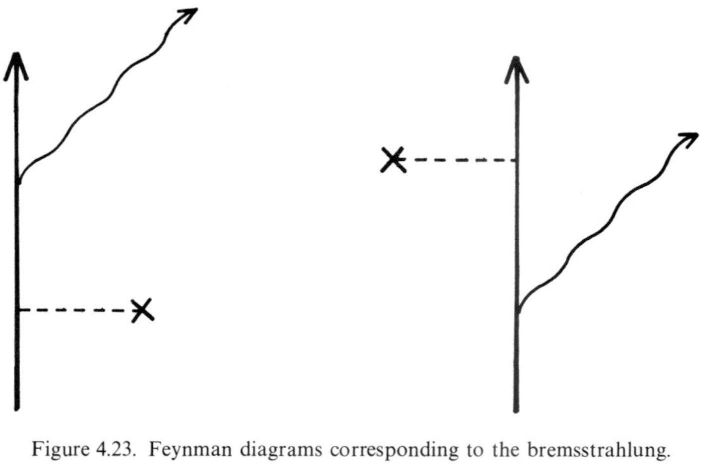

Figure 4.23. Feynman diagrams corresponding to the bremsstrahlung.

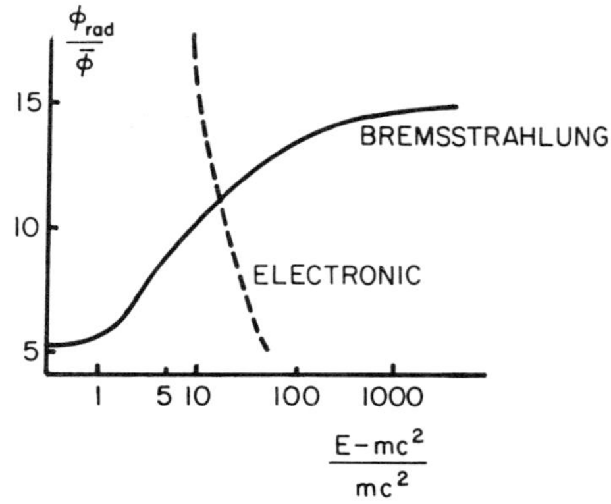

Figure 4.24. Energy dependence of the cross-section of electron excitation and bremsstrahlung emission.

for $E \ll mc^2$ and

$$-\frac{dE}{dz} = 4NE\left(\log\frac{2E}{mc^2} - \frac{1}{3}\right)\bar{\phi} \tag{4.9.2}$$

for $E \gg mc^2$, where $\bar{\phi} = Z^2 r_0^2/137$, r_0 being the classical electron radius and N the atom density. Figure 4.24 shows the calculated results of $\phi_{rad} = \left(-\frac{dE}{dz}\right)\Big/ NE$ for Pb as a function of beam energy E. Note that the stopping power due to bremsstrahlung becomes larger than that due to single electron excitation for energy values higher than $\sim 20\,mc^2$, or ~ 10 MeV.

4.9.2. *Coherent bremsstrahlung (Überall effect)*

An interference phenomenon for bremsstrahlung occurs when high energy electron and positron beams pass through a single crystal, and this is called the 'Überall effect'.[106,107] Consider an electron beam moving along a straight line with velocity v making an angle α with the atomic axis, as shown in figure 4.25. If we take the axial spacing to be d, the electron collides with atoms once in a path length $D = d/\sin\alpha$. As is clear from figure 4.25, the observer who measures the photon at an angle θ to the electron beam gets a signal of period $c(D/v) - D\cos\theta$, since the photon propagates $c(D/v)$ during the collision time D/v. If the period coincides with the photon wavelength λ satisfying the relation

$$c(D/v) - D\cos\theta = n\lambda \tag{4.9.3}$$

where $n = 1, 2, 3, \ldots$, there is interference.

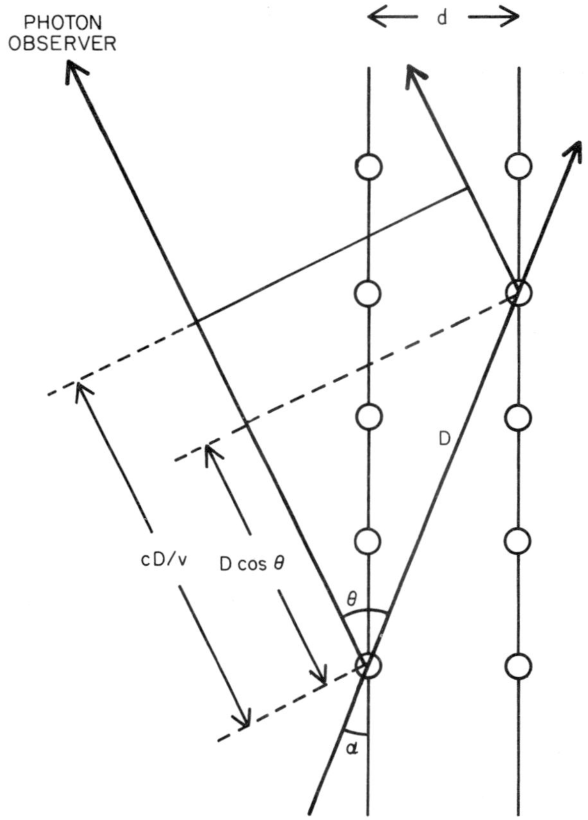

Figure 4.25. Electrons moving along a straight line with velocity v and an angle α with the atomic axis. Signal detection is of period $c(D/v) - D\cos\theta$ at the observer with an angle θ to the electron beam.

Condition 4.9.3 is rewritten as

$$\omega = n\omega_{1\theta}$$
$$\omega_{1\theta} = \frac{\omega_0}{1 - \dfrac{v}{c}\cos\theta} \tag{4.9.4}$$
$$(\omega_0 = 2\pi v/D)$$

We may interpret $\omega_{1\theta}$ as the Doppler shift of angular frequency ω_0.

Since the electron beam is relativistic (high energy), we get the classical trajectory denoted by $\mathbf{r}(t)$. The radiation intensity in the solid angle $d\Omega$ around the direction $\mathbf{n}$ in the frequency interval $(\omega, \omega + d\omega)$ takes the following form according to ordinary electrodynamics[108]

$$\frac{d^2W}{d\Omega\, d\omega} = \frac{e^2}{4\pi^2 c}\left|\int_{-\infty}^{\infty} \frac{\mathbf{n}\times[(\mathbf{n}-\boldsymbol{\beta})\times\dot{\boldsymbol{\beta}}]}{(1-\boldsymbol{\beta}\cdot\mathbf{n})^2}\exp\{i\omega[t-\mathbf{n}\cdot\mathbf{r}(t)/c]\}\,dt\right|^2$$
$$\simeq \frac{e^2\omega^2}{4\pi^2 c}\left|\int_{-\infty}^{\infty}[(\mathbf{n}\times(\mathbf{n}\times\boldsymbol{\beta})\exp\{i\omega[t-\mathbf{n}\cdot\mathbf{r}(t)/c]\}\,dt\right|^2 \tag{4.9.5}$$

We may approximate the electron trajectory near the atom by a small circle with angular velocity ω_c. For the observer in the zy plane of figure 4.26, we obtain

$$t-\mathbf{n}\cdot\mathbf{r}(t)/c = t - \frac{v}{c\omega_c}\sin\omega_c t\cos\theta \tag{4.9.6}$$

and

$$\mathbf{n}\times(\mathbf{n}\times\boldsymbol{\beta}) = \beta(-\boldsymbol{\varepsilon}_{\parallel}\sin\omega_c t + \boldsymbol{\varepsilon}_{\perp}\cos\omega_c t\sin\theta) \tag{4.9.7}$$

where $\boldsymbol{\varepsilon}_{\parallel}$ and $\boldsymbol{\varepsilon}_{\perp}$ are the unit vector in the centrifugal direction (x axis) and the unit vector perpendicular to $\mathbf{n}$ and $\boldsymbol{\varepsilon}_{\parallel}$ in figure 4.26, respectively. These small circular motions appear with every atomic collision with period $t_0 = D/v$. Therefore, equations 4.9.6 and 4.9.7 are replaced, respectively, by

$$t\left(1-\frac{v}{c}\cos\theta\right) - \frac{v}{c\omega_c}\sin\omega_c t\cos\theta\sum_m \delta_{t,mt_0} \tag{4.9.8}$$

and

$$\beta(-\boldsymbol{\varepsilon}_{\parallel}\sin\omega_c t + \boldsymbol{\varepsilon}_{\perp}\cos\omega_c t\sin\theta)\sum_m \delta_{t,mt_0} \tag{4.9.9}$$

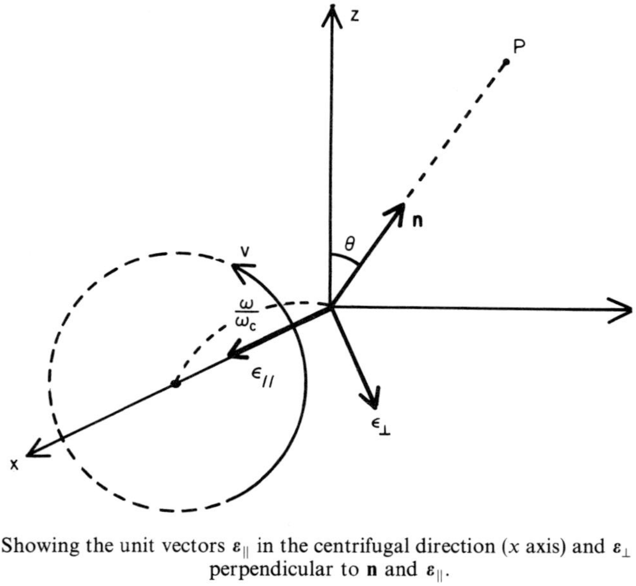

Figure 4.26. Showing the unit vectors $\boldsymbol{\varepsilon}_{\parallel}$ in the centrifugal direction (x axis) and $\boldsymbol{\varepsilon}_{\perp}$ in the direction perpendicular to $\mathbf{n}$ and $\boldsymbol{\varepsilon}_{\parallel}$.

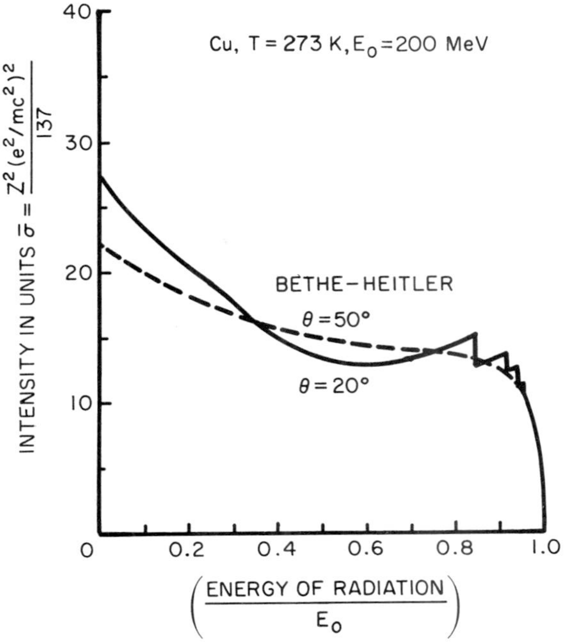

Figure 4.27. Theoretical results of the radiation power of coherent bremsstrahlung, due to Überall's analytical expression. The ordinary bremsstrahlung power due to Bethe and Heitler is also shown as the dashed line.

Inserting equations 4.9.8 and 4.9.9 into equation 4.9.5, we obtain the 'structure factor' defined by

$$\sum_m \exp(i2\pi m\omega/\omega_{1\theta})$$

which indicates the coherent condition 4.9.4.

Überall first derived the radiation power of coherent bremsstrahlung by making use of the works of Bethe and Heitler.[104] His results are represented in analytical form too complicated to show here. Typical numerical results are shown in figure 4.27, where several peaks appear in the high energy region, corresponding to coherent bremsstrahlung. For comparison, the ordinary Bethe–Heitler equation is also graphed in the same figure with the dashed line. Figure 4.28 shows the experimental result of Barbiellini *et al.*,[109] for 4.8 Ge Ve^- in diamond $\langle 110 \rangle$. The angle α was taken to be 2·48 mrad ($\theta \approx 0$). Examples of other experimental works of coherent bremsstrahlung are given in references 110 and 111.

We now consider coherent bremsstrahlung for $\alpha = 0$. The periodic field due to the atomic row with lattice constant a as seen by the electron of velocity v has the frequency v/a. Accordingly, the coherence condition takes the form

$$\omega = n\omega^*_{1\theta} \qquad (4.9.10)$$

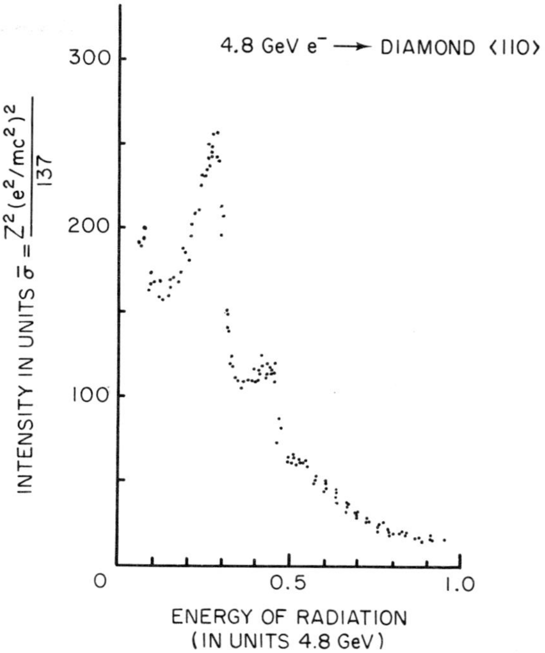

Figure. 4.28. Experimental curve of coherent bremsstrahlung for 4·8 GeV electrons in diamond ⟨110⟩ with $\alpha = 2{\cdot}48$ mrad ($\theta \approx 0$), according to Bologna *et al.*[111]

$$\omega^*_{1\theta} = \frac{2\pi v/a}{1 - \dfrac{v}{c}\cos\theta} \tag{4.9.11}$$

However, when $\alpha = 0$ the electron beam may be trapped in the atomic axis or plane, i.e., the channelling trajectories, which results in a slowly varying periodic motion. Kumakhov[112,113] demonstrated another coherent radiation from the channelling trajectories, which we call 'channelling radiation'. For an electron beam which oscillates with angular frequency $\bar{\omega}$ in the laboratory system, the coherence condition takes the form

$$\omega = n\omega_m$$

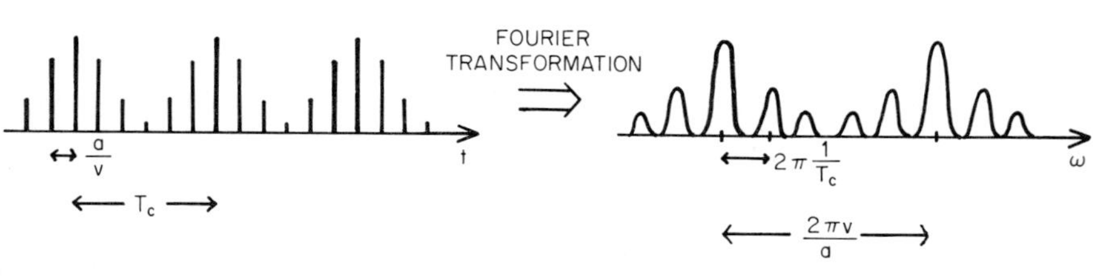

Figure 4.29. Relation between coherent bremsstrahlung and channelling radiation. The values a/v and T_c indicate the field periodicity and the channelling periodicity, respectively.

$$\omega_m = \frac{\bar{\omega}}{1 - \dfrac{v}{c}\cos\theta} \simeq 2\gamma^2\bar{\omega} \tag{4.9.12}$$

from which we can see that the channelling radiation peaks appear in the low energy region. Figure 4.29 shows the schematic relationship between the coherent-bremsstrahlung spectrum and the channelling radiation spectrum, caused by the double periodicity (the field periodicity a/v and the channelling periodicity T_c) of the acceleration of the electron beam.

4.9.3. *Channelling radiation*

Kumakhov first calculated the planar channelling radiation spectrum for very high energy positrons where the quantum effect is negligible. The channelling trajectories are represented by the simple sine function

$$y = y_m \sin(\bar{\omega}t + \delta_0) \tag{4.9.13}$$

from which we obtain the acceleration

$$\ddot{y} = -\bar{\omega}^2 y_m \sin(\bar{\omega}t + \delta_0) \tag{4.9.14}$$

Where $\bar{\omega}$ is the relativistically modified angular frequency ω_0 determined by the channelling condition

$$\bar{\omega}^2 = \omega_0^2/\gamma \tag{4.9.15}$$

The origin of the modification 4.9.15 is the use of the relativistic mass in the equation for angular frequency ω_0.

If the positron (electron) trajectory is represented by the position vector $\mathbf{r}(t)$, the radiation intensity emitted in the solid angle $\mathrm{d}\Omega$ in the direction $\mathbf{n}$ in the frequency interval $(\omega, \omega + \mathrm{d}\omega)$ takes the form

$$\frac{\mathrm{d}^2 W}{\mathrm{d}\omega\,\mathrm{d}\Omega} = 2|A(\omega)|^2$$

where

$$A(\omega) = \sqrt{\frac{e^2}{8\pi^2 c}} \int_{-\infty}^{\infty} \frac{\mathbf{n} \times [(\mathbf{n} - \boldsymbol{\beta}) \times \dot{\boldsymbol{\beta}}]}{(1 - \boldsymbol{\beta}\cdot\mathbf{n})^2} \exp\{i\omega[t - \mathbf{n}\cdot\mathbf{r}(t)/c]\}\,\mathrm{d}t \tag{4.9.16}$$

in which

$$\boldsymbol{\beta} = \dot{\mathbf{r}}(t)/c$$

If we define ω' by

$$\omega' = \omega\left(1 - \frac{\mathbf{n}\cdot\mathbf{v}}{c}\right)$$

$$= \omega\left(1 - \frac{v\cos\theta}{c}\right) \tag{4.9.17}$$

$$\doteq \omega\left(1-\frac{v}{c}+\frac{v}{2c}\theta^2\right) \tag{4.9.18}$$

the integral 4.9.16 may be reduced to the evaluation of

$$\dot{\boldsymbol{\beta}}_{\omega'} = \int_{-\infty}^{\infty} \dot{\boldsymbol{\beta}} \exp(i\omega' t)\,\mathrm{d}t \tag{4.9.19}$$

since the velocity $c\boldsymbol{\beta}$ is considered to be constant for very high energy positrons.

For the planar channelled positrons, the acceleration vector $c\dot{\boldsymbol{\beta}}$ takes the form

$$c\dot{\boldsymbol{\beta}} = c\dot{\boldsymbol{\beta}}_y = -\bar{\omega}^2 y_m \sin(\bar{\omega}t+\delta_0)$$

and its Fourier transform is

$$c\dot{\boldsymbol{\beta}}_{\omega'} = -\bar{\omega}^2 y_m \delta_{\omega',\bar{\omega}} \tag{4.9.20}$$

Taking into account equation 4.9.20, we obtain

$$\begin{aligned}\frac{\mathrm{d}W}{\mathrm{d}\omega} &= \int 2|A(\omega)|^2\,\mathrm{d}\Omega = 2\int\int |A(\omega)|^2\,\mathrm{d}\varphi\sin\theta\,\mathrm{d}\theta \\ &= \frac{e^2}{4\pi^2 c}\int \mathrm{d}\varphi \int \sin\theta\,\mathrm{d}\theta \frac{|\mathbf{n}\times[(\mathbf{n}-\boldsymbol{\beta})\times\dot{\boldsymbol{\beta}}_{\omega'}]|^2}{(1-\boldsymbol{\beta}\cdot\mathbf{n})^4}\end{aligned} \tag{4.9.21}$$

Noting that $\sin\theta\,\mathrm{d}\theta \doteq \frac{c}{v\omega}\mathrm{d}\omega'$ according to equation 4.9.18, and inserting equation 4.9.19 into 4.9.21, we get the final form

$$\begin{aligned}\frac{\mathrm{d}W}{\mathrm{d}\omega} &= \frac{e^2}{4\pi^2 v\omega}\int \mathrm{d}\varphi \int_{\omega m}^{\infty} \frac{|\mathbf{n}\times[(\mathbf{n}-\boldsymbol{\beta})\times\dot{\boldsymbol{\beta}}_{\omega'}]|^2}{(1-\boldsymbol{\beta}\cdot\mathbf{n})^4}\mathrm{d}\omega' \\ &= I_0\frac{3\omega}{\omega_m^2}\left[1-2\frac{\omega}{\omega_m}+2\left(\frac{\omega}{\omega_m}\right)^2\right]\end{aligned} \tag{4.9.22}$$

where I_0 is the total intensity defined by

$$I_0 = \bar{\omega}^4 y_m^2 \gamma^4 e^2/3c^3 \tag{4.9.23}$$

In fact, we can get I_0 from equation 4.9.22

$$\int_0^{\omega_m}\frac{\mathrm{d}W}{\mathrm{d}\omega}\mathrm{d}\omega = I_0\frac{3}{\omega_m^2}\left(\tfrac{1}{2}\omega_m^2 - \tfrac{2}{3}\omega_m^2 + \tfrac{1}{2}\omega_m^2\right) = I_0$$

The intensity spectrum $\frac{\mathrm{d}W}{\mathrm{d}\omega}$ is plotted in figure 4.30 for the 5 GeV electrons in the $\langle 110\rangle$ direction in diamond.[115]

To extend the classical theory of channelling radiation to the quantum mechanical one is not so difficult. Our starting point is the Dirac equation

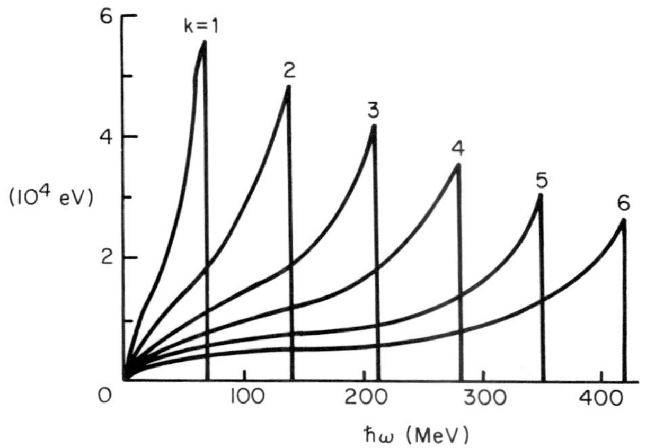

Figure 4.30. The intensity spectrum of channelling radiation for 5 GeV electrons in diamond ⟨110⟩.

$$(E - U - mc^2)\varphi = c(\hat{\sigma}\hat{p})\chi$$

$$(E - U + mc^2)\chi = c(\hat{\sigma}\hat{p})\varphi \tag{4.9.24}$$

where φ and χ are the two component spinor functions and $\hat{\sigma}$ is the Pauli matrix. U is the continuum potential for the electron or positron beam. For a harmonic potential we can solve equation 4.9.24 easily and obtain the quantum levels, which correspond to standing waves in waveguides. The channelling radiation is emitted by transitions between these levels.[116] The transition probability $W_{n_1n_2}$ resulting in a photon emission with polarization σ and momentum $\hbar k$ takes the form[117]

$$W_{n_1n_2} = \frac{e^2}{\hbar}\frac{4\pi^2}{L^3}\sum_k \frac{\delta(\omega_{n_1n_2} - \omega)}{k}\Phi_\sigma \tag{4.9.25}$$

and

$$\Phi_\sigma = (\alpha^*_{n_1n_2}\beta_\sigma)(\alpha_{n_1n_2}\beta_\sigma) \tag{4.9.26}$$

In the above, L is the normalization constant, $\hbar\omega_{n_1n_2}$ is the transition energy, and $\alpha_{n_1n_2}$ is the transition matrix element.

Integrating over the solid angle we obtain the radiation intensity spectrum

$$\frac{\mathrm{d}I}{\mathrm{d}\omega} \approx \frac{3I_{n_1n_2}\omega}{\omega_m^2}\left[1 - 2\frac{\omega}{\omega_m} + 2\left(\frac{\omega}{\omega_m}\right)^2\right] \tag{4.9.27}$$

where

$$I_{n_1n_2} = \frac{4e^2x^2_{n_1n_2}\omega^4_{n_1n_2}\gamma^4}{3c^3} \tag{4.9.28}$$

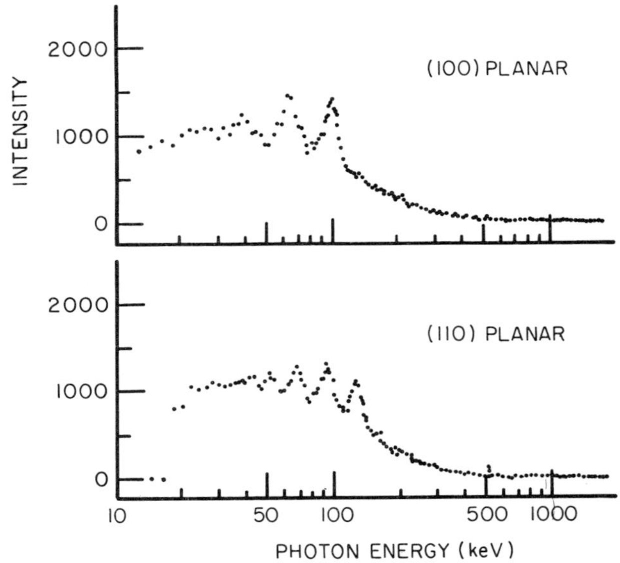

Figure 4.31. Experimental results of planar channelling radiation for 56 MeV electrons in Si by Swent *et al.*[118] (*a*) (100) planar, (*b*) (110) planar.

In the above, $\omega_m = \omega_{n_1 n_2}[1-(v/c)\cos\theta]$, which is the maximum frequency, and $x_{n_1 n_2}$ is the oscillator strength of the transition.

Figure 4.31 shows the experimental results of the planar channelling radiation of 56 MeV electrons in Silicon (Si) by Swent *et al.*,[118] (*a*) for (100) planar channelling and (*b*) for (110) planar channelling. The intensities due to randomly directed beams were subtracted from those due to planar channelling radiation. 'Random direction' refers to an incident angle that is slightly deviated

Table 4.2. Theoretical and experimental peak profiles for (100) and (110) planar channelling.

Surface	Theoretical results[119] (keV)	Experimental results[118] (keV)
(100)	109·1	99
	72·3	64
	50·3	39
(110)	139·1	128
	101·4	94
	75·4	68
	58·2	52

from the channelling condition, which also results in forward radiation. The peak positions for the (100) planar case are 99 keV, 64 keV and 39 keV; and those for the (110) planar case are 128 keV, 94 keV and 68 keV. Calculations by Komaki and Fujimoto[119] of the peak profile using quantum mechanics also correspond to the experiments of Swent *et al.*[118] Their results are listed in table 4.2.

In contrast to planar channelling, the calculation of axial channelling is complicated whether using classical or quantum mechanics, since the trajectories in axial channelling are difficult to obtain. There is no rigorous theory for these cases. However, for very high energy, computer simulation is useful. The acceleration vector $c\dot{\boldsymbol{\beta}}$ has been computed by simulated trajectories.

As an example, figure 4.32 shows the results of a computer simulation[120,121] (open circles) for 56 MeV electrons in Si⟨110⟩ with an initial angular beam spread of $\Delta\theta = 0{\cdot}114°$, compared to the experimental results (black circles) of Swent *et al.*[118] Figure 4.33 shows results for 600 MeV electrons in diamond ⟨110⟩ (open circles), compared with the experimental ones (black circles) of Adishchev *et al.*[122] with an initial angular beam spread of $\Delta\theta = 0{\cdot}0172°$.

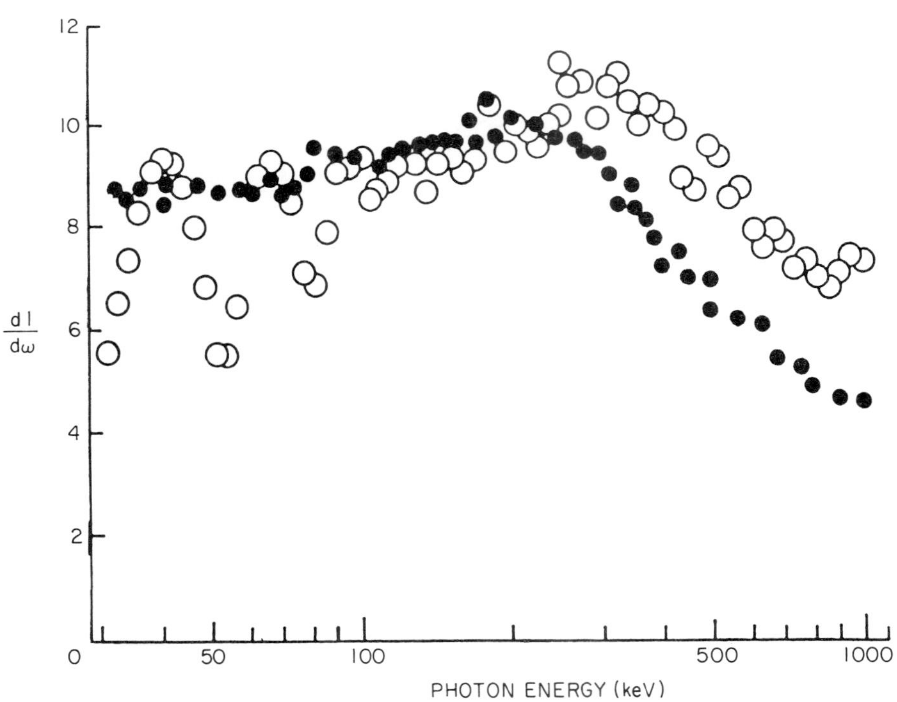

Figure 4.32. Comparison between the results of computer simulation[120] (open circles) for 56 MeV electrons in Si⟨110⟩ and the experimental results (black circles) of Swent *et al.*[118]

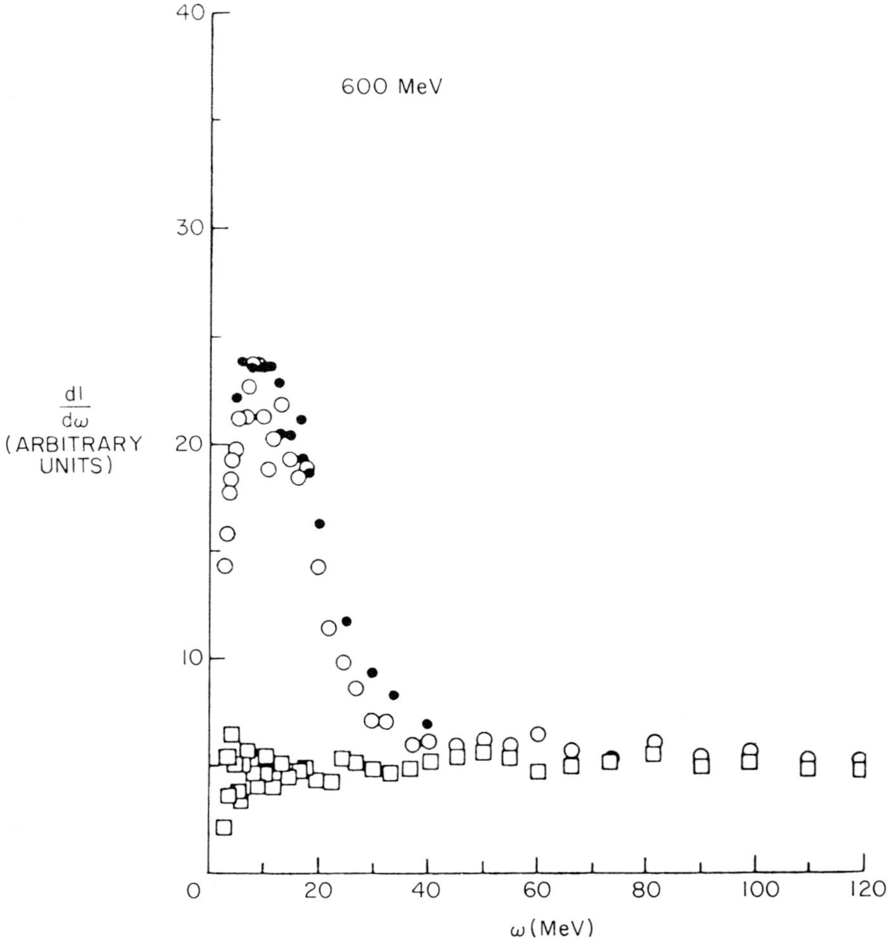

Figure 4.33. Channelling radiation intensity obtained by computer simulation for 600 MeV electrons in the diamond ⟨110⟩ (open circles) compared to the experimental intensity (black circles) of Adishchev *et al.*[122] The squares describe the random case.

Table 4.3. Various calculation methods for planar and axial channelling at low and high energy.

	Low energy (quantum mechanical)	High energy (classical)
Planar channelling	Analytical + computer (Dirac equation)	Analytical (sine-type trajectories)
Axial channelling	Computer (*N*-beam diffraction theory)	Computer simulation

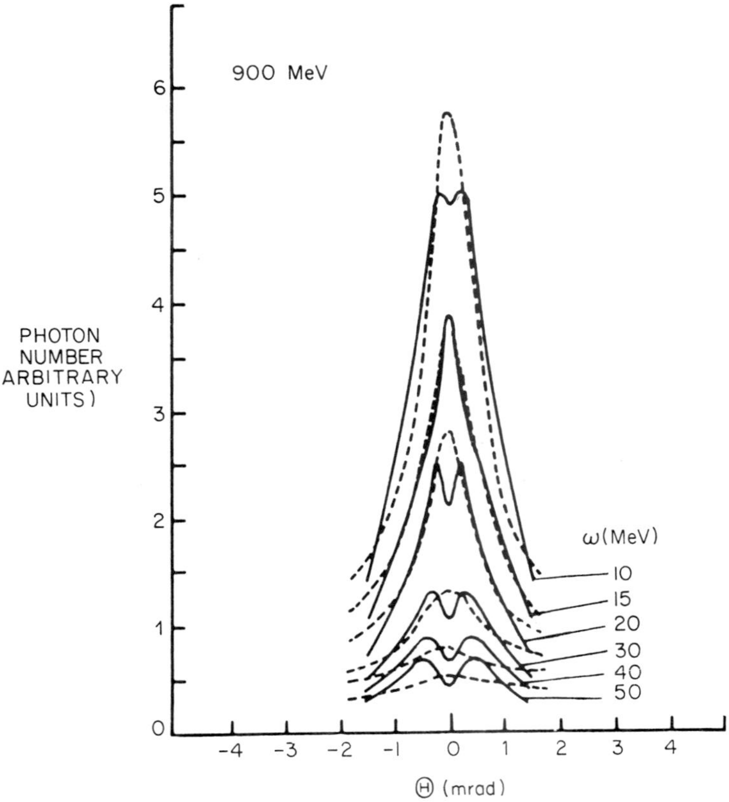

Figure 4.34. Emission-angle dependence of the photon intensity for 900 MeV electrons in diamond ⟨110⟩. The solid and dotted lines show the results of computer simulations and of experiments, respectively.

The rectangles depict the intensity spectrum for the random beam with an incident angle of $\theta = 3^\circ$ to the ⟨110⟩ axis. It should be noted that we can see in the simulation a fine structure near $\hbar\omega = 20$ MeV. In figure 4.34 comparisons are made between the computer simulation (solid lines) for the dependence of the channelling radiation of 900 MeV electrons on the emission angle for diamond ⟨110⟩ and the experimental results (dashed lines) by Adishchev *et al.*[122] On the whole, the computer simulations of Yamamura and Ohtsuki[120,121] agree well with the experimental results.

It is more difficult to make quantum-mechanical calculations for axial channelling radiation due to low energy electrons or positrons.[123] Table 4.3 shows the various calculation methods for planar and axial cases divided into two energy regions. Kalashnikov and Olchack have recently proposed and calculated[124] the Mössbauer-type recoil-less channelling radiation.

References

1. Ohtsuki, Y.H., *J. Phys. Soc. Japan* **24**, 555 (1968).
2. Borrmann, G., *Phys. Z.* **42**, 157 (1940).
3. Borrmann, G., *Z. f. Phys.* **127**, 297 (1950).
4. Borrmann, G., *Naturwiss.* **38**, 330 (1951).
5. Yoshioka, H., *J. Phys. Soc. Japan* **12**, 618 (1957).
6. Hashimoto, H., Howie, A. and Whelan, M.J., *Proc. Roy. Soc. London* **A269**, 80 (1962).
7. Pines, D., *Elementary Excitations in Solids* (New York: Benjamin, 1964).
8. Lindhard, J., *Kgl. Dan. Viden. Sels. Mat-Fys. Medd.* **28**, 8 (1954).
9. Ferrell, R.A., *Phys. Rev.* **101**, 554 (1956).
10. Gjøness, J., *J. Phys. Soc. Japan* **17**, Supplement BII, 137 (1962).
11. *Int. Table for X-ray Crystallography*, III (The Kynoch Press, 1962).
12. Freeman, A.J. *Acta Cryst.* **12**, 929 (1959).
13. Whelan, M.J., *J. Appl. Phys.* **36**, 2099 (1965).
14. Radi, G., *Acta Cryst.* **A26**, 41 (1970).
15. Ohtsuki, Y.H., *J. Phys. Soc. Japan* **21**, 2300 (1966).
16. Horton, G.K. and Maradudin, A.A. (eds.), *Dynamical Properties of Solids*, Vol. 1 (Amsterdam: North Holland, 1974).
17. Hall, C.R. and Hirsch, P.B., *Proc. Roy. Soc. London* **286**, 158 (1965).
18. Hall, C.R., *Phil. Mag.* **12**, 815 (1965).
19. Yoshioka, H. and Kainuma, Y., *J. Phys. Soc. Japan* 17 Supplement BII, 134 (1962).
20. Mott, N.F. and Massey, H.S.W., *The Theory of Atomic Collisions* (Oxford University Press, 1965).
21. Goldberger, M.L. and Watson, K.M., *Collision Theory* (New York, 1964).
22. Kohra, K. and Watanabe, H., *J. Phys. Soc. Japan* **16**, 580 (1961).
23. Watanabe, H., *Japan J. Appl. Phys.* **3**, 480 (1964).
24. Dupouy, G., Perrier, F., Uyeda, R., Ayroles, R. and Mazel, A., *J. Microscopie* **4**, 429 (1965).
25. Castaing, R., El Hili, A. and Henry, L., *C.R. Acad. Sci.* **262B**, 169 (1966).
26. Castaing, R., El Hili, A. and Henry, L., *C.R. Acad. Sci.* **262B**, 1051 (1966).
27. Uyeda, R. and Nonoyama, M., *Japan J. Appl. Phys.* **4**, 498 (1965).
28. Fukuhara, A., *J. Phys. Soc. Japan* **18**, 964 (1963).
29. Lehmpfuhl, G. and Molière, K., *Z. Phys.* **164**, 389 (1961).
30. Lehmpfuhl, G. and Molière, K., *J. Phys. Soc. Japan* **17**, Supplement BII, 130 (1962).
31. Goodman, P. and Lehmpfuhl, G., *Acta Cryst.* **22**, 14 (1967).
32. Hashimoto, M., Howie, A. and Whelan, M.J., *Proc. Roy. Soc. London* **A269**, 80 (1962).
33. Hashimoto, H., *J. Appl. Phys.* **35**, 277 (1964).
34. Meyer, G., *Phys. Letters* **20**, 240 (1966).
35. Meyer, G., *Z. Naturforsch.* **21a**, 1524 (1966).
36. Horstmann, M. and Meyer, G., *Acta Cryst.* **15**, 271 (1962).
37. Boersch, H., Bostanjoglo, O. and Niedrig H., *Z. Phys.* **180**, 407 (1964).
38. Kuwabara, S., *J. Phys. Soc. Japan* **22**, 1245 (1967).
39. Hansen-Schmidt, J. and Horstmann, M., *Z. Naturforsch.* **20a**, 1239 (1965).
40. Hansen-Schmidt, J., *Z. Phys.* **198**, 433 (1967).
41. Goringe, M.J. and Whelan, M.J., *Electron Microscopy* (Tokyo: Maruzen, 1966).
42. Fujimoto, F. and Howie, A., *Phil. Mag.* **13**, 1131 (1966).
43. Takagi, S., *J. Phys. Soc. Japan* **13**, 287 (1958).
44. Howie, A., *Proc. Roy. Soc. London* **A271**, 268 (1963).
45. Fukuhara, F., *J. Phys. Soc. Japan* **18**, 496 (1963).

46. Gjønness, J., *Acta Cryst.* **20**, 240 (1966).
47. Cowley, J.M. and Pogany, A.P., *Acta Cryst.* **A24**, 109 (1968).
48. Doyle, P.A., *Acta Cryst.* **A25**, 569 (1969).
49. Kikuchi, S., *Proc. Imp. Acad. Japan* **4**, 271 (1928).
50. Kikuchi, S., *Proc. Imp. Acad. Japan* **4**, 275 (1928).
51. Nishikawa, S. and Kikuchi, S., *Nature* **121**, 1019 (1928).
52. Kikuchi, S., *Proc. Imp. Acad. Japan* **4**, 354 (1928).
53. Kikuchi, S., *Proc. Imp. Acad. Japan* **4**, 471 (1928).
54. Shinohara, K., *I.P.C.R.* **18**, 223 (1932).
55. Shinohara, K., *I.P.C.R.* **20**, 39 (1932).
56. Hayashi, T., *Sci. Rep. Tohoku Imp. Univ.* **23**, 491 (1934).
57. Shinohara, K., *Phys. Rev.* **47**, 730 (1935).
58. Kainuma, Y., *J. Phys. Soc. Japan* **8**, 685 (1953).
59. Takagi, S., *J. Phys. Soc. Japan* **13**, 278 (1958).
60. Takagi, S., *J. Phys. Soc. Japan* **13**, 287 (1958).
61. Okamoto, K., Ichinokawa, T. and Ohtsuki, Y.H., *Phys. Letters* **31A**, 570 (1970).
62. Okamoto, K., Ichinokawa, T. and Ohtsuki, Y.H. *J. Phys. Soc. Japan* **30**, 1690 (1971).
63. Ichinokawa, T., Kamigaki, Y. and Ohtsuki, Y.H., *Japan J. Appl. Phys.* **8**, 969 (1969).
64. Kogiso, M. and Kainuma, Y., *J. Phys. Soc. Japan* **25**, 498 (1968).
65. Kainuma, Y. and Kogiso, M., *Acta Cryst.* **A24** 81 (1968).
66. Gjønness, J. and Høier, R., *Acta Cryst.* **A25**, 300 (1969).
67. Hall, C.R., *Phil. Mag.* **22**, 63 and 175 (1970).
68. Ishida, K., *J. Phys. Soc. Japan* **30**, No. 5 (1971).
69. Mosser, A., Burggraf, C., Goldsztaub, S. and Ohtsuki, Y.H., *Surf. Sci.* **54**, 580 (1976).
70. Bohr, N., *Phil. Mag.* **25**, 10 (1913).
71. Bohr, N., *Phil. Mag.* **30**, 581 (1915).
72. Bohr, N., *Kgl. Dan. Viden. Sels. Mat-Fys. Medd.* **18**, No. 8 (1948).
73. Bethe, H., *Ann. Phys. (Germany)* **5**, 325 (1930).
74. Bloch, F., *Ann. Phys. (Germany)* **16**, 285 (1933).
75. Lindhard, J., *Kgl. Dan. Viden. Sels. Mat-Fys. Medd.* **28**, No. 8 (1954).
76. Lindhard, J. and Winter, A., *Kgl. Dan. Viden. Sels. Mat-Fys. Medd.* **34**, No. 4 (1964).
77. Firsov, O.B., *Zh. Eksp. Teor. Fiz.* **36**, 1517 (1959).
78. Bøttiger, J. and Bason, F., *Rad. Effects* **2**, 105 (1969).
79. Eisen, F.H., *Can. J. Phys.* **46**, 561 (1968).
80. Land, D.J. and Brennen, J.G., *Nucl. Instr. Meth.* **132**, 89 (1976).
81. Cheshire, I.M. and Poate, J.M., in *Atomic Collision Phenomena in Solids* (North Holland, 1970), p. 351.
82. Bhalla, C.P., Bradford, J.N. and Reese, G., in *Atomic Collision Phenomena in Solids* (North Holland, 1970), p. 361.
83. Lindhard, J. and Scharff, M., *Phys. Rev.* **124**, (1961).
84. Lindhard, J., Scharff, M., and Schiøtt, S., *Kgl. Dan. Vid. Sels. Mat-Fys. Medd.* **33**, No. 14 (1963).
85. Kitagawa, M. and Ohtsuki, Y.H., *Phys. Rev.* **B**9, 4719 (1974).
86. Briggs, J.S. and Pathak, A.P., *J. Phys. C* **6**, L153 (1973).
87. Nelson, R.S., *The Observation of Atomic Collisions in Crystalline Solids* (North Holland, 1968).
88. Datz, S., Moak, C.D., Noggle, T.S., Appleton, B.R. and Lutz, H.O., *Phys. Rev.* **179**, 315 (1969).
89. Ohtsuki, Y.H., *J. Phys. Soc. Japan* **34**, 129 (1973).
90. Kitagawa, M. and Ohtsuki, Y.H., *Phys. Rev.* **B5**, 3418 (1972).
91. Esbensen, H. and Golovchenko, J.A., *Nucl. Physics* **A298**, 382 (1978).
92. Burenkov, A.F., Komarov, F.F. and Kumakhov, M.A., *Phys. Stat. Sol.* **99**, 417 (1980).

93. Ohtsuki, Y.H. and Kitagawa, M., *Phys. Letters* **A40**, 313 (1972).
94. Dettmann, K. and Robinson, M.T., *Phys. Rev.* **B10**, 1 (1974).
95. Robinson, M.T., *Phys. Rev.* **B4**, 1461 (1971).
96. Fano, U., *Ann. Rev. Nucl. Sci.* **13**, 1 (1963).
97. Andersen, H.H. and Ziegler, J.F., in *The Stopping Power and Ranges of Ions in Matter* edited by J.F. Ziegler (Pergamon, 1977).
98. Barkas, W.H., Dyer, N.J. and Heckman, H.H., *Phys. Rev. Lett.* **11**, 26 (1963).
99. Ashley, J.C., Brandt, W. and Ritchie, R.H., *Phys. Rev.* **B5**, 2393 (1972).
100. Ashley, J.C., Brandt, W. and Ritchie, R.H., *Phys. Rev.* **A8**, 2402 (1973).
101. Ritchie, R.H. and Brandt, W., *Phys. Rev.* **A17**, 2102 (1978).
102. Lindhard, J., *Nucl. Inst. Methods* **132**, 1 (1976).
103. Ishiwari, R., Shiomi, N. and Sakamoto, N., *Phys. Lett.* **A75**, 112 (1979).
104. Bethe, H. and Heitler, W., *Proc. Roy. Soc. London* **A146**, 83 (1934).
105. Racah, G., *Nuovo Cimento* **11**, No. 7 (1934).
106. Überall, H., *Phys. Rev.* **103**, 1055 (1956).
107. Diambrini Palazzi, G., *Rev. Mod. Phys.* **46**, 129 (1974).
108. Jackson, J.D., *Classical Electrodynamics* (New York: Wiley, 1965).
109. Barbiellini, G., Bologna, G., Diambrini, G. and Murtas, G.P., *Phys. Rev. Lett.* **9**, 396 (1962).
110. Kifune, T., Kimura, Y., Kobayashi, M., Kondo, K. and Nishikawa, T., *J. Phys. Soc. Japan* **21**, 1905 (1966).
111. Bologna, G., Lutz, G., Schulz, H.D., Timm, U. and Zimmermann, W., *Nuovo Cimento* **A42**, 844 (1966).
112. Kumakhov, M.A., *Phys. Lett.* **D57**, 17 (1976).
113. Kumakhov, M.A., *Phys. Stat. Sol.* **B84**, 41 (1977).
114. Bair, V.N., *et al.*, *Sov. Phys. Dokl.* **24** (6) (1979).
115. Kumakhov, M.A., and Trikalinos, C. *Zh. Eksp. Teor. Fiz.* **78**, No. 4 (1980).
116. Terhune, R.W. and Pantell, R.H., *Appl. Phys. Lett.* **30**, 265 (1977).
117. Beloshitsky, V.V. and Kumakhov, M.A., *Zh. Eksp. Teor. Fiz.* **74**, 1244 (1978).
118. Swent, R.L., Pantell, R.H., Alguard, M.J., Berman, B.L., Bloom, S.D. and Datz, S., *Phys. Rev. Lett.* **43**, 1723 (1979).
119. Komaki, K. and Fujimoto, F., *Rad, Effects* **56**, 13 (1981).
120. Yamamura, Y. and Ohtsuki, Y.H. *Phys. Rev.* **B24**, 3430 (1981).
121. Yamamura, Y. and Ohtsuki, Y.H., *Rad. Effects* **56**, 1 (1981).
122. Adishchev, Yu.N., Kaplin, V.V., Potalitsin, A.P. and Vorobiev, S.A., *Phys. Lett*, **A75**, 316 (1980).
123. Andersen, J.U. and Laegsgard, E., *Phys. Rev. Lett.* **44**, 1079 (1980).
124. Kalashnikov, N.P. and Olchak, A.S., *Nuovo Cimento*, to be published.

CHAPTER 5

dechannelling

5.1. Phenomenological theory

Channelled beams escape from the channelling path in various ways; this is called 'dechannelling'. For an ideal crystal dechannelling may be attributed to single electron excitation, plasmon excitation and lattice vibrations due to phonon excitation. Lindhard[1] first presented the phenomenological theory of dechannelling in 1965. As an introduction to dechannelling we shall start with Lindhard's theory.

We assume a continuum potential which exerts a force

$$\mathbf{F} = -\operatorname{grad} U(\mathbf{r}) \tag{5.1.1}$$

on the channelled particles. The force $\mathbf{F}$ may deviate from $\mathbf{F}$ to $\mathbf{F} + \delta\mathbf{F}$ due to lattice vibrations, for example. The channelled path is changed by this $\delta\mathbf{F}$ to a random path. The transverse momentum change $\delta\mathbf{p}_\perp$ of a charged particle of mass M_1 and velocity v while traversing a path length of one atomic spacing d is given by

$$\delta\mathbf{p}_\perp = \frac{d}{v}\delta\mathbf{F}(\mathbf{r}) \tag{5.1.2}$$

Thus the transverse energy change of the channelled ions during the path length δz will be

$$\delta\langle E_\perp\rangle_{\mathrm{ch}} = \sum_{\mathrm{atoms}}^{\delta N} \left\langle \frac{\delta p_\perp^2}{2M_1} \right\rangle_{\mathrm{ch}} = \frac{(\delta N)d^2}{4E}\langle \delta\mathbf{F}(\mathbf{r})^2\rangle_{\mathrm{ch}}$$

$$= \frac{d\delta z}{4E}\langle \delta\mathbf{F}(\mathbf{r})^2\rangle_{\mathrm{ch}} \tag{5.1.3}$$

where $\delta N = \delta z/d$ is the number of atoms within the path length δz, and $E = M_1 v^2/2$ is the kinetic energy of the incident ion beam. We use $\langle \ldots \rangle_{\mathrm{ch}}$ for the average over the channelling path.

We now take into account the effect of lattice vibrations on the force deviation $\delta\mathbf{F}(\mathbf{r})$. If we represent the deformation of the lattice by the vector $\boldsymbol{\rho}$, the interaction potential ΔU between the ion beam and the lattice deviation is given by

$$\Delta U = -\boldsymbol{\rho}\cdot \mathrm{grad}\, U(\mathbf{r}) \tag{5.1.4}$$

from which we obtain

$$\delta \mathbf{F}(\mathbf{r}) = -\frac{\partial(\Delta U)}{\mathrm{d}\mathbf{r}} = \frac{\partial}{\partial \mathbf{r}}(\boldsymbol{\rho}\cdot \mathrm{grad}\, U(\mathbf{r})) \tag{5.1.5}$$

We can represent expression 5.1.5 in polar co-ordinates

$$\delta \mathbf{F}(\mathbf{r})^2 = \left[\frac{\partial}{\partial r}(\boldsymbol{\rho}\cdot \mathrm{grad}\, U(\mathbf{r}))\right]^2 + (1/r^2)\left[\frac{\partial}{\partial \theta}[\boldsymbol{\rho}\cdot \mathrm{grad}\, U(\mathbf{r})]\right]^2$$

$$= \left[\frac{\partial}{\partial r}(\boldsymbol{\rho}\cdot \mathbf{F})\right]^2 + (1/r^2)\left[\frac{\partial}{\partial \theta}(\boldsymbol{\rho}\cdot \mathbf{F})\right]^2$$

The terms including $(\boldsymbol{\rho}\cdot \mathbf{F}) = \rho \mathbf{F} \cos\theta$ should be replaced by their average values, because of axial symmetry. Thus we obtain

$$\delta \mathbf{F}(\mathbf{r})^2 = \rho^2(\mathbf{F}'^2 \cos^2\theta + (1/r^2)\mathbf{F}^2 \sin^2\theta$$

$$\rightarrow \frac{\rho^2}{2}\left(\mathbf{F}'^2 + \frac{\mathbf{F}^2}{r^2}\right) \tag{5.1.6}$$

Inserting equation 5.1.6 into equation 5.1.3, we obtain

$$\frac{\delta \langle E_\perp \rangle_{\mathrm{ch}}}{\delta z} = \frac{\rho^2 d}{8E}\left\langle \mathbf{F}'^2 + \frac{\mathbf{F}^2}{r^2}\right\rangle_{\mathrm{ch}} \tag{5.1.7}$$

which is the expression derived by Lindhard for the 'diffusion power' and represents the transverse diffusion of energy over path length δz.

On the other hand, we can easily derive the diffusion power for single electron excitation, since the derivation is the same as that of Bohr[2] for the diffusion of random beams.

We assume that an ion with mass M_1 collides with an electron with mass m, and acquires a transverse velocity $v_\perp$. The resulting change in kinetic energy of the ion is

$$\delta E_\perp = \tfrac{1}{2} M_1 v_\perp^2$$

The electron may be scattered by the collision with the ion and acquire a transverse velocity $V_\perp$ and a velocity component in the channelling direction V_z. We obtain from transverse momentum conservation

$$M_1 v_\perp = m V_\perp$$

Thus, we have

$$\delta E_\perp = \tfrac{1}{2} M_1 v_\perp^2 = \tfrac{1}{2} M_1 \cdot (m/M_1)^2\, V_\perp^2$$
$$= (m/M_1)\tfrac{1}{2} m V_\perp^2$$

If we consider the conservation of energy for an ion and an electron, the energy loss of the ion over path length δz, $\delta z\left(-\dfrac{\mathrm{d}E}{\mathrm{d}z}\right)$, equals the gain in kinetic energy by the electron. Thus

$$\left(-\frac{dE}{dz}\right)\delta z = \tfrac{1}{2}mV_{\perp}^{2}$$

where $\left(-\frac{dE}{dz}\right)$ is the stopping power for ions for single electron excitation (close collision). As shown in the previous chapter the close collision stopping power is half of the total electron stopping power $\left(-\frac{dE}{dz}\right)_{e}$

$$\delta E_{\perp} = \frac{m}{M_1}\cdot\tfrac{1}{2}\left(-\frac{dE}{dz}\right)_{e} dz$$

or

$$\frac{\delta E_{\perp}}{\delta z} = \frac{m}{2M_1}\left(-\frac{dE}{dz}\right)_{e} \tag{5.1.8}$$

Since the stopping power $\left(-\frac{dE}{dz}\right)_{e}$ is proportional to the electron density n, following to Bethe–Bloch theory, equation 5.1.8 leads to the following result, if we replace the electron density n by the position dependent density $n(r)$

$$\frac{\delta\langle E_{\perp}\rangle_{ch}}{\delta z} = \frac{m}{2M_1}\left\langle\left(-\frac{dE}{dz}\right)_{e}\right\rangle_{ch}$$

$$= \frac{m}{2M_1} Se\langle n(r)\rangle_{ch} \tag{5.1.9}$$

Two important characteristics of the diffusion power $\frac{\delta\langle E_{\perp}\rangle_{ch}}{\delta z}$ in the Lindhard theory are that it is (*a*) proportional to the local electron density n for single electron excitation and (*b*) proportional to the squared mean amplitude ρ^2 of the lattice vibration for phonon excitation. At room temperature, ρ^2 is proportional to the temperature, which corresponds to the case of single phonon excitation (see Section 5.2, p. 135 for an example).

Let us define the diffusion coefficient of the beam for the transverse momentum $p_{\perp}$ by

$$D(p_{\perp}) = \tfrac{1}{2}\frac{\delta\langle p_{\perp}^{2}\rangle_{ch}}{\delta z}$$

$$= M_1\frac{\delta\langle E_{\perp}\rangle_{ch}}{\delta z} \tag{5.1.10}$$

from which we may write down the diffusion equation, expressing the diffusion process in the beam direction.

Let $g(\mathbf{p}_\perp, z)\, d\mathbf{p}_\perp$ be the number of ions in the transverse momentum region $d\mathbf{p}_\perp$. Then the diffusion equation can be represented in the form

$$\frac{\partial}{\partial z} g(\mathbf{p}_\perp, z) = \mathrm{div}[D(\mathbf{p}_\perp)\, \mathrm{grad}_{\mathbf{p}_\perp} g(\mathbf{p}_\perp, z)] \qquad (5.1.11)$$

For axial channelling with a single atomic row the problem will be solved in the axially symmetric manner, in which equation 5.1.11 becomes the much simpler one-dimensional diffusion equation. Using the transverse energy $E_\perp$ instead of the transverse momentum, the one-dimensional diffusion equation can be expressed in the form

$$\frac{\partial}{\partial z} g(E_\perp, z) = \frac{\partial}{\partial E_\perp}\left[D(E_\perp)\frac{\partial}{\partial E_\perp} g(E_\perp, z)\right] \qquad (5.1.12)$$

The derivation of equations 5.1.11 and 5.1.12, and its physical meaning, will be presented in sections 5.3 and 5.4.

Almost all calculations of dechannelling are based on the above Lindhard theory. For example, Kumakhov *et al.*,[3,4,5] Morita and Itoh[6] and Bonderup *et al.*[7] tried to compare their calculations of the axial dechannelling intensity with experimental results. Figure 5.1 shows a comparison by Bonderup *et al.*[7] of theoretical results of the dechannelling fraction for 1·6 MeV H^+ into Si$\langle 110\rangle$ at different temperatures. The dechannelling fraction is defined by

$$\chi(z) = \int_{E_\perp^c}^{\infty} g(E_\perp, z)\, \mathrm{d}E_\perp \qquad (5.1.13)$$

which represents the total yield of diffused beams with transverse energies larger

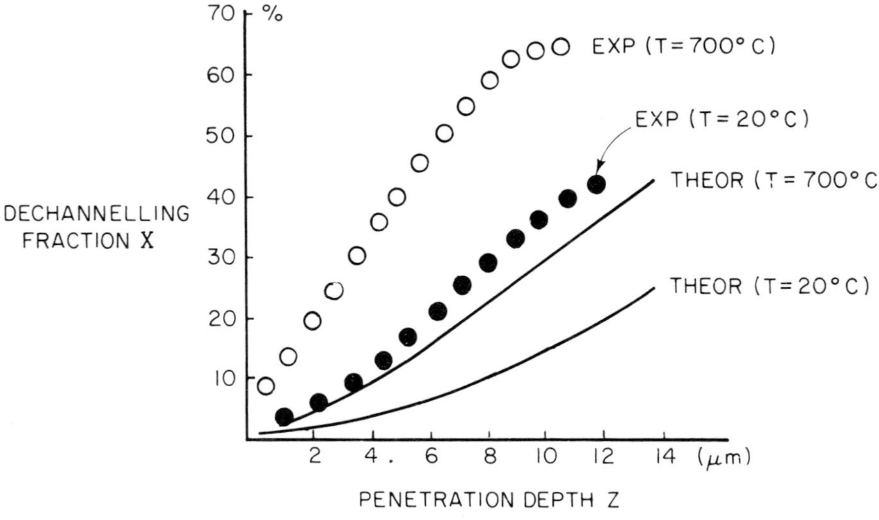

Figure 5.1. Dechannelling fraction for 1·6 MeV H^+ in Si$\langle 110\rangle$ according to Bonderup *et al.*[7]

than the critical energy $E^c (= E_0\theta_c)$. In figure 5.1 the dechannelling fractions are shown as a function of the penetration length of the beam. The experimental data of Pendersen *et al.*[8] are also included, from which we can see that the theoretical results are far from agreement with the experimental ones.

For planar channelling, many authors (e.g., Feldman and Appleton,[9] Morita,[10] Davies *et al.*,[11] Campisano *et al.*[12]) also used the Lindhard theory. Their results do not agree well with experimental ones. Consequently, we need to refine the Lindhard theory in a very fundamental way.

5.2. Calculation of diffusion coefficients by the inelastic scattering function

It is necessary to study beam diffusion by dividing the process into (*a*) single electron excitation, (*b*) lattice vibration, and (*c*) plasmon excitation for both (i) high energy and (ii) low energy ions. In addition, our theory takes special account of (*a*) the fact that the scattering centres are localized in the lattice and (*b*) the quantum mechanical effects on beam diffusion.

5.2.1. *Diffusion coefficients for fast ions*

We introduce the method for calculating diffusion coefficients for fast ions by making use of the inelastic scattering function introduced and explained in the previous chapter.

The diffusion coefficient for axial channelling should be represented quantum mechanically,† making use of the transition probability function P_n introduced in equation 4.8.42

$$D_i(\mathbf{r}_\perp) = \left\langle \frac{\delta p_i^2}{\delta z} \right\rangle = \frac{1}{v} \int \sum_{n(\neq 0)} (\hbar q_i)^2 \, P_n(\mathbf{K}', \mathbf{K}; \mathbf{r}_\perp) \, \mathrm{d}\mathbf{K}' \qquad (5.2.1)$$

where $i = x$ or y, and where $\hbar q_i$ is the momentum transfer in the direction of the ith component. Similarly, for planar channelling, we have

$$D_i(y) \equiv \left\langle \frac{\delta p_i^2}{\delta z} \right\rangle = \frac{1}{v} \int \sum_{n(\neq 0)} (\hbar q_i)^2 \, P_n(\mathbf{K}', \mathbf{K}; y) \, \mathrm{d}\mathbf{K}' \qquad (5.2.2)$$

In equations 5.2.1 and 5.2.2 the inelastic scattering function $P_n(\mathbf{K}', \mathbf{K}; \mathbf{r}_\perp)$ is given by equation 4.8.42, and related to the inelastic scattering factor $S(\mathbf{q}, \mathbf{q}')$ introduced in equation 4.7.10

$$\sum_{n(\neq 0)} P_n(\mathbf{K}', \mathbf{K}; \mathbf{r}_\perp) = \frac{1}{4\pi^2 \hbar^2 v} \sum_{\mathbf{h}\perp} S(\mathbf{q}, \mathbf{q} + \mathbf{h}_\perp) \exp(i\mathbf{h}_\perp \cdot \mathbf{r}_\perp)$$

---

†The diffusion coefficient defined here differs from the one introduced by equation 5.1.10 by a factor of $\frac{1}{2}$.

Inserting this expression into equation 5.2.1, we obtain

$$D_i(\mathbf{r}_\perp) = \frac{1}{4\pi^2 v^2} \int \sum_{\mathbf{h}\perp} (q_i)^2 S(\mathbf{q}, \mathbf{q} + \mathbf{h}_\perp) \exp(i\mathbf{h}_\perp \cdot \mathbf{r}) \, d\mathbf{q} \tag{5.2.1'}$$

Electron contribution

First, we shall study the contribution of plasmon excitation to the diffusion coefficient. The plasmon is excited in the nearly free electron gas which is independent of the crystal lattice. In this case, the inelastic scattering function $P_n(\mathbf{K}', \mathbf{K}; \mathbf{r}_\perp)$ could be represented in terms of the plasmon frequency ω_p, where the position dependence of the inelastic scattering function disappears.

The inelastic scattering factor $S(\mathbf{q}, \mathbf{q})$ takes the form[13]

$$S(\mathbf{q}, \mathbf{q}) = \frac{4Z_1^2 e^2 \hbar v}{q_\perp^2 + q_z^2} \cdot \frac{\pi \omega_p}{2v} \delta\left(q_z - \frac{\omega_p}{v} \right)$$

from we which we get

$$D_i = \frac{Z_1^2 e^2 \hbar \omega_p}{2v^2} \left\{ k_c^2 - \left(\frac{\omega_p}{v} \right)^2 \ln\left[\left(\frac{k_c v}{\omega_p} \right)^2 + 1 \right] \right\} \tag{5.2.3}$$

For high energies, where $v^2 \gg (\omega_p/k_c)^2$, the above equation becomes much simpler and reduces to

$$D_i = \frac{Z_1^2 e^2 \hbar \omega_p}{2v^2} k_c^2 \tag{5.2.4}$$

It should be noted that the diffusion coefficient is proportional to $n^{7/6}$ since ω_p and k_c are proportional to $n^{1/2}$ and $n^{1/3}$, respectively. This is very nearly proportional to n^1. Numerical studies of equation 5.2.4 will be discussed later.

For single electron excitations the inelastic scattering factor $S(\mathbf{q}, \mathbf{q} + \mathbf{h}_\perp)$ is represented by the Compton scattering factor f_{ij} in the form

$$S(\mathbf{q}, \mathbf{q} + \mathbf{h}_\perp) = \frac{1}{V_{\text{cell}}} v_{\mathbf{q}} v_{\mathbf{q}+\mathbf{h}_\perp} \left[\sum_i f_{ii}(\mathbf{h}_\perp) - \sum_{ij} f_{ij}(\mathbf{q} + \mathbf{h}_\perp) f_{ij}(\mathbf{q}) \right]$$
$$\times \delta(|\mathbf{K}_z + \mathbf{q}| - K_z + I/\hbar v) \tag{5.2.5}$$

where I is the averaged excitation energy. In the previous chapter the Compton scattering factor f_{ij} was calculated in detail and we can use it here.

For high energies we can neglect the second term in square brackets, in equation 5.2.5 and the diffusion coefficient is in the form

$$D_i(\mathbf{r}_\perp) = \frac{1}{4\pi^2 v^2 V_{\text{cell}}} \int \sum_{\mathbf{h}_\perp} q_i^2 v_{\mathbf{q}} v_{\mathbf{q}+\mathbf{h}_\perp} \sum_j f_{jj}(\mathbf{h}_\perp) \exp(i\mathbf{h}_\perp \cdot \mathbf{r}_\perp)$$
$$\times \delta(|\mathbf{K}_z + \mathbf{q}| - K_z + I/\hbar v) \, d\mathbf{q} \tag{5.2.6}$$

for axial channelling, and

$$D_i(y) = \frac{1}{4\pi^2 v^2 V_{\text{cell}}} \int \sum_{h_y} q_i^2 v_{\mathbf{q}} v_{\mathbf{q}+\mathbf{h}_y} \sum_j f_{jj}(h_y) \exp(i\mathbf{h}_y \cdot y) \times \delta(|\mathbf{K}_z + \mathbf{q}| - K_z + I/\hbar v) \mathrm{d}\mathbf{q} \tag{5.2.7}$$

for planar channelling.

We shall now make a rough approximation of the above quantities. We neglect the reciprocal lattice ($\mathbf{h}_\perp$ or $\mathbf{h}_y$) dependence of the factors $v_{\mathbf{q}+\mathbf{h}_\perp}$ and $v_{\mathbf{q}+\mathbf{h}_y}$ in equations 5.2.6 and 5.2.7, respectively. According to the definition of the Compton scattering factor we get

$$\frac{1}{V_{\text{cell}}} \sum_j f_{jj}(\mathbf{h}_\perp) = n(\mathbf{h}_\perp)$$

where $n(\mathbf{h}_\perp)$ is the Fourier component of the electron density $n(\mathbf{r}_\perp)$

$$\sum_{\mathbf{h}_\perp} n(\mathbf{h}_\perp) \exp(i\mathbf{h}_\perp \cdot \mathbf{r}_\perp) = n(\mathbf{r}_\perp)$$

Then, we can conclude that the diffusion coefficient $D_i(\mathbf{r}_\perp)$ or $D_i(y)$ is proportional to the electron density $n(\mathbf{r}_\perp)$ or $n(y)$, respectively, as in the Lindhard theory.

In order to derive the Lindhard expression, we have to perform an integration of the form

$$\int_{q_{\min}}^{q_{\max}} q_i^2 v_q^2 \delta(|\mathbf{K}_z + \mathbf{q}| - K_z + I/\hbar v) \mathrm{d}\mathbf{q}$$

where the upper and lower limits of the integration are taken as $q_{\max} = 2mv/\hbar$ and $q_{\min} = I/\hbar v$, respectively, as in the calulation for the stopping power. For the isotropic case we note that $D_x(\mathbf{r}_\perp) = D_y(\mathbf{r}_\perp) = \frac{1}{2} D_\perp(\mathbf{r}_\perp)$, since $q_x^2 + q_y^2 = q_\perp^2$. $D_\perp(\mathbf{r})$ is the diffusion coefficient obtained from equation 5.2.6 by replacing q_i^2 with $q_\perp^2$ in the integral. The integral can be performed over the plane perpendicular to the axis, since the incident energy is much larger than the ionization energy. Then we arrive at the form

$$\begin{aligned} D_\perp(\mathbf{r}_\perp) &= \frac{n(\mathbf{r}_\perp)}{4\pi^2 v^2} \int_{I/\hbar v}^{2mv/\hbar} q_\perp^2 \left(\frac{4\pi Z_1 e^2}{q_\perp^2} \right)^2 2\pi q_\perp \mathrm{d}q_\perp \\ &= \frac{8\pi Z_1^2 e^4}{v^2} L_e n(\mathbf{r}_\perp) \end{aligned} \tag{5.2.8}$$

where

$$L_e = \ln\left(\frac{2mv^2}{I} \right) \tag{5.2.9}$$

The result 5.2.8 is the Lindhard expression 5.1.10 multiplied by 2.

However, it should be noted that there is no reason to neglect the scattering factor's dependence on the reciprocal lattice; i.e., to neglect the Umklapp (U) process in deriving equation 5.2.8 from equation 5.2.6. For example, the ratio of the minimum values of q and $h_\perp$ is given by

$$\frac{(h_\perp)_{min}}{I/\hbar v} \frac{Mv^2}{I} \cdot \frac{\lambda_{ion}}{d}$$

where d and λ_{ion} are the lattice constant and the de Broglie ion wavelength, respectively. For 1·5 MeV protons, λ_{ion} is of the order of 10^{-5} nm and the above ratio is of the order of 10^2. Thus we may conclude that the $h_\perp$ dependence cannot be neglected when compared with the q dependence. This fact becomes more dominant with increasing ion energy where the Lindhard expression 5.1.9 has to be greatly modified.

Waho and Ohtsuki[14] tried to perform the integration in equation 5.2.6 with the U process. For the planar case, it can be shown† that $D_x(y) < D_y(y)$. The result of Waho and Ohtsuki for the diffusion coefficient $D_y(y)$ is expressed by

$$D_y(y) = \frac{4\pi(Z_1 e^2)^2}{v^2} \left\{ (L_e - \tfrac{1}{2})\, n(y) + \sum_{h_y} [I_1(h_y) + I_2(h_y) + I_3(h_y)] \times f(h_y) \cos(h_y y) \right\} \tag{5.2.10}$$

Where $f(h_y)$ is given by

$$f(h_y) = N \sum f_{jj}(h_y) = N \int n(\mathbf{r}) \exp(-i\mathbf{h}_y \cdot \mathbf{r})\, d\mathbf{r} \tag{5.2.11}$$

In the above, N is the number of atoms, and the functions $I_1(h_y), I_2(h_y)$ and $I_3(h_y)$ are given by‡

$$I_1(h_y) = \frac{1}{4R} \ln \left| \left(\frac{q_{max}^2 + h_y^2 R - \beta^2}{q_{max}^2 - h_y^2 R - \beta^2} \right) \times \left(\frac{R+1}{R-1} \right) \right| \tag{5.2.12}$$

$$I_2(h_y) = \frac{1}{4h_y^2} (\beta^2 + h_y^2 - q_{max}^2) \tag{5.2.13}$$

and

$$I_3(h_y) = \tfrac{3}{4} \ln \left| \frac{q_{max}^2 - h_y^2 + \beta^2}{2q_{max}^2} \right| \tag{5.2.14}$$

where

$$R = \sqrt{1 + 4(q_{min}^2/h_y^2)}$$

and

$$\beta^2 = [(q_{max}^2 - h_y^2)^2 + 4q_{min}^2 h_y^2]^{1/2}$$

† We shall discuss this property again in section 5.5, p. 149.

‡ It should be noted that the sign in $(\beta^2 + h_y^2 - g_{max}^2)$ of equation 5.2.13 was misprinted in the original paper.[14]

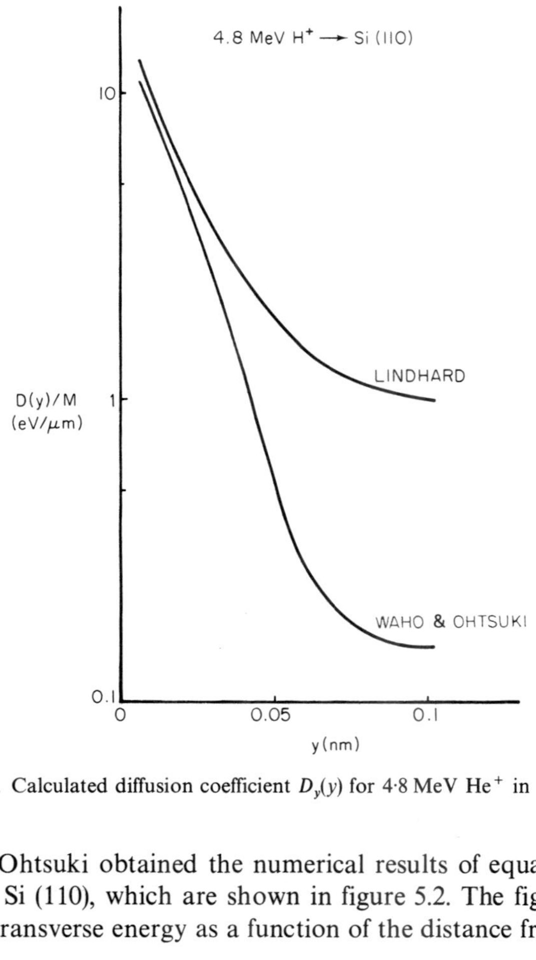

Figure 5.2. Calculated diffusion coefficient $D_y(y)$ for 4·8 MeV He^+ in Si (110).

Waho and Ohtsuki obtained the numerical results of equation 5.2.10 for 10 MeV He^+ in Si (110), which are shown in figure 5.2. The figure shows the diffusion of the transverse energy as a function of the distance from the atomic plane

$$\left\langle \frac{\delta E_\perp}{\delta z} \right\rangle = \frac{D_y(y)}{2M}$$

In the figure the diffusion coefficient calculated by the Lindhard theory is also shown. It should be noted that the dependence on position is much stronger and the value of the diffusion coefficient much smaller than those of the Lindhard theory. This may mean that the U process carries much more momentum in the y direction than the Lindhard theory assumed.

It is complicated to perform the numerical calculation of equation 5.2.10 for each case. So, Ohtsuki *et al.*[15] tried to find a simpler formula from equation 5.2.10. The numerical calculation for many cases of equation 5.2.10 leads to the result that the second term in the curly brackets in equation 5.2.10 is also a linear function of the electron density. Thus, the diffusion coefficient can be expressed in the form

$$D_y(y) = \frac{8\pi(Z_1 e^2)^2}{v^2}[L_e^* n(y) + C] \tag{5.2.15}$$

where L_e^* is the factor obtained by modifying L_e, and C is a constant which does not depend on the position y.

We can prove equation 5.2.15 analytically as follows. The scattering factor $f(h_y)$ approaches zero rapidly if h_y increases beyond h_y^{max}, as seen in equation 5.2.10. It should also be noted that $h_y^{max} \ll q_{max}$. Therefore equations 5.2.12, 5.2.13 and 5.2.14 become

$$I_1(h_y) \approx \frac{|h_y|}{2\sqrt{h_y^2 + 4q_{min}^2}} \ln\left|\frac{\sqrt{h_y^2 + 4q_{min}^2} + h_y}{\sqrt{h_y^2 + 4q_{min}^2} - h_y}\right|$$

$$I_2(h_y) \approx 0$$

and

$$I_3(h_y) \approx \tfrac{3}{4}\ln|1 - (h_y^2/q_{max}^2)|$$

since $\beta^2 \approx q_{max}^2(1 - h_y^2/q_{max}^2)$.

Accordingly, we obtain for $h_y \gg q_{min}$

$$I(h_y) = I_1(h_y) + I_2(h_y) + I_3(h_y)$$

$$\simeq \frac{|h_y|}{2\sqrt{h_y^2 + 4q_{min}^2}} \ln\left(\frac{\sqrt{h_y^2 + 4q_{min}^2} + h_y}{\sqrt{h_y^2 + 4q_{min}^2} - h_y}\right)$$

In equation 5.2.10 the term for $h_y = 0$ takes the form

$$I_1(0) \to 0$$

$$I_2(0) \to 0$$

$$I_3(0) \to 0$$

and we can therefore, also obtain

$$D_y(y) = \frac{4\pi(Z_1 e^2)^2}{v^2}\left[(L_e - \tfrac{1}{2})n(y) + \sum_{h_y \neq 0} I(h_y) f(h_y) \cos h_y y\right]$$

where

$$I(h_y) = \frac{|h_y|}{2\sqrt{h_y^2 + 4q_{min}^2}} \ln\left|\frac{\sqrt{h_y^2 + 4q_{min}^2} + h_y}{\sqrt{h_y^2 + 4q_{min}^2} - h_y}\right| \tag{5.2.16}$$

Since the factor $I(h_y)$ is a slowly varying function and q_{min} is very small, the above equation becomes

$$\frac{|\bar{h}_y|}{2\sqrt{\bar{h}_y^2+4q_{\min}^2}}\ln\left|\frac{\sqrt{\bar{h}_y^2+4q_{\min}^2}+\bar{h}_y}{\sqrt{\bar{h}_y^2+4q_{\min}^2}-\bar{h}_y}\right|\sum_{h_y\neq 0} f(h_y)\cos h_y y$$

$$=\frac{|\bar{h}_y|}{2\sqrt{\bar{h}_y^2+4q_{\min}^2}}\ln\left|\frac{\sqrt{\bar{h}_y^2+4q_{\min}^2}+\bar{h}_y}{\sqrt{\bar{h}_y^2+4q_{\min}^2}-\bar{h}_y}\right|\left[\sum_{h_y} f(h_y)\cos h_y y - f(0)\right]$$

$$=\frac{|\bar{h}_y|}{2\sqrt{\bar{h}_y^2+4q_{\min}^2}}\ln\left|\frac{\sqrt{\bar{h}_y^2+4q_{\min}^2}+\bar{h}_y}{\sqrt{\bar{h}_y^2+4q_{\min}^2}-\bar{h}_y}\right|[n(y)-\bar{n}]$$

where $\bar{n}$ is the mean electron density, and $\bar{h}_y$ is the mean value of the reciprocal lattice vector which contributes effectively to the scattering.

Thus we arrive at the following expression for the constants in equation 5.2.15

$$L_e^* = L_e - \frac{1}{2} + \frac{|\bar{h}_y|}{2\sqrt{\bar{h}_y^2+4q_{\min}^2}}\ln\left|\frac{\sqrt{\bar{h}_y^2+4q_{\min}^2}+\bar{h}_y}{\sqrt{\bar{h}_y^2+4q_{\min}^2}-\bar{h}_y}\right|$$

$$C = -\left|\frac{|\bar{h}_y|}{2\sqrt{\bar{h}_y^2+4q_{\min}^2}}\right|\bar{n}\ln\left|\frac{\sqrt{\bar{h}_y^2+4q_{\min}^2}+h_y}{\sqrt{h_y^2+4q_{\min}^2}-h_y}\right| \qquad (5.2.16)'$$

In these expressions the only uncertainty is in selecting the average value $\bar{h}_y$. In ordinary cases we can take $\bar{h}_y \approx C(2\pi/d)$, where C has a value of $5 \sim 10$ and d is the spacing of the channelling planes. When $q_{\min} = I/\hbar v$ is very small, $I(h_y)$ becomes $\ln(\bar{h}_y/q_{\min})$, and hence we may simply take $\bar{h}_y \cong (2\pi/d)$.

Phonon excitation

Ion beams under channelling conditions collide with atoms with large impact parameters and recoil, exciting lattice vibrations. In hyperchannelling the recoil amplitude is very small and a single phonon is excited. However, in general channelling the recoil may be expressed by multi-phonon excitation. At the same time, the U process, which was considered for single electron excitation, is also important in the calculation of the diffusion coefficient in the nuclear (multi-phonon) excitation.

Here we show in detail the derivation of the nuclear diffusion coefficient, according to Kitagawa and Ohtsuki.[16] In Chapter 4 the inelastic scattering factor $S(\mathbf{q},\mathbf{q}+\mathbf{h}_\perp)$ due to multi-phonon excitation was given as

$$S(\mathbf{q},\mathbf{q}+\mathbf{h}_\perp) = \frac{1}{V_{\text{cell}}} v_{\mathbf{q}} v_{\mathbf{q}+\mathbf{h}_\perp} \{\exp[-M(\mathbf{h}_\perp)] - \exp[-M(\mathbf{q}) - \mathbf{M}(\mathbf{q}+\mathbf{h}_\perp)]\}$$
$$\times\, \delta(|\mathbf{K}_z + \mathbf{q}| - K_z)$$

Inserting this expression into equation 5.2.1′ we obtain

$$D_i(\mathbf{r}_\perp) = \frac{1}{4\pi^2 v^2 V_{\text{cell}}} \int \sum_{h_\perp} q_i^2 v_{\mathbf{q}} v_{\mathbf{q}+\mathbf{h}_\perp} \{\exp[-M(\mathbf{h}_\perp)] - \exp[-M(\mathbf{q}) - M(\mathbf{q}+\mathbf{h}_\perp)]\}$$

$$\times \exp(i\mathbf{h}_\perp \cdot \mathbf{r}_\perp)\, \delta(|\mathbf{K}_z + \mathbf{q}| - K_z)\, \mathrm{d}\mathbf{q} \qquad (5.2.17)$$

We now consider the approximation necessary to derive the Lindhard expression 5.1.7 from equation 5.2.17. In equation 5.2.17 we expand the Debye–Waller factor as

$$\exp(-M) = 1 - M + \ldots.$$

Since the Debye–Waller coefficient M is represented by

$$M = \tfrac{1}{4}\rho_\perp^2 q^2$$

$\rho_\perp$ being the transverse component of the thermal displacement, as introduced in Chapter 3, we obtain

$$D_i(\mathbf{r}_\perp) = \frac{\rho_\perp^2}{16\pi^2 v^2 V_{\text{cell}}} \int \sum_{h_\perp} q_i^2 v_{\mathbf{q}} v_{\mathbf{q}+\mathbf{h}_\perp} [q^2 + (\mathbf{q}+\mathbf{h}_\perp)^2 - h_\perp^2]$$

$$\times \exp(i\mathbf{h}_\perp \cdot \mathbf{r}_\perp)\, \delta(|\mathbf{K}_z + \mathbf{q}| - K_z)\, \mathrm{d}\mathbf{q} \qquad (5.2.18)$$

For the isotropic case the integral is performed over the two-dimensional plane perpendicular to the axis

$$D_\perp(\mathbf{r}_\perp) = \frac{\rho_\perp^2}{8\pi^2 v^2 V_{\text{cell}}} \int \sum_{\mathbf{h}_\perp} q_\perp^2 v_{\mathbf{q}} v_{\mathbf{q}+\mathbf{h}_\perp} (\mathbf{q}_\perp^2 + \mathbf{q}_\perp \cdot \mathbf{h}_\perp)$$

$$\times \exp(i\mathbf{h}_\perp \cdot \mathbf{r}_\perp)\, \delta(|\mathbf{K}_z + \mathbf{q}| - K_z)\, 2\pi q_\perp\, \mathrm{d}q_\perp \qquad (5.2.19)$$

Now we compare equation 5.2.19 with the Lindhard formula, which is proportional to

$$F = [\mathrm{grad}(\boldsymbol{\rho}_\perp \cdot \mathrm{grad}\, V)]^2$$

where V is the Lindhard continuum potential, which is expanded in the Fourier series as

$$V = \sum_{\mathbf{h}} V_{\mathbf{h}} \exp(i\mathbf{h}\cdot\mathbf{r})$$

Therefore F takes the form

$$F = \sum_{\mathbf{h},\mathbf{g}} [(\boldsymbol{\rho}_\perp \cdot \mathbf{h})(\boldsymbol{\rho}_\perp \cdot \mathbf{g}) + (\boldsymbol{\rho}_\perp \cdot \mathbf{g})^2](\mathbf{h}\cdot\mathbf{g} + g^2)\, V_{\mathbf{h}+\mathbf{g}} V_{\mathbf{g}}^* \exp(i\mathbf{h}\cdot\mathbf{r})$$

The first term in the brackets becomes zero by the summation over all directions of the vectors **h** and **g**. The second term, on the other hand, takes the value $\frac{1}{2}$ by the summation over all directions. Accordingly, we get

$$F = \tfrac{1}{2}\rho_\perp^2 \sum_{\mathbf{h},\mathbf{g}} \mathbf{g}^2(\mathbf{h}\cdot\mathbf{g} + \mathbf{g}^2)\, V_{\mathbf{h}+\mathbf{g}} V_{\mathbf{g}}^* \exp(i\mathbf{h}\cdot\mathbf{r})$$

which is equivalent to equation 5.2.19.

Thus we may conclude that we arrive at the Lindhard expression from the expression of Kitagawa and Ohtsuki[16] if we assume single-phonon excitation. By numerical calculation we can show that the single-phonon approximation is correct only for the hyperchannelling of very low energy ions.

Now we shall try to simplify equation 5.2.17, but still include the effects of multi-phonon excitation. Instead of using the expansion of the Debye–Waller factor $\exp(-M)$, we can neglect the second term in the brackets in equations 5.2.17, which is valid for the region $|\mathbf{q}| \gg h_\perp$. We should note that equation 5.2.17 vanishes for $|\mathbf{q}| \ll h_\perp$, since

$$v_{\mathbf{q}} v_{\mathbf{q}+\mathbf{h}_\perp} \{\exp[-M(\mathbf{h}_\perp)] - \exp[-M(\mathbf{q}) - M(\mathbf{q}+\mathbf{h}_\perp)]\}$$
$$\approx v_0 v_{h_\perp} \{\exp[-M(\mathbf{h}_\perp)] - \exp[-M(\mathbf{h}_\perp)]\} = 0$$

When is the condition $|\mathbf{q}| \gg h_\perp$ satisfied? In general, for ions whose energy is a few MeV the fundamental reciprocal lattice vector $2\pi/d$ is in the range of q such that $q_{\max} < 2\pi/d < q_{\max}$. However, the range where q is small does not contribute to the integral, since the integral includes the factor q_i^2. Thus the range $2\pi/d < q < q_{\max}$ contributes to the integral. For increasing beam energy the contribution range becomes larger, and the condition $q \gg h_\perp$ is easily satisfied. It is known that for an electron beam of energy greater than 1 MeV, the term $\exp[-M(q) - M(\mathbf{q}+\mathbf{h}_\perp)]$ is neglected.

Thus, according to equation 5.2.17, for the isotropic axial case we obtain†

$$D_\perp(\mathbf{r}_\perp) = \frac{1}{4\pi^2 v^2 V_{\text{cell}}} \int \sum_{h_\perp} q_\perp^2 v_{\mathbf{q}} v_{\mathbf{q}+\mathbf{h}_\perp} \exp[-M(\mathbf{h}_\perp)] \exp(i\mathbf{h}_\perp \cdot \mathbf{r}_\perp) \times \delta(|\mathbf{K}_z + \mathbf{q}| - K_z)\, \mathrm{d}\mathbf{q} \qquad (5.2.20)$$

Since $v_{\mathbf{q}+\mathbf{h}_\perp} \sim v_{\mathbf{q}}$ for the region $|q| \gg h_\perp$, we may perform‡ the summation over $\mathbf{h}_\perp$

$$(P(r_\perp) \equiv) \sum_{h_\perp} \exp[(-M(h_\perp)] \exp[i\mathbf{h}_\perp \cdot \mathbf{r}] = \frac{r_0^2}{\rho_\perp^2} \exp(-r_\perp^2/\rho_\perp^2) \qquad (5.2.21)$$

where $2r_0$ is the distance between atomic rows and $P(\mathbf{r}_\perp)$ is the probability of finding the atom at the point $\mathbf{r}_\perp$ from its equilibrium position due to lattice vibration.§

---

†Kagan and Kononets (*Zh. Eksp. Teor. Fiz.* **64**, 1042 (1973)) also tried to derive a general expression of $D_\perp(\mathbf{r}_\perp)$ due to many-phonon excitations. Their general expression is same as equation 5.2.20 if we neglect the U process. Equation 5.2.20 is derived for the Einstein model (see equation (4.3.26)). Sirotinin *et al.* (*Phys. Lett.* **88A**, 199 (1982)) have recently discussed the correlation effect, which is not included here.

‡Numerical integration of equation 5.2.20 without the assumption $v_{q+h_\perp} \sim v_q$ has been performed by Waho (Thesis, Waseda Univ., Tokyo, 1978).

§The normalization of $P(\mathbf{r})$ is taken by $\frac{1}{\pi r_0^2}\int P(r_\perp) 2\pi r_\perp \, \mathrm{d}r_\perp = 1$.

Finally, equation 5.2.20 becomes

$$D_\perp(\mathbf{r}_\perp) = D_{\text{random}} P(\mathbf{r}_\perp) \tag{5.2.22}$$

where D_{random} is given by

$$D_{\text{random}} = \frac{1}{4\pi^2 v^2 V_{\text{cell}}} \int_{q_{\min}}^{q_{\max}} q_\perp^2 \, |v_{q_\perp}|^2 \, \delta(|\mathbf{K}_z - \mathbf{q}| - K_z) \, \mathrm{d}\mathbf{q} \tag{5.2.23}$$

which is the normal diffusion coefficient first derived by Bohr[2] for a random system (gas target). In fact, equation 5.2.23 can be integrated

$$D_{\text{random}} = \frac{32\pi^2 Z_1^2 Z_2^2 e^4 N}{4\pi^2 v^2} \cdot \pi \ln\left|\frac{q_{\max}}{q_{\min}}\right|$$

$$= \frac{8\pi Z_1^2 Z_2^2 e^4 N}{v^2} L_n \tag{5.2.24}$$

where

$$L_n = \ln\left|\frac{q_{\max}}{q_{\min}}\right| = \tfrac{1}{2}\ln\left|\frac{T_{\max}}{T_{\min}}\right| = \ln\left|\frac{2EM_2}{\kappa(a_{\text{TF}})(M_1+M_2)}\right|$$

$$= \ln\left(1{\cdot}29 \frac{a_{\text{TF}} E}{Z_1 Z_2 e^2} \frac{M_2}{M_1 + M_2}\right) \tag{5.2.25}$$

where we employed the simple screened r^{-1} potential (see Section 1.3, p. 5).

For planar channelling we obtain

$$D_y(y) = \left\langle \frac{\delta p_y^2}{\delta z} \right\rangle = \left\langle \frac{\delta p_y^2}{\delta z} \right\rangle_{\text{random}} P(y)$$

where

$$P(y) = \frac{d_y}{\sqrt{2\pi}\, \rho_y} \exp(-y^2/2\rho_y^2) \tag{5.2.21$'$}$$

where d_y is the distance between atomic planes and ρ_y is the mean squared amplitude of the transverse component of the lattice vibration. The factor $\left\langle \frac{\delta p_y^2}{\delta z} \right\rangle_{\text{random}}$ is defined by

$$\left\langle \frac{\delta p_y^2}{\delta z} \right\rangle_{\text{random}} = \frac{1}{4\pi^2 v^2 V_{\text{cell}}} \int_{q_{\min}}^{q_{\max}} q_y^2 \, |v_q|^2 \, \delta(|\mathbf{K}_z - \mathbf{q}| - K_z) \mathrm{d}\mathbf{q}$$

$$= \tfrac{1}{2} D_{\text{random}}$$

It was Ellegard and Lassen[17] who first pointed out that the Gaussian type diffusion coefficient is more reasonable than the Lindhard type. Our results for $D_\perp(\mathbf{r}_\perp)$ or $D_y(y)$ will now be compared to those of Ellegard and Lassen.

We have derived the position-dependent diffusion coefficient. However, as seen in sections 5.3 and 5.4, it is more convenient to get the transverse momentum $p_\perp$ or transverse energy $E_\perp$ dependent diffusion coefficient $D_\perp(E_\perp)$ or $D_y(E_y)$, which is calculated by averaging over the accessible region (see Chapter 2)

$$\langle D_\perp(\mathbf{r}_\perp)\rangle_A = \frac{1}{\pi(r_0^2 - r_1^2)}\int D_\perp(\mathbf{r}_\perp)\,2\pi r_\perp\,dr_\perp$$

Now, we compare three forms of the axial diffusion coefficient

Lindhard

$$\langle D_\perp(\mathbf{r}_\perp)\rangle_A = D_{\text{random}}\gamma_n$$

$$\gamma_n \cong (1/L_n)(\rho_\perp^2/3a^2)\exp\{(2E_\perp/E_c)[1-\exp(-2E_\perp/E_c)]^3\}$$

Ellegard and Lassen

$$\gamma_n = \exp(-a^2/\rho_\perp^2) \tag{5.2.26}$$

Kitagawa and Ohtsuki

$$\gamma_n = \exp\{-3a^2/\rho_\perp^2[\exp[(E_\perp/2E_c)-1]^{-1}]\} \tag{5.2.27}$$

It is worth noting that the expression of Kitagawa and Ohtsuki agrees with Ellegard and Lassen's if the factor $\exp[(E_\perp/2E_c)-1]$ in the former is replaced by 3. Since Ellegard and Lassen obtained their expression phenomenologically, the above replacement has no reasonable basis.

Figure 5.3 shows[18] the average value of γ_n over $E_\perp$ as a function of temperature (or $\rho_\perp^2$) for 5 MeV H^+ in bismuth (Bi), comparing the theoretical results of Kitagawa and Ohtsuki and Ellegard and Lassen with the experimental results of Ellegard and Lassen.[17]

5.2.2. *Diffusion coefficient for slow ions*

We cannot use perturbation theory for the calculation of the electron diffusion coefficient for slow ions, as shown in the calculation for the stopping power in the previous chapter. Therefore, we must try to derive the electron diffusion coefficient for slow ions in the electron gas by making use of the Green function method.[19]

According to linear response theory[20] the inelastic scattering probability per unit length $P(\mathbf{q},\omega)$ is represented by the Fourier component, $K^R(\mathbf{q},\omega)$, of the retarded Green function, $G^R(\mathbf{q},t)$

$$P(\mathbf{q},\omega) = -\frac{4Z_1^2e^4}{\pi v q^4}\,Im\,K^R(\mathbf{q},\omega)\,\delta\left(\hbar\mathbf{q}\cdot\mathbf{v} + \frac{\hbar^2q^2}{2M} + \hbar\omega\right)$$

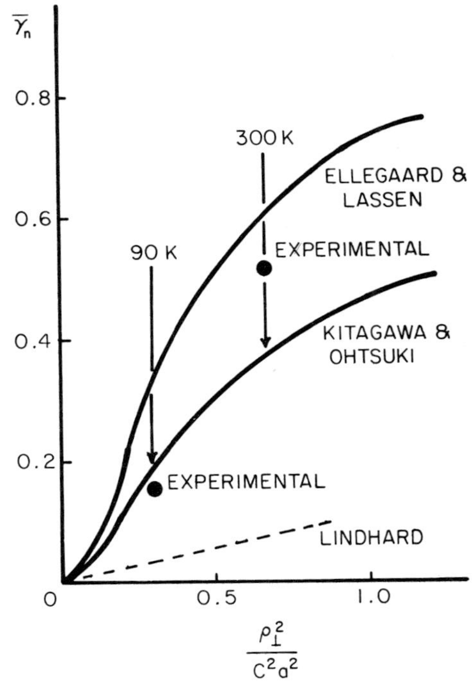

Figure 5.3. Temperature dependence of γ_n for 5 MeV H^+ in Bi, comparing Kitagawa and Ohtsuki's, Ellegard and Lassen's, and Lindhard's expressions with experimental results.[17]

Therefore we get the following expression for the diffusion coefficient

$$D_y = -\frac{4Z_1^2e^4}{\pi v}\int_0^\infty \mathrm{d}\omega \int \mathrm{d}\mathbf{q}\,(\hbar q_y)^2\, Im\, K^R(\mathbf{q},\omega)\frac{\delta(\hbar\mathbf{q}\cdot\mathbf{v} + \hbar q^2/2M + \hbar\omega)}{q^4}$$

If we take into account the relation

$$\delta(\omega) = \int_{-\infty}^{\infty} \exp(-i\omega t)\,\mathrm{d}t$$

and perform the integration over the angle in $\mathbf{q}$, we get

$$D_y = -i\frac{2Z_1^2e^4\hbar}{\pi}\int\frac{\mathrm{d}q}{q}\int_{-\infty}^{\infty}\mathrm{d}\omega\, Im\, K^R(q,\omega)$$
$$\times\int_{-\infty}^{\infty}\mathrm{d}t\,\exp(-i\omega t)\,I_1(qvt) \qquad (5.2.28)$$

where

$$I_1(qvt) = \frac{q}{v}\left[\frac{2n_1(qvt)}{qvt} + \frac{1}{qvt} + \frac{1}{(qvt)^3}\right] \qquad (5.2.29)$$

$n_1(x)$ being the spherical Neuman function. For very high energy ions the function $I_1(qvt)$ is approximated by

$$I_1(qvt) \approx \frac{1}{v^2}\mathrm{P}\left(\frac{1}{t}\right) + \frac{2}{q^2v^4}\mathrm{P}\left(\frac{1}{t^3}\right) \tag{5.2.30}$$

and for low energy ions by

$$I_1(qvt) \approx \tfrac{1}{4}q^2 t \tag{5.2.31}$$

where P signifies 'to take the principal value'.

Inserting equation 5.2.31 into equation 5.2.28 we get

$$\begin{aligned} D_y &= \frac{-2iZ_1^2e^4\hbar}{4\pi}\int q\,\mathrm{d}q\int_{-\infty}^{\infty}\mathrm{d}\omega\, Im\,K^{\mathrm{R}}(q,\omega)\int_{-\infty}^{\infty}\mathrm{d}t\,t\exp(-i\omega t) \\ &= \frac{\hbar Z_1^2e^4}{2\pi}\int q\,\mathrm{d}q\int \mathrm{d}\omega\, Im\,K^{\mathrm{R}}(q,\omega)\frac{\mathrm{d}}{\mathrm{d}\omega}\int_{-\infty}^{\infty}dt\exp(-i\omega t) \\ &= -Z_1^2e^4\hbar\int q\,\mathrm{d}q\frac{\mathrm{d}}{\mathrm{d}\omega}Im\,K^{\mathrm{R}}(q,\omega)\bigg|_{\omega=0} \end{aligned} \tag{5.2.32}$$

Note that the diffusion coefficient for slow ions D_y does not depend on the ion velocity v_0.

Now we can use the Green function expression $K^{\mathrm{R}}(q,\omega)$ derived by Hubbard[20]

$$\frac{\mathrm{d}}{\mathrm{d}\omega}Im\,K^{\mathrm{R}}(q,\omega)\bigg|_{\omega=0} = -\frac{m^{*2}}{2\pi\hbar^3}\frac{q^3}{[q^2+q_{\mathrm{T}}^2 f(q)]^2} \tag{5.2.33}$$

where m^* is the effective mass of the electron in the electron gas, and

$$q_{\mathrm{T}}^2 = 4e^2m^*(q_{\mathrm{F}}/\pi\hbar^2)$$

$$f(q) = 1 - \frac{q^2}{2(q^2+q_{\mathrm{F}}^2)}$$

Inserting equation 5.2.33 into equation 5.2.32 we get

$$D_y = \frac{Z_1^2e^4m^{*2}}{\pi\hbar^2}q_{\mathrm{F}}\int_0^1 \mathrm{d}\eta\frac{\eta^4}{[\eta^2+(m^*/m)\,\alpha f(\eta)]^2} \tag{5.2.34}$$

where $\eta = q/2q_{\mathrm{F}}$, $f(\eta) = (2\eta^2+1)/(4\eta^2+1)$ and $\alpha = e^2/\pi\hbar v_{\mathrm{F}}$.

For a very dense gas, where $\alpha \ll 1$, equation 5.2.34 becomes

$$D_y = \frac{1{\cdot}44Z_1^2e^4m^2}{\pi^{1/3}\hbar^2}n^{1/3} \tag{5.2.35}$$

This shows that the diffusion coefficient for slow ions in a very dense electron gas does not depend on the ion velocity, and is proportional to $n^{1/3}$, n being the mean electron density, i.e.

$$q_F = (3\pi^2 n)^{1/3}$$

For a dilute gas ($\alpha \gg 1$) we obtain

$$m^* \approx 3\pi q_F \hbar^2/4e^2$$
$$(m^*/m)\alpha \approx \tfrac{3}{4}$$

and

$$D_y = 0{\cdot}26\pi^3 Z_1^2 \hbar^2 n \tag{5.2.36}$$

The important difference between equations 5.2.35 and 5.2.36 is the screening effect.

There are no exact experimental studies of the diffusion coefficient or dechannelling for slow ions. Davies *et al.*[11] tried to scale the dechannelling fraction as a function of (depth)/(ion energy). It was possible to scale up the data for 2·0 MeV H^+, 3·5 MeV H^+, 8 MeV H^+ and 12·0 MeV H^+ into W(110), but not the data for 0·5 MeV H^+. The deviation of the data for the slow ion is consistent with the theory proposed here.

5.3. Planar dechannelling

Dechannelling rates or dechannelling fractions are derived by solving the diffusion equation or the Fokker–Planck equation analytically or numerically by making use of the diffusion coefficient obtained in the previous section.

The Fokker–Planck equation for the distribution function $f(\mathbf{r},\mathbf{p},t)$ takes the form

$$\frac{\partial f}{\partial t} + \mathbf{v}\cdot\mathrm{grad}_r f - \mathrm{grad}\, U(\mathbf{r})\,\mathrm{grad}_p f$$
$$= -f\,\mathrm{div}_p\left\langle\frac{\Delta\mathbf{p}}{\Delta t}\right\rangle - \left\langle\frac{\Delta\mathbf{p}}{\Delta t}\right\rangle \mathrm{grad}_p f$$
$$+\frac{1}{2}\sum_{i=x,y,z}\frac{\partial^2}{\partial p_i^2}\left(\left\langle\frac{\Delta p_i^2}{\Delta t}\right\rangle f + \ldots\right) \tag{5.3.1}$$

where $\langle\ldots\rangle$ is the expectation value of the momentum transfer rates, and $U(r)$ is the crystal potential, which we may take as the continuum potential $U^{\mathrm{pl}}(r_\perp)$ in this case.

We should note that $\langle\Delta p_x\rangle = \langle\Delta p_y\rangle = 0$, since this system is axially symmetric. The term $\left\langle\frac{\Delta p_z}{\Delta t}\right\rangle$ is related to the stopping power. The contribution of the term $\left\langle\frac{\Delta p_z}{\Delta t}\right\rangle$ is called the 'damping effect'. This term will be discussed in section 5.4.

Therefore, equation 5.3.1 becomes

$$\frac{\partial f}{\partial t} + \mathbf{v}\cdot \mathrm{grad}\, f - \mathrm{grad}\, U(r_\perp)\, \mathrm{grad}_\mathrm{p}\, f$$
$$= \frac{1}{2} \sum_{i=x,y} \frac{\partial^2}{\partial p_i^2} \left(\left\langle \frac{\Delta p_i^2}{\Delta t} \right\rangle f \right) \tag{5.3.2}$$

Making use of the relation

$$E_\perp = \frac{p^2 \theta^2}{2M_1} + U(r_\perp)$$
$$p^2 = p_x^2 + p_y^2$$

the right-hand side of equation 5.3.2 is changed to

$$+\frac{1}{2} \frac{\partial}{\partial E_\perp} \left\{ [E_\perp - U(r_\perp)] \frac{1}{M_1} \left\langle \frac{\Delta p_\perp^2}{\Delta t} \right\rangle \frac{\partial f}{\partial E_\perp} \right\}$$

Here we assume the equilibrium distribution $F(p, E_\perp, t)$ for $E_\perp \geq U(r_\perp)$

$$f(\mathbf{r}_\perp, \mathbf{p}, E_\perp, t) = \begin{cases} F(p, E_\perp, t) & \text{when } E_\perp \geq U(r_\perp) \\ 0 & \text{when } E_\perp < U(r_\perp) \end{cases}$$

Considering this assumption and integrating over the accessible region (see Chapter 2) of the minimum distance r_1 to the string, we obtain

$$\pi(r_0^2 - r_1^2) \frac{\partial F}{\partial t} = \int \left\{ \frac{1}{2} \frac{\partial}{\partial E_\perp} [E_\perp - U(r_\perp)] \frac{1}{M_1} \left\langle \frac{\Delta p_\perp^2}{\Delta t} \right\rangle \frac{\partial F}{\partial E_\perp} \right\} 2\pi r_\perp \, \mathrm{d}r_\perp$$

or

$$\frac{\partial F}{\partial z} = \frac{1}{2S(E_\perp)} \frac{\partial}{\partial E_\perp} \int \left\{ [E_\perp - U(r_\perp)] \frac{1}{M_1} \left\langle \frac{\Delta p_\perp^2}{\Delta z} \right\rangle \frac{\partial F}{\partial E_\perp} \right\} 2\pi r_\perp \, \mathrm{d}r_\perp$$
$$= \frac{1}{2S(E_\perp)} \frac{\partial}{\partial E_\perp} \left[S(E_\perp) D_\perp(E_\perp) \frac{\partial F}{\partial E_\perp} \right] \tag{5.3.3}$$

$$D_\perp(E_\perp) = \frac{1}{S(E_\perp)} \int [E_\perp - U(r_\perp)] \frac{1}{M_1} \left\langle \frac{\Delta p_\perp^2}{\Delta z} \right\rangle 2\pi r_\perp \, \mathrm{d}r_\perp \tag{5.3.4}$$

$$S(E_\perp) = \pi(r_0^2 - r_1^2)$$

where πr_0^2 is the area per string. The diffusion type equation 5.3.3 was derived by Beloshitsky and Kumakhov.[5]

If we neglect the $E_\perp$ dependence of $S(E_\perp)$, or if we assume $r_0 \gg r_1$, equation 5.3.3 turns out to be

$$\frac{\partial F}{\partial z} = \frac{1}{2} \frac{\partial}{\partial E_\perp} \left[D(E_\perp) \frac{\partial F}{\partial E_\perp} \right] \tag{5.3.5}$$

In the same manner, we may obtain the planar diffusion equation where the distribution function f is represented by the transverse energy E_y at depth z in the crystal

$$\frac{\partial f}{\partial z} = \frac{1}{2}\frac{\partial}{\partial E_y}\left[D_y(E_y)\frac{\partial f}{\partial E_y}\right] \tag{5.3.6}$$

$$D_y(E_y) = \frac{1}{(y_0 - y_1)}\int [E_y - U(y)]\frac{1}{M_1}\left\langle\frac{\Delta p_y^2}{\Delta z}\right\rangle \mathrm{d}y$$

$$= \frac{1}{(y_0 - y_1)}\int [E_y - U(y)]\frac{1}{M_1}D_y\,\mathrm{d}y \tag{5.3.7}$$

where $y_0 - y_1$ is the accessible area (see Chapter 2). This equation was employed by many authors,[1,12] to study planar dechannelling.

If we consider that the transverse momentum $p_\perp$ is changed into $p_\perp + \Delta p_\perp$ due to inelastic scatterings, we obtain

$$\Delta E_\perp = \frac{(\mathbf{p}_\perp + \Delta\mathbf{p}_\perp)^2}{2M_1} - \frac{p_\perp^2}{2M_1} \simeq \frac{\mathbf{p}_\perp \cdot \Delta\mathbf{p}_\perp}{M_1}$$

Therefore, equation 5.3.4 is rewritten in the form

$$D_\perp(E_\perp) = \frac{1}{2S(E_\perp)}\int \left\langle\frac{\Delta E_\perp^2}{\Delta z}\right\rangle 2\pi r_\perp\,\mathrm{d}r_\perp$$

$$\equiv \frac{1}{2}\left\langle\left\langle\frac{\Delta E_\perp^2}{\Delta z}\right\rangle\right\rangle_A \tag{5.3.8}$$

since $\dfrac{p_\perp^2}{2M_1} = E_\perp - U(r_\perp)$.

If we neglect the $E_\perp$ dependence of the accessible region in equations 5.3.4 or 5.3.7, we obtain

$$\frac{\partial D_\perp(E_\perp)}{\partial E_\perp} = \frac{\partial}{\partial E_\perp}\left(\frac{1}{\pi r_0^2}\int E_\perp \frac{1}{M_1}\left\langle\frac{\Delta p_\perp^2}{\Delta z}\right\rangle 2\pi r_\perp\,\mathrm{d}r_\perp\right)$$

$$= \frac{1}{\pi r_0^2}\int \frac{1}{M_1}\left\langle\frac{\Delta p_\perp^2}{\Delta z}\right\rangle 2\pi r_\perp\,\mathrm{d}r_\perp$$

$$= \frac{1}{M_1}\langle D_\perp(r_\perp)\rangle_A \tag{5.3.9}$$

or †

$$\frac{\partial D_y(E_y)}{\partial E_y} = \frac{1}{M_1}\langle D_y(y)\rangle_A \tag{5.3.10}$$

---

†We can also derive this equation[7] by taking the relation

$$\frac{\mathrm{d}(\Delta E_y^2)}{\mathrm{d}(\Delta E_y)} = 2\Delta E_y = \Delta p_y^2/M_1$$

and fixing the value ΔE_y in the integrand of equation 5.3.4.

This is the simplified relation between $D_{\perp}(r_{\perp})$ and $D_{\perp}(E_{\perp})$, or $D_y(y)$ and $D_y(E_y)$. However, this relation is too simplified to obtain $D(E_{\perp})$ from $D_{\perp}(r_{\perp})$.

Waho has shown[21] that the diffusion equation in the form of 5.3.6 can be successfully applied to the calculation of planar dechannelling. If $D_y(E_y)$ takes a simple form, the diffusion equation 5.3.6 can be solved easily. If $D_y(E_y)$ does not depend on E_y, then equation 5.3.6 is just the heat transfer equation which can be solved analytically with the following conditions

$$\text{initial condition: } F(E_y, 0) = F^0(E_y) = \text{(initial distribution)}$$
$$\text{final condition: } F(E_y^c, t) = 0 \text{ for } t > 0$$

In most cases the initial distribution function $F^0(E_y)$ is assumed to be the delta function. The final condition means that a beam with transverse energy E_y larger than the critical energy $E_y^c = E\theta_c^2$ deviates completely from the channelled trajectories.

However, as stated in the previous section, we cannot assume that $D_y(E_y)$ is a constant. We should take the form

$$D_y(E_y) = D_0 E_y^l \tag{5.3.11}$$

where D_0 and l are unknown constants. We can determine D_0 and using figure 5.4,

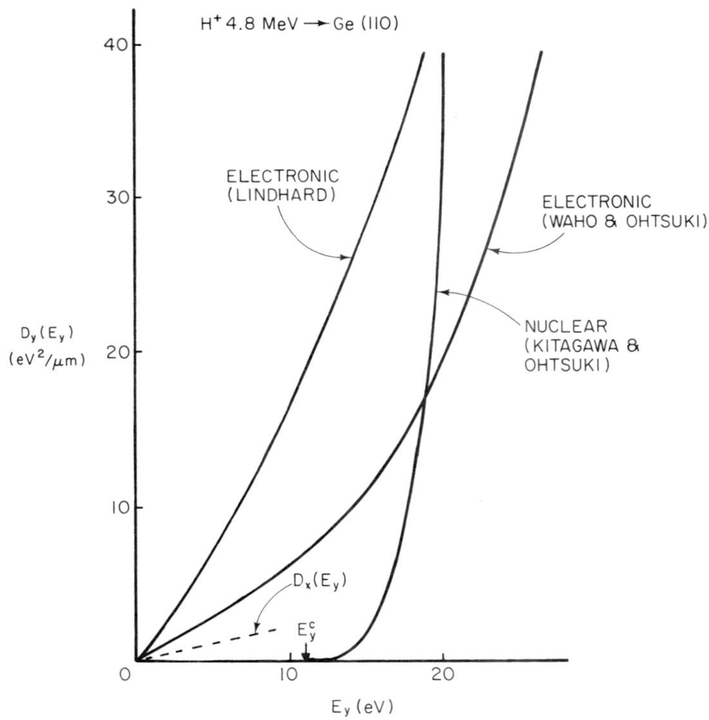

Figure 5.4. Detailed profile of $D_y(E_y)$ for 4·8 MeV H^+ in Ge (110) according to Waho and Ohtsuki,[14] and Kitagawa and Ohtsuki.[16]

which shows the detailed profile of $D_y(E_y)$ for 4·8 MeV H^+ into Ge(110) for planar channelling by making use of the formulae of Waho and Ohtsuki[14] (equation 5.2.10 or 5.2.15), and Kitagawa and Ohtsuki[16] (e.g., equation 5.2.27). This shows:

(*a*) that for the region $E_y \leq E_y^c$ the nuclear contribution is very small, and

(*b*) that $D_y(E_y)$ is proportional to E_y for $E_y \leq E_y^c$.

Therefore we determine l to be 1 for this case.

Inserting equation 5.3.11 into equation 5.3.6, we obtain[21]

$$F(E_y, t) = \sum_{k=1}^{\infty} C_k E_y^{(1-l)/2} J_p[j_{p,k}(E_y/E_y^c)^{(2-l)/2}] \times \exp\left[-D_0(E_y^c)^{l-2}\left(\frac{2-l}{2} j_{p,k}\right)^2 t\right] \tag{5.3.12}$$

where

$$C_k = \frac{2-l}{(E_y^c)^{2-l} J_{p+1}^2(j_{p,k})} \int_0^{E_y^c} E_y^{(1-l)/2} F^0(E_y) J_p[j_{p,k}(E_y/E_y^c)^{(2-l)/2}]\, dE_y$$

$$p = |(1-l)/(2-l)|$$

In the above $j_{p,k}$ is the kth zero of the Bessel function.

Here we define the channelled fraction $N(t)$ by

$$N(t) = \int_0^{E_y^c} F(E_y, t)\, dE_y$$

and the half thickness $t_{1/2}$ by

$$N(t_{1/2}) = \tfrac{1}{2} N(0).$$

The channelled fraction $N(t)$ gives the intensity of the channelled beam which has not escaped from the channelling trajectories, and the half thickness $t_{1/2}$ is that thickness where $N(t)$ has decreased to one-half of its initial value.

As seen for 4·8 MeV H^+ into Ge(110) we may take $l=1$ and $k=1$ in equation 5.3.12, since the terms for $k \geq 2$ are very small. We then obtain

$$F(E_y, t) = C_1 J_0[j_{0,1}(E_y/E_y^c)^{1/2}] \exp\left[-\frac{D_0}{E_y^c}(\tfrac{1}{2}j_{0,1})^2 t\right]$$

from which

$$t_{1/2} = \frac{4 \ln 2 (E_y^c)^2}{(j_{0,1})^2 D(E_y^c)} = 4 \ln 2 (E_y^c)/(j_{0,1})^2 D_0 \tag{5.3.13}$$

This formula for $t_{1/2}$ has a very simple and clear physical interpretation. Since the half-thickness $t_{1/2}$ is proportional to $E_y^c = E\theta_c^2$, and $\theta_c^2 \propto E^{-1}$, E_y^c does not depend on the energy E. However, D_0 is proportional to E^{-1} in the high energy region, and therefore $t_{1/2}$ is proportional to E. In figure 5.5 we

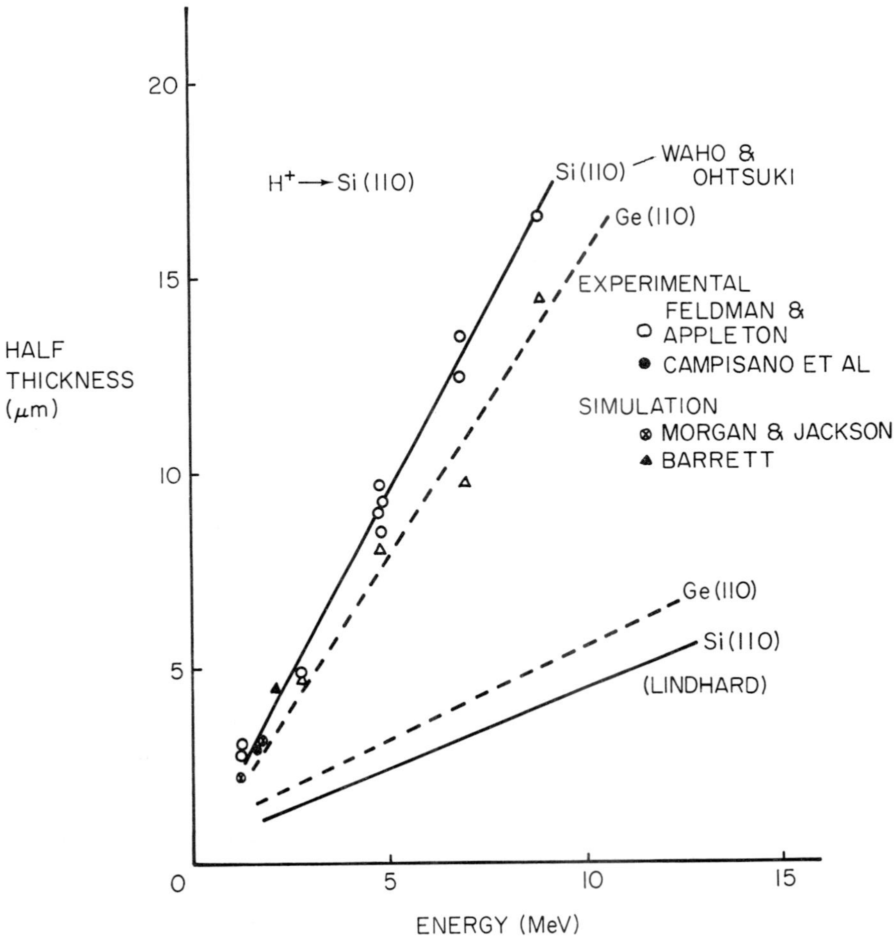

Figure 5.5. Calculated half-thickness $t_{1/2}$ for H^+ in Si as a function of the ion energy compared with experimental results.[9] Lindhard-type calculation and computer simulation by Barrett,[22] and Morgan and Jackson[23] are also plotted.

compare the calculation[21] of $t_{1/2}$ using equation 5.2.12 for H^+ into Si with experimental results.[9] Comparisons are also made between the Lindhard diffusion coefficient and computer simulations by Barrett,[22] and Morgan and Jackson.[23] The calculated results agree well with the experimental and computed ones.†

† Grasso showed in *Channelling Theory, Observation and Applications* (edited by D.V. Morgan, New York, Wiley, 1973) that the calculation of $t_{1/2}$ using the Lindhard diffusion coefficient agrees well with experimental results. However, in the calculation it was assumed that $D_y = D_x = \frac{1}{2}D$. As was pointed out in solution 5.2 and will be pointed out again in section 5.5, the assumption is not correct.

5.4. Axial dechannelling

In the previous section we showed that the solution of the simple diffusion equation using the diffusion coefficient given by Waho and Ohtsuki gives reasonable results for planar dechannelling phenomena. However, as shown by many workers (for reviews see reference 24), the same method cannot produce good results for *axial* dechannelling. So, Kumakhov and his coworkers,[3,4,5] and Kitagawa and Ohtsuki[16] tried to modify the axial dechannelling theory by making use of the Fokker–Planck equation.

If we denote the normal stopping power $W(E_{(z)})$ of a beam of energy $E_{(z)}$ at position z, we obtain the simple relation

$$z=\int_{E_z}^{E_0}\frac{\mathrm{d}E_{(z)}}{W(E_{(z)})} \tag{5.4.1}$$

where E_0 is the initial energy. Changing the distribution function from f *to* u as follows

$$u=\frac{E_{(z)}W(E_z)}{E_0W(E_0)}f \tag{5.4.2}$$

and taking into account equation 5.4.1, equation 5.3.2 takes the form

$$-\frac{\partial u}{\partial E_{(z)}}=\frac{1}{2E_{(z)}}(\mathbf{p}_\perp\cdot\mathrm{grad}_{p\perp}u)+\frac{1}{W(E_{(z)})\,v_{(z)}}\sum_{i=x,y}\frac{\partial^2}{\partial p_i^2}(\sigma_{ii}\,u) \tag{5.4.3}$$

where $\sigma_{ii}=\dfrac{1}{2}\left\langle\dfrac{\Delta p_i^2}{\Delta t}\right\rangle$. Equation 5.4.3 is the simplified Fokker–Planck equation in the energy–momentum space $(p_x, p_y, E_{(z)})$.

It is more convenient if we write the Fokker–Planck equation in cylindrical co-ordinates and use the variables defined by

$$\mu_\theta=\frac{D_\theta}{2W(E_{(z)})}=\frac{1}{2W(E_{(z)})}\left\langle\frac{\Delta p_\theta^2}{\Delta z}\right\rangle \tag{5.4.4}$$

and

$$\mu_\perp=\frac{D_\perp}{2W(E_{(z)})}=\frac{1}{2W(E_{(z)})}\left\langle\frac{\Delta p_\perp^2}{\Delta z}\right\rangle \tag{5.4.5}$$

Accordingly we arrive at the equation

$$-\frac{\partial u}{\partial E_{(z)}}=\frac{p_\perp}{2E_{(z)}}\frac{\partial u}{\partial p_\perp}+\frac{\partial^2}{\partial p_\perp^2}(\mu_\perp u)+\frac{1}{p_\perp}\frac{\partial}{\partial p_\perp}[(2\mu_\perp-\mu_\theta)u] \tag{5.4.6}$$

When the beam trajectories are sufficiently mixed and arrive at a statistical equilibrium the modified diffusion coefficient μ_θ can be neglected. Furthermore, we note that the first term of equation 5.4.6 is negligible since $p_\perp$ is very small under channelling conditions; whence we obtain

$$-\frac{\partial u}{\partial E_{(z)}} = \frac{\partial^2}{\partial p_\perp^2}(\mu_\perp u) + \frac{2}{p_\perp} \cdot \frac{\partial(\mu_\perp u)}{\partial p_\perp} \tag{5.4.7}$$

The equations 5.4.6 and 5.4.7 were derived by Kitagawa and Ohtsuki,[16] and solved numerically by Ohtsuki *et al.*[15] The results obtained approach the experimental ones more nearly than those given by the diffusion equation. However, agreement of theory with experiments is not sufficient. This may be attributed to the arbitrary solving of the differential equation.

5.5. Diffusion of electron beams

There are no fundamental differences between ion beam diffusion and electron beam diffusion. The diffusion coefficients can also be obtained from the formulae given by Kitagawa and Ohtsuki,[16] and Waho and Ohtsuki,[14] if we replace the scattering factor v_q for ions with that for electrons, and then take the maximum and minimum momentum transfers for single electron excitations in the forms

$$\hbar q_{\max} = mv \tag{5.5.1}$$

and

$$\hbar q_{\min} = \frac{I}{v} \tag{5.5.2}$$

where I is the mean excitation energy. The minimum momentum transfer $\hbar q_{\min}$ is the same as for ions (see equation 5.2.8), but the maximum momentum transfer $\hbar q_{\max}$ does not include the factor 2 in equation 5.5.1. This means that the maximum momentum transfer should be mv for a head-on collision between two electrons.†

Figure 5.6 shows the calculated diffusion coefficients (Fourier components) obtained[25] in this way for electron beams with an incident energy of 3 MeV into Si as a function of (*a*) the reciprocal lattice vector $\mathbf{h} = (2n, 2n, 0)$, and (*b*) the temperature. It should be noted that the y component (which is perpendicular to the atomic plane) of the diffusion coefficient D_{yh} is about twice as large as the x component D_{xh} at the reciprocal lattice point $\mathbf{h} = (4, 4, 0)$. It is easily understood that the diffusion coefficient due to phonon excitation for $h = 0$ does not depend on temperature, as shown in figure 5.6(*b*). For $\mathbf{h} = (4, 4, 0)$, we see that $D_{yh} > D_{xh}$, but the values for $D_{yh} = D_{xh}$ are not very different. Comparing the diffusion coefficient due to single electron excitation with that

† In actual calculations there are no important differences between the two types of maximum momentum transfer.

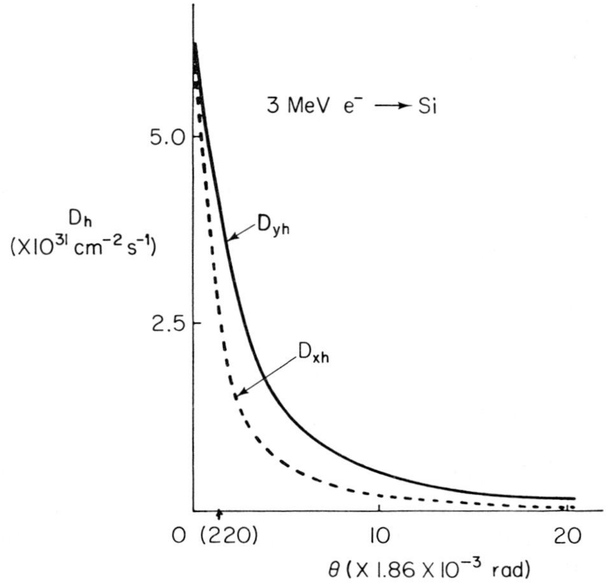

(*a*)

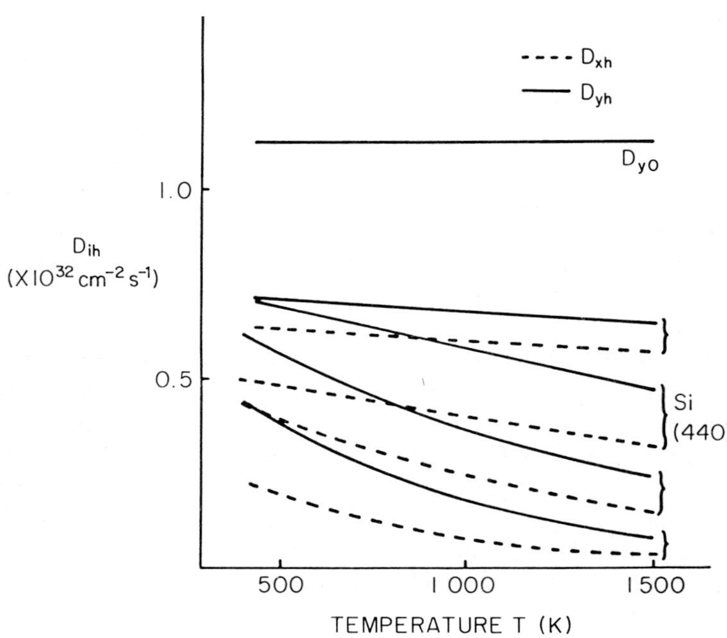

(*b*)

Figure 5.6. Diffusion coefficients for 3 MeV electrons into Si as a function of (*a*) the reciprocal lattice vector $\mathbf{h} = (2n, 2n, 0)$ and (*b*) the temperature.

of phonon excitation at $\mathbf{h} = (4, 4, 0)$, the latter is twice as large as the former. We also note that the diffusion coefficient due to phonon excitation decreases with increasing temperature. This is obtained by the Debye–Waller factor, which is the origin of the temperature dependence of the diffusion coefficient D_y introduced in section 5.2.

By making use of the diffusion coefficient D_{ih}, we can calculate the diffusion of the diffraction spots. We use the two-wave approximation and, for simplicity, take a case where the intensities of the 0 and **h** waves are the same (Chapter 3). The change of the intensity corresponding to the jth branch $\chi^{(j)}$ is represented by the diffusion equation in the form

$$\frac{\partial \chi^{(j)}}{\partial t} = \frac{\partial^2}{\partial x^2}(D_x^{(j)} \chi^{(j)}) + \frac{\partial^2}{\partial y^2}(D_y^{(j)} \chi^{(j)}) \tag{5.5.3}$$

which was first derived by Ohtsuki[26] using quantum mechanics. In equation 5.5.3 the diffusion coefficient of the jth branch $D_i^{(j)}$ is given in the form

$$D_i^{(j)} = D_{i0} \pm C\, D_{ih} \tag{5.5.4}$$

being + for $j = 1$ and − for $j = 2$. The constant C depends on the method of calculation. Ohtsuki originally gave $C = \sqrt{2}$, but recently Tamura[27] proved $C = 1$ by using a different model.

Since the simple solution of the diffusion equation 5.5.3 is the Gaussian distribution, the intensity $I(\theta_x, \theta_y)$, as a function of the angle $\theta = (\theta_x, \theta_y)$, takes the form

$$I(\theta_x, \theta_y) = \sum_{j=1,2} \frac{1}{2\pi \bar{\theta}_x^{(j)} \bar{\theta}_y^{(j)}} \exp\left[-\left(\frac{\theta_x}{\bar{\theta}_x^{(j)}}\right)^2 - \left(\frac{\theta_y}{\bar{\theta}_y^{(j)}}\right)^2 \right] \tag{5.5.5}$$

where

$$\bar{\theta}_i^{(j)2} = \frac{4\hbar^2 z}{m^2 v^3} D_i^{(j)} \tag{5.5.6}$$

z being the crystal thickness and $i = x, y$. Figure 5.7 shows the equi-intensity lines around the Si (0, 0, 0) and (2, 2, 0) spots for a 3 MeV electron beam passing through 500 nm, calculated by Tamura and Ohtsuki.[25] Also shown in figure 5.7 are the experimental diffraction spots found by Nakai.[28] We can see that the calculated asymmetrical intensity profile agrees with the experimental one.

5.6. Applications of dechannelling

Dechannelling is successfully applied in research into defects in crystals. We can detect the backscattering yield, which is the random beam fraction $R_d(z)$, where z is the depth from the surface. When the defect density is small, we may use the single scattering model.[29] In this model the channelled fraction

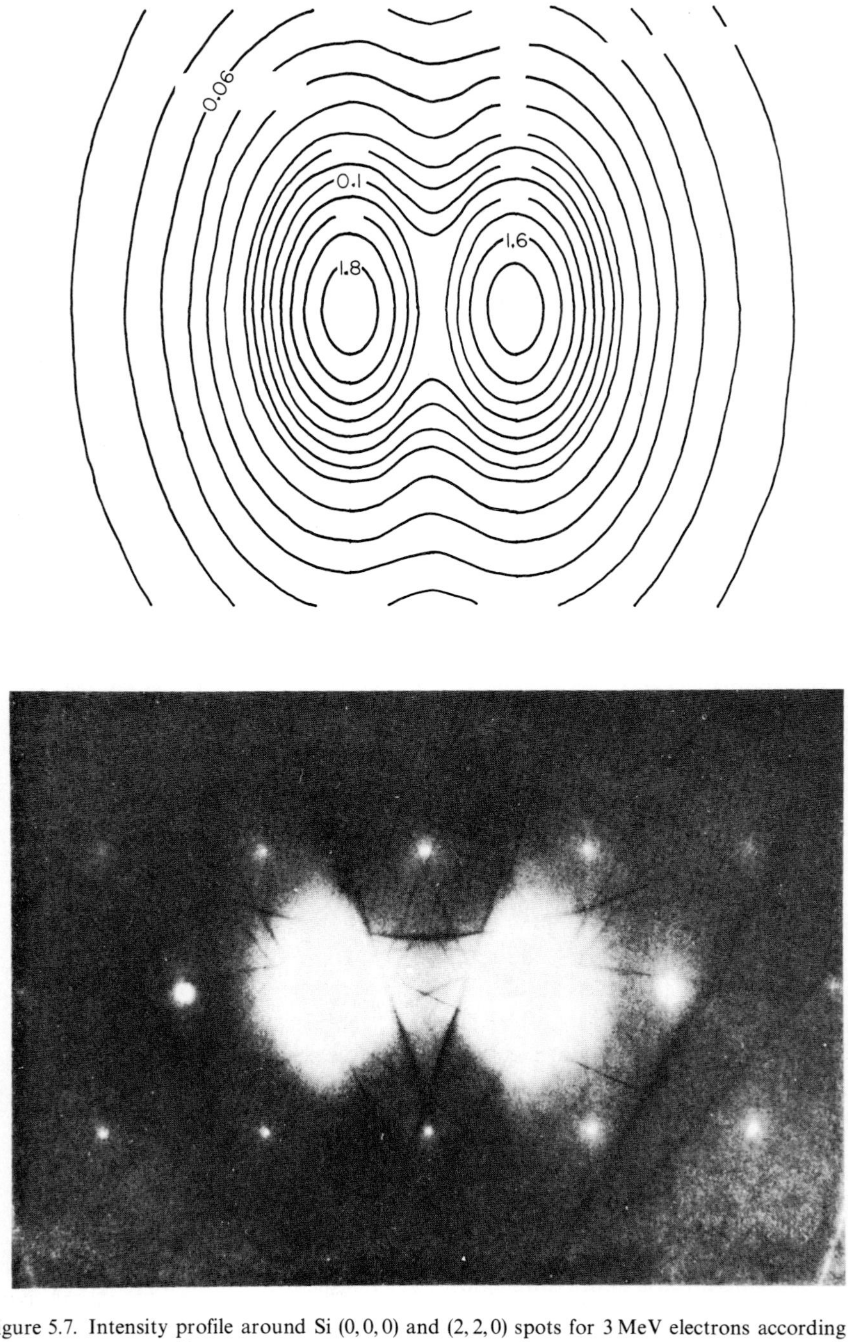

Figure 5.7. Intensity profile around Si (0, 0, 0) and (2, 2, 0) spots for 3 MeV electrons according to Tamura and Ohtsuki,[25] compared with the photograph obtained by Nakai.[28] (The author is indebted to Dr Nakai for the photograph.)

$1 - R_d(z)$ decreases simply in the form

$$\frac{d}{dz}[1 - R_d(z)] = -[n_d(z)A(z) + \chi(z)][1 - R_d(z)] \qquad (5.6.1)$$

where $n_d(z)$ is the defect density, and $A(z)$ represents the cross-section of dechannelling due to defects. $\chi(z)$ is the dechannelling fraction in a crystal with no defects.

When the defect is in the centre of the channel the channelled beam is scattered by the defect. Therefore $\chi(z)$ is calculated by Rutherford scattering with an angle greater than the critical angle.[29] Morita and Sizmann[30] worked out this process by computer simulation.

Quéré[31] derived an analytical formula of $\chi(z)$ for dislocations by considering the centrifugal force induced by distorted lattice planes due to dislocations. A few experiments for this process have been reported (see, e.g., reference 32).

For a large defect density, we have to consider[33,34] multiple scattering in order to obtain the dechannelled fraction. Matsunami and Itoh[35,36] have developed a diffusion-type theory in which they take into account diffusion coefficients due to defects besides electronic and phonon excitations.

References

1. Lindhard, J., *Kgl. Dan. Viden. Sels. Mat-Fys. Medd.* **34**, No. 14 (1965).
2. Bohr, N., *Kgl. Dan. Viden. Sels. Mat-Fys. Medd.* **18**, No. 8 (1948).
3. Kumakhov, M.A., *Phys. Lett.* **A33**, 133 (1970)
4. Beloshitsky, V.V., Kumakhov, M.A. and Muralev, V.A., *Rad. Effects* **13**, 9 (1972).
5. Beloshitsky, V.V. and Kumakhov, M.A., *J. Exp. Theor. Physics* **35**, 605 (1972).
6. Morita, K. and Itoh, N., *J. Phys. Soc. Jap.* **30**, 1430 (1971).
7. Bonderup, E., Esbensen, H., Andersen, J.U. and Schiøtt, H.E., *Rad. Effects* **12**, 261 (1972).
8. Pendersen, M.J., Andersen, J.U., Elliott, D.J. and Laegsgard, F., in *Atomic Collision in Solids II* (Plenum, 1975), p. 863.
9. Feldman, L.C. and Appleton, B.R., *Phys. Rev.* **B8**, 935 (1973).
10. Morita, K., *Rad. Effects* **14**, 195 (1972).
11. Davies, J.A., Howe, L.M., Marsden, D.A. and Whitton, J.L., *Rad. Effects* **12**, 247 (1972).
12. Campisano, S.U., Foti, G., Grasso, F., LoSavio, M. and Rimini, E., *Rad. Effects* **13**, 157 (1972).
13. Pines, D., *Elementary Excitations in Solids* (New York: Benjamin, 1964).
14. Waho, T. and Ohtsuki, Y.H., *Rad. Effects* **21**, 217 (1974).
15. Ohtsuki, Y.H., Omura, T., Tanaka, H. and Kitagawa, M., *Nucl. Inst. Methods* **149**, 361 (1978).
16. Kitagawa, M. and Ohtsuki, Y.H., *Phys. Rev.* **B8**, 3117 (1973).
17. Ellegard, C. and Lassen, N.O., *Kgl. Dan. Viden. Sels. Mat-Fys. Medd.* **35**, No. 16 (1967).
18. Kitagawa, M., *Phys. Lett.* **A58**, 487 (1976).
19. Kitagawa, M. and Ohtsuki, Y.H., *Phys. Rev.* **B9**, 4719 (1974).

20. Hubbard, J., *Proc. Roy. Soc. London* **A243**, 336 (1957).
21. Waho, T., *Phys. Rev.* **B14**, 4830 (1976).
22. Barrett, J.H., *Phys. Rev.* **B3**, 1527 (1971).
23. Morgan, D.V. and Jackson, D.P., *Nucl. Inst. Methods* **132**, 153 (1976).
24. Gemmell, D.S., *Rev. Mod. Physics* **46**, No. 1 (1974).
25. Tamura, E. and Ohtsuki, Y.H., *Phys. Stat. Sol.* **B93**, 167 (1979).
26. Ohtsuki, Y.H., *Phys. Stat. Sol.* **B67**, 495 (1975).
27. Tamura, E., *J. Phys. Soc. Japan* (to be published).
28. Nakai, Y., private communication.
29. Bøgh, E., *Canad. J. Phys.* **46**, 653 (1968).
30. Morita, K. and Sizmann, R., *Rad. Effects* **24**, 281 (1975).
31. Quéré, Y., *Phys. Stat. Sol.* **30**, 713 (1968).
32. Merkle, K.L., Pronko, P.P., Gemmell, D.S., Mikkelson, R.C. and Wrobel, J.R., *Phys. Rev.* **B8**, 1002 (1973).
33. Westmoreland, J.E., Mayer, J.W., Eisen, F.H. and Welch, B., *Rad. Effects* **6**, 161 (1970).
34. Feldman, L.C. and Rodgers, J.W., *J. Appl. Phys.* **41**, 3776 (1976).
35. Matsunami, N. and Itoh, N., *Phys. Lett.* **A45**, 453 (1973).
36. Matsunami, N., *J. Phys. Soc. Japan* **38**, 848 (1975).

CHAPTER 6

charge states, screening and wake

6.1. Introduction

The charge states of an ion and the Coulomb potential of the ion's charge are greatly modified when the ion moves in a solid, because of the virtual ionizations (and electron captures) and polarizations of the solid caused by the ion's motion. The polarizations are well described by plasmon excitation. These effects are also discussed in the field of atomic collisions in gases, but they are much more pronounced in dense media like solids. There is, however, a new phenomenon of the charge states which is found only in crystals, known as the resonance effect.

We shall first study the charge states of ions moving through a solid. The problem is very difficult theoretically, and, except for simple cases, there are only a few phenomenological or semi-empirical theories available. In section 6.3 we show that the charge states are modified under channelling conditions, and we discuss coherent phenomena (Okorokov effect) as new topics. We shall study dynamical screening and wake potentials, as related to recent experimental results, in the last section.

6.2. Equilibrium charge states and effective charge

The charge states change rapidly by the exchange of electrons between ions and the solid during ion penetration. As a result, in a short time the charge state becomes an equilibrium charge state. Bohr showed interest in this equilibrium charge state in an excellent review article published in 1948.[1] The interaction between ions and gaseous targets (especially the charge state) is one of the most important subjects in the field of atomic collisions. Recently, the charge state of ions in solids has been of interest to both theoreticians and experimentalists in this field.

The effective charge is defined as the charge which contributes to the energy loss during the penetration of ions in matter as defined in the Bethe–

Bloch expression for the stopping power. With the increasing interest in ion implantation and ion beam fusion the effective charge has been studied by many workers. What is the relation between the equilibrium charge states and the effective charge? We shall try to answer this question in the following subsections.

6.2.1. *Equilibrium charge*

Let us consider an ion which has Z_1 electrons in its orbit before penetrating into a solid. Within a time interval t after impinging on the surface, the number of orbital electrons is Q (i.e., Z_1 Q electrons were captured by the solid), and the quantum state of the ion is characterized by s. We represent the probability that the ion is in the above charge and quantum state at time t by the function $N_{Z_1}(Q,s;t)$. We assume that the ion state (Q,s) changes to another state (Q',s') with transition probability $Pz_1(Q',s';Q,s;t)$. It is evident that there are inverse processes $(Q'',s'')\rightarrow(Q,s)$. Therefore, the rate of change from ion state (Q,s) during a short time dt is given by

$$\frac{\mathrm{d}N_{Z_1}(Q,s;t)}{\mathrm{d}t}=\sum_{Q's'}{}'[P_{Z_1}(Q,s;Q',s';t)N_{Z_1}(Q',s';t)-P_{Z_1}(Q',s';Q,s;t)N_{Z_1}(Q,s;t)] \tag{6.2.1}$$

where $\sum'$ denotes sum over all (Q',s') except $(Q',s')=(Q,s)$.

For gaseous targets the ion is in the ground state immediately before the collision, and the ion state is completely characterized by Q; therefore, for gases we neglect the s terms in equation 6.2.1. Since the collision cross-section σ is defined in terms of the transition probability P as $\sigma=P/(nv)$, n and v being the number density of the target atom and the ion velocity, respectively, equation 6.2.1 becomes

$$\frac{\mathrm{d}}{\mathrm{d}t}N_{Z_1}(Qt)=nv\sum_{Q'}{}'[\sigma_{Z_1}(Q,Q')\,N_{Z_1}(Q't)-\sigma_{Z_1}(Q';Q)\,N_{Z_1}(Qt)] \tag{6.2.2}$$

where $\sigma_{Z_1}(Q';Q)$ is the capture cross-section σ_c when $Q'<Q$ and the loss cross-section σ_L when $Q'>Q$. The calculation of these cross-sections is very complicated, except for a proton in gaseous hydrogen.

For large t, $N_{Z_1}(Q;t)$ becomes stationary, independent of time, and is called the equilibrium charge distribution, denoted by $\bar{N}_{Z_1}(Q)$. Furthermore, the average equilibrium charge $\bar{Q}$ is defined by

$$\bar{Q}=\sum_{Q}Q\,\bar{N}_{Z_1}(Q) \tag{6.2.3}$$

The rate of change of the fraction of neutral hydrogen atoms ϕ_0 in the beam is given by

$$\frac{\mathrm{d}\phi_0}{\mathrm{d}t}=nv(\phi^+\sigma_c-\phi_0\sigma_L)$$

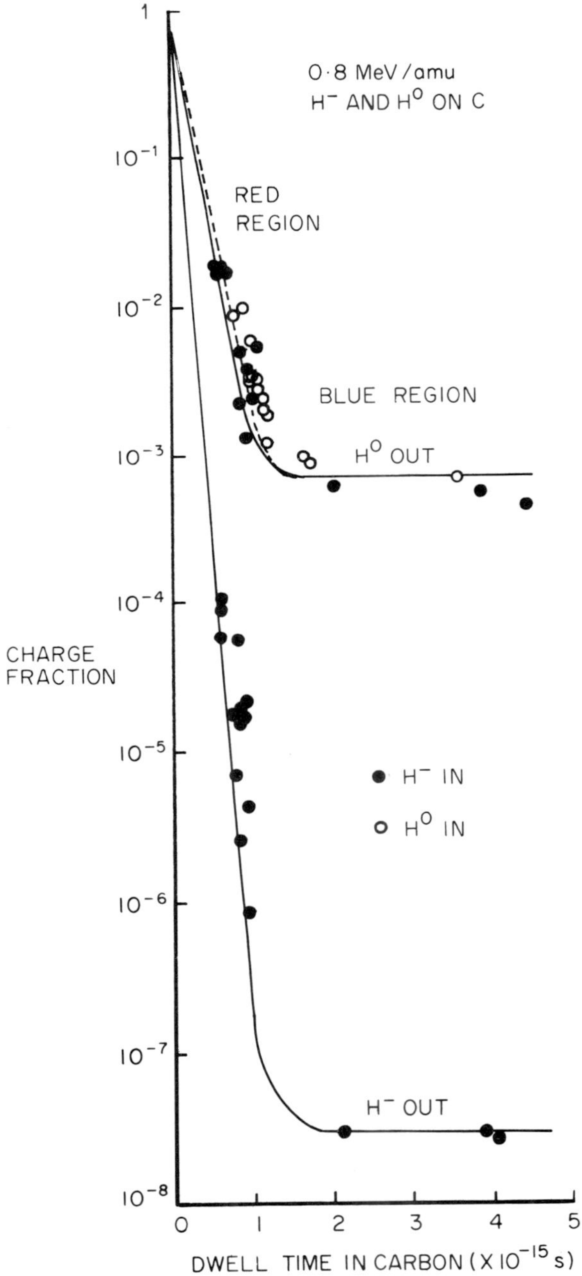

Figure 6.1. Charge fraction of the ion as a function of time, showing the first stage in a short time to be the 'red region' and the second stage to be the 'blue region', according to the Lyon group.[2]

Inserting the relation $\phi^+ = 1 - \phi_0$, we obtain

$$\frac{d\phi_0}{dt} = -nv(\sigma_c + \sigma_L)\phi_0 + nv\sigma_c$$

We can easily solve for the initial condition $\phi_0 = 0$ when $t = 0$

$$\phi_0 = \{1 - \exp[-nv(\sigma_c + \sigma_L)t]\} + (1 + \sigma_L/\sigma_c)^{-1}$$

Under the stationary condition we obtain

$$\frac{d\phi_0}{dt} = 0$$

or

$$\phi_0 = (1 + \sigma_L/\sigma_c)^{-1} \tag{6.2.4}$$

The Lyon group[2] called the time region where the time-dependent non-stationary term $\exp[-nv(\sigma_c + \sigma_L)t]$ is dominant the 'red region', and the stationary region the 'blue region'. Experimentally the red and blue regions are well separated. However, there are cases where the boundary between the two regions is not so clear. Figure 6.1 shows an example of the time dependence of ϕ_0, indicating the red and blue regions.

For He ions, equation 6.2.2 becomes simply

$$\frac{dN_2(0,t)}{dt} = nv[\sigma_{10}N_2(1,t) + \sigma_{20}N_2(2,t)] - nv(\sigma_{01} + \sigma_{02})N_2(0,t)$$

$$\frac{dN_2(1,t)}{dt} = nv[\sigma_{01}N_2(0,t) + \sigma_{21}N_2(2,t)] - nv(\sigma_{10} + \sigma_{12})N_2(1,t)$$

$$\frac{dN_2(2,t)}{dt} = nv[\sigma_{02}N_2(0,t) + \sigma_{12}N_2(1,t)] - nv(\sigma_{20} + \sigma_{21})N_2(2,t)$$

here σ_{ij} denotes the cross-section of charge transfer $i \rightarrow j$. At equilibrium ($t \rightarrow \infty$), we obtain

$$\sigma_{10}N_2(1,\infty) + \sigma_{20}N_2(2,\infty) - (\sigma_{01} + \sigma_{02})N_2(0,\infty) = 0$$
$$\sigma_{01}N_2(0,\infty) + \sigma_{21}N_2(2,\infty) - (\sigma_{10} + \sigma_{12})N_2(1,\infty) = 0$$
$$\sigma_{02}N_2(0,\infty) + \sigma_{12}N_2(1,\infty) - (\sigma_{20} + \sigma_{21})N_2(2,\infty) = 0$$

With the normalization condition

$$N_2(0,\infty) + N_2(1,\infty) + N_2(2,\infty) = 1 \tag{6.2.5}$$

and, neglecting σ_{20} and σ_{02}, we obtain

$$\sigma_{10}N_2(1,\infty) - \sigma_{01}N_2(0,\infty) = 0$$
$$\sigma_{01}N_2(0,\infty) + \sigma_{21}N_2(2,\infty) - (\sigma_{10} + \sigma_{12})N_2(1,\infty) = 0$$
$$\sigma_{12}N_2(1,\infty) - \sigma_{21}N_2(2,\infty) = 0$$

In general, we can show that $\sigma_{01} \ll \sigma_{10}$, and that $N_2(1, \infty) \ll N_2(0, \infty)$. Therefore we obtain

$$\begin{aligned}\bar{Q} &= \sum_i i N_2(i, \infty) = 2 - \frac{\sigma_{12}}{\sigma_{12} + \sigma_{21}} \\ &= 2 - \frac{\sigma_c}{\sigma_c + \sigma_L}\end{aligned} \tag{6.2.6}$$

where $\sigma_c = \sigma_{12}, \sigma_L = \sigma_{21}$.

Bohr[1] gave the phenomenological formula for $\sigma_{Z_1}(Q; Q')$ without using quantum mechanics. First, we shall discuss the ionization (or electron loss) cross-section σ_L. Let us consider an atom in a solid moving with velocity $-\mathbf{v}$ colliding with a static ion, rather than an ion moving with velocity $\mathbf{v}$ colliding with a stationary atom in the solid. We can then consider that the electron cloud and the nucleus collide with the ion with a velocity $-\mathbf{v}$ at the same time, and are scattered according to Rutherford scattering theory.

The differential cross-section for Rutherford scattering by an ion with charge Ze in which energy is transferred to the electron over the energy interval $[T, T+\mathrm{d}T]$ is given by (see Chapter 1)

$$\mathrm{d}\sigma = \frac{2\pi Z^2 e^4}{mv^2} \frac{\mathrm{d}T}{T^2}$$

Accordingly, the total ionization cross-section σ_L is calculated by summing up the contributions of the electron cloud (containing Z_2 electrons) and the nucleus

$$\sigma_L = \frac{2\pi}{mv^2} [(Z_2 e)^2 e^2 + Z_2 \times (e^2 \cdot e^2)] \frac{1}{I}$$

where I is the ionization energy. The second term corresponds to the collision of electrons of the atoms of the solid (their number being Z_2) with the electron of the ion. For inner shell electrons in light ions $I = mv_B^2 Z_1^2/2$, and we obtain

$$\sigma_L = 4\pi a_B^2 \frac{Z_2^2 + Z_2}{Z_1^2} \left(\frac{v_B}{v}\right)^2$$

When Z_2 and the screening effect become large, Bohr indicated the formula

$$\sigma_L = \pi a_B^2 \frac{Z_2^{2/3}}{Z_1} \left(\frac{v_B}{v}\right)$$

We can calculate σ_L from first principles. For example, Kaneko and Ohtsuki[3] calculated σ_L by considering the ionization probability in the Born approximation. Their starting Hamiltonian in atomic units takes the form

$$H = -\frac{\hbar^2}{2M} \Delta_R + \frac{Z_1 Z_2 e^2}{R} + H_t + H_i + H_{int} \tag{6.2.7}$$

where

$$H_t = \sum_i \left(-\frac{\hbar^2}{2m}\Delta_{2i} - \frac{Z_2 e^2}{r_{2i}} \right) + \frac{1}{2} \sum_{i \neq j} \frac{e^2}{|\mathbf{r}_{2i} - \mathbf{r}_{2j}|} \tag{6.2.8}$$

$$H_i = -\frac{\hbar^2}{2m}\Delta_1 - \frac{Z_1 e^2}{r_1} \tag{6.2.9}$$

and

$$H_{int} = \sum_i \left(-\frac{Z_1 e^2}{|\mathbf{r}_{2i} - \mathbf{R}|} \right) - \frac{Z_2 e^2}{|\mathbf{r}_1 + \mathbf{R}|} + \sum_i \frac{e^2}{|\mathbf{r}_{2i} - \mathbf{r}_1 - \mathbf{R}|} \tag{6.2.10}$$

M is the reduced mass, $\mathbf{r}_{2i}(i = 1, \ldots, Z_2)$ is the position vector of the target atom's electrons measured from the target atom's nucleus, and H_t is the Hamiltonian for the target atom's electrons. The vector $\mathbf{r}$ is the position vector of the ion's (extra) electron (projectile's electron) measured from the ion's nucleus. Therefore, H_i is the Hamiltonian for the ion's electron. H_{int} is the interaction Hamiltonian between the ion's nucleus and the target atom's electrons (first term), between the target atom's nucleus and the ion's electron (second term), and between the ion's electron and the target atom's electron (third term). $\mathbf{R}$ is the relative position vector of the ion's nucleus and the target atom's nucleus.

Making use of the above Hamiltonian we can derive the ionization cross-section σ_{ion} in which the target atom remains in the ground state

$$\sigma_{ion} = 8\pi(Z_2/Z_1^2\eta)^2 a_B^2 \int_q^\infty \frac{dq}{q^3} \langle |1 - (1/Z_2) f^t_{00}(-Z_1\mathbf{q})|^2 \rangle_{angle} \times \int_0^{k_{max}(q)} I(q,k)\, dk\, (a_B/Z_1) \tag{6.2.11}$$

where $k_{max}(q) = (Z_1/a_B)[2q\eta(a_B/Z_1) - 1]^{1/2}$, $q_{min} = Z_1/2\eta a_B$ and $\mathbf{q}$ is the momentum transfer. The parameter η is the ratio of the projectile velocity and the orbital velocity. The function $f^t_{00}(-Z_1\mathbf{q})$ is the atomic form factor.

In the above

$$I(q,k) = \frac{2^8 \bar{q}^2 \bar{k}[\bar{q}^2 + \frac{1}{3}(1 + \bar{k}^2)]}{[1 + (\bar{k} + \bar{q})^2]^3 [1 + (\bar{k} - \bar{q})^2]^3} \times \frac{\exp[-(2/\bar{k})\tan^{-1}(2\bar{k}/1 + q^2 - k^2)]}{1 - \exp(-2\pi/k)} \tag{6.2.12}$$

where $\bar{q} = a_B q/Z_1$, $k = a_B k/Z_1$ and $\langle \ldots \rangle_{angle}$ implies the average taken over the direction of the momentum transfer $\mathbf{q}$. k indicates the electron wave number ionized from the bound state. Kaneko and Ohtsuki calculated equation 6.2.11 in detail making use of the Hartree–Fock wave functions[4] for the target atom.

Next we consider electron capture. Thomas[5] adopted the following model for electron capture: electrons emitted from the target atom by the incident ion are scattered by the target nucleus; as a result, they begin to move with the same velocity as the incident ion, and are captured immediately by the ion. With the above model we can get the following expression for the capture cross-section

$$\sigma_c = \frac{64\sqrt{2}}{3}\left(\frac{v_B}{v}\right)^{11}(\pi a_B^2)\frac{Z_1^{7/2}Z_2^2}{(r/a_B)^{7/2}} \tag{6.2.13}$$

where r is the radius of the orbit of the captured electron. Thomas's model is reasonable for high energy beams since forward scattering is assumed.

Bohr,[1] on the other hand, obtained the capture cross-section for an ion with a velocity comparable to the velocity of the atomic orbit electron. He assumed that an electron with a velocity the same as that of an ion should be captured by that ion. In order to be captured by an ion with velocity v it is necessary that the kinetic energy $\frac{1}{2}mv^2$ be given to the electron. The cross-section of the kinetic energy transfer by collision is obtained easily by classical mechanics

$$\sigma = 4\pi Z_1^2 \frac{e^4}{m^2v^4} = 4\pi a_B^2 Z_1^2\left(\frac{v_B}{v}\right)^4$$

Next we have to calculate the capture probability of electrons with the above kinetic energy. This probability may be considered as the ratio of the volume in the velocity space of the orbit electron to that of the ion

$$P = \left(\frac{Z_1 v_B}{v}\right)^3$$

Since the number of electrons with orbit velocity v is

$$n \cong Z_2^{1/3}\left(\frac{v}{v_B}\right)$$

the total capture cross-section will be obtained by multiplying the above three equations

$$\sigma_c = 4\pi a_B^2 Z_1^5 Z_2^{1/3}\left(\frac{v_B}{v}\right)^6 \tag{6.2.14}$$

Bohr and Lindhard,[6] by further developing the above reasoning, arrived at another expression

$$\sigma_c = \pi a_B^2 Z_1^2 Z_2^{1/3}\left(\frac{v_B}{v}\right)^3 \tag{6.2.15}$$

Recently, a few authors have tried to derive σ_c from first principles by calculating the matrix elements quantum mechanically, which is very difficult. Cross[7] derived the following matrix element for the capture cross-section

$$\langle f|V|i\rangle = \iint \exp\left\{-\frac{i}{\hbar}\left[\left(\frac{E_i - E_f}{v} + \frac{mv}{2}\right)z_b + \mathbf{p}_\perp\cdot\boldsymbol{\rho}_b\right]\right\}\psi_f^*(\mathbf{r}_b)V(\mathbf{r}_a, \mathbf{r}_b)$$

$$\times\,\psi_i(\mathbf{r}_a)\exp\left\{\frac{i}{\hbar}\left[\left(\frac{E_i - E_f}{v} - \frac{mv}{2}\right)z_a + \mathbf{p}_\perp\cdot\boldsymbol{\rho}_a\right]\right\}\mathrm{d}\mathbf{r}_a\,\mathrm{d}\mathbf{r}_b \tag{6.2.16}$$

where ψ_f and ψ_i are the wave functions of the final and initial states of the electron, respectively, and $\mathbf{r}_a = (\boldsymbol{\rho}_a, z_a)$ is the position vector of the electron measured from the nucleus A, z being the direction of the ion velocity $\mathbf{v}$. The ion velocity v is included in the factor exp [. . .] of equation 6.2.16, which means that the orbit electron is moving with the ion. The Hamiltonian $V(\mathbf{r}_a, \mathbf{r}_b)$ is the interaction potential for the electron transition. Let us take $V \doteqdot -e^2/r_b$ (Brinkman–Kramers approximation[8]) and $E_i - E_f \doteqdot \frac{1}{2}mv^2 \pm \hbar v/R_i$, R_i being $n'a_B/Z_B$(n' is the quantum number of the electron in the target atom which will be captured by the ion). We then get the capture cross-section from the target atomic state n' to the ion state n

$$\sigma_c = (2\pi\hbar^2 v)^{-2} \sum_i \int d^2 p_\perp |\langle f|V|i\rangle|^2$$

$$= \frac{2^{18}}{5}\pi a_B^2 \frac{1}{n^3}\frac{Z_2^5}{n'^3}\left(\frac{v}{v_B}\right)^8 \left[\frac{(\frac{1}{2}mv_B^2)^2}{\left(E_i - E_f - \dfrac{mv^2}{2}\right)^2 + (\hbar v/R_i)^2}\right]^5 \tag{6.2.17}$$

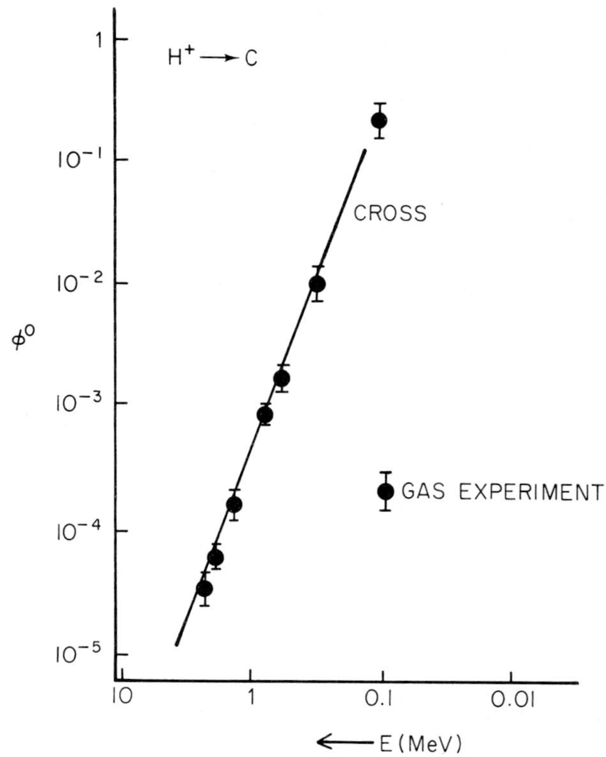

Figure 6.2. Calculated neutral fraction ϕ_0 for 100 keV–10 MeV H^+ into carbon foil according to Cross.[7]

Cross calculated in detail the neutral fraction $\phi_0 = (1 + \sigma_L/\sigma_c)^{-1}$ making use of expression 6.2.16 for 100 keV to 10 MeV protons going into carbon foil, and compared their values with experimental results, which are shown in figure 6.2. The agreement is quite good.

Kaneko and Ohtsuki[3], on the other hand, calculated the capture cross-section in more detail by making use of the Landau–Zener method[8] with the Brinkman–Kramers approximation. They considered two states and expanded the total wave function with the eigenfunctions ϕ_A and φ_B of the target atom's and ion's electrons, respectively

$$\psi = a(t)\phi_A(\mathbf{r}_2)\exp\left[-\frac{i}{\hbar}\left(\frac{m\mathbf{v}\cdot\mathbf{r}}{2} + \frac{m}{8}v^2t + \varepsilon_A t\right)\right] + b(t)\,\varphi_B(\mathbf{r}_1)$$
$$\exp\left[-\frac{i}{\hbar}\left(-\frac{m\mathbf{v}\cdot\mathbf{r}}{2} + \frac{m}{8}v^2t + \varepsilon_B t\right)\right] \tag{6.2.18}$$

Inserting this wave function into the Schrödinger equation for ψ, we get a coupled equation for the amplitudes $a(t)$ and $b(t)$, from which we get

$$\sigma_c = 2\pi N_A \int_0^{\rho_{\max}} d\rho\, \rho\, |b(\infty)|^2 \tag{6.2.19}$$

where ρ is the impact parameter and N_A is the number of electrons in the target atom.

The calculation of $b(\infty)$ is not easy, but Kaneko and Ohtsuki obtained (for *s* state capture; in atomic units)

$$b(\infty) = iZ_1^{5/2}\frac{4v_B}{v}\sum_i C_i N_i \left[\frac{\partial}{\partial(-\zeta_i)}\right]^i I_1 \tag{6.2.20}$$

where

$$I_1 = [K_0(\alpha\rho) - K_0(\beta_i\rho)]/X$$
$$X = \beta_i^2 - \alpha^2$$
$$\alpha^2 = \frac{Z_1^2}{a_B^2} + \left(\frac{mv}{2} - \frac{\omega}{v}\right)^2\Big/\hbar^2$$
$$\beta_i^2 = \zeta_i^2 + \left(\frac{mv}{2} + \frac{\omega}{v}\right)^2\Big/\hbar^2$$
$$\omega = \varepsilon_A - \varepsilon_B \tag{6.2.21}$$

and $C_i N_i$ represents the expansion coefficient[4] of ϕ_A with respect to the spherical harmonic function $Y_{00}(\theta, \varphi)$.

For He ions, Kaneko and Ohtsuki calculated in detail the mean charge $\bar{Q} = 2 - \sigma_c/(\sigma_c + \sigma_L)$ from equations 6.2.11 and 6.2.19, varying Z_2 from 4 to 45. Figure 6.3 shows their results as a function of Z_2 and the experimental results

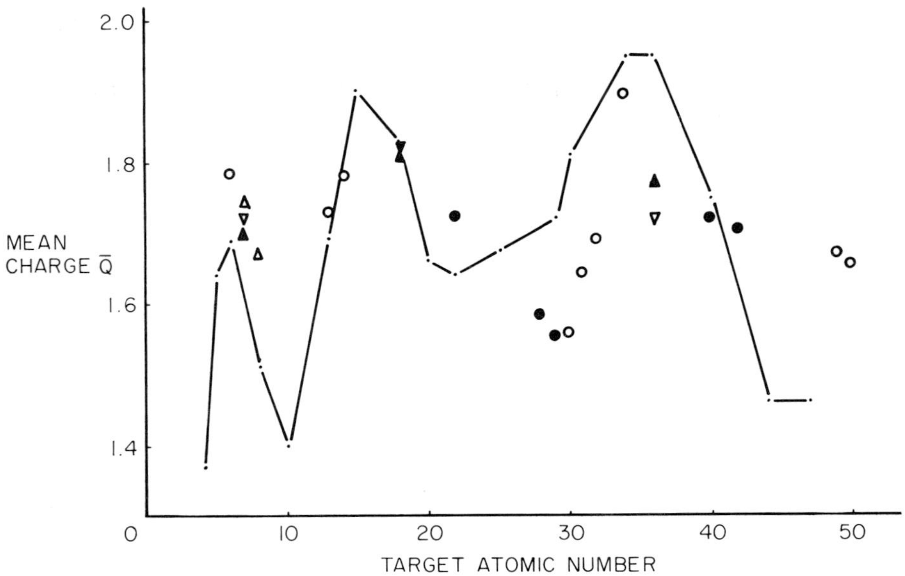

Figure 6.3. Calculated mean charge $\bar{Q}$ as a function of Z_2 by Kaneko and Ohtsuki,[3] compared with the experimental data of Kido *et al.*[9]

of Kido *et al.*,[9] in which we can see oscillatory behaviour. This oscillation is due to the periodic shell electrons contributing dominantly to σ_c.

Without getting detailed capture and loss cross-sections, there are some empirical and/or phenomenological formulae to give the mean equilibrium charge $\bar{Q}$. Bohr[1] assumed that $\bar{Q}/Z_1$ becomes the ratio of the ion velocity v to the orbit velocity $v_0(\cong v_B Z_1^{2/3})$

$$\frac{\bar{Q}}{Z_1} = \frac{v}{v_0} \cong \frac{v}{v_B Z_1^{2/3}} \tag{6.2.22}$$

This assumption is based on the fact that all the electrons whose velocities are less than that of the heavy ions, v, are stripped (Bohr's velocity criterion). The electrons whose orbital velocities are large are tightly bound to the atom and are not stripped by the ion. Equation 6.2.22 is correct for $1 < v/v_B < Z_1^{2/3}$.

A much more precise estimate was proposed by Knipp and Teller,[10] and Bruning *et al.*,[11] making use of the Thomas–Fermi model. Lamb[12] obtained a formula for $\bar{Q}/Z_1$ assuming that all the electrons whose ionization energy is smaller than $\frac{1}{2}mv^2$ are stripped (Lamb's energy criterion). The above theories are compared with experiments by Betz in figure 6.4. The agreement is not very good.

On the other hand, Betz[13] proposed an empirical formula in the form

$$(\bar{Q}/Z_1) = 1 - C\exp[-(v/v_B Z_1^{\gamma})] \tag{6.2.23}$$

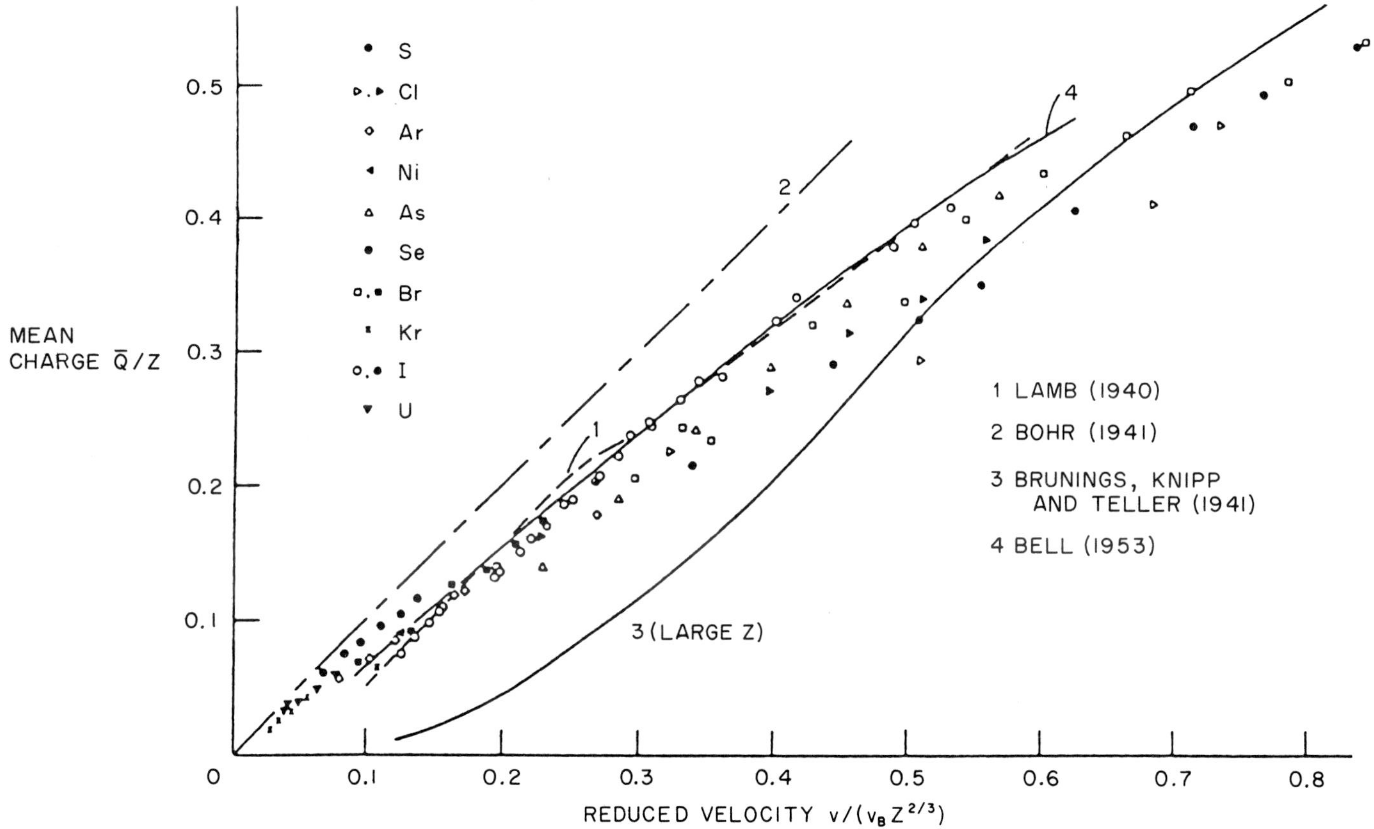

Figure 6.4. Several theoretical (phenomenological) expressions for the mean charge $\bar{Q}$ compared with experiments by Betz.[13]

where $C \doteqdot 1{\cdot}1$ and $\gamma = 0{\cdot}64$ give a good fit with experimental results. Another empirical formula was introduced by Nikolaev and Dmitriev[14] for $Z_1 \geq 16$

$$\frac{\bar{Q}}{Z_1} = \left[1 + \left(Z_1^{-\alpha}\frac{v}{v'}\right)^{-1/k}\right]^{-k} \tag{6.2.24}$$

where $v' = 3{\cdot}6 \times 10^8\,\mathrm{cm\,s^{-1}}$, $\alpha = 0{\cdot}45$, and $k = 0{\cdot}6$. Brandt *et al.*[15,16,17] and Kitagawa[18] derived an expression for $\bar{Q}$ using statistical approximation.

6.2.2. *Effective charge*

Since the stopping power for an energetic ion is proportional to the square of its charge (Bethe–Bloch formula), we can renormalize the experimental deviation from the Bethe–Bloch formula to the charge, which is called the effective charge Q_{eff}. The physical meaning of the effective charge may be considered to be a charge which contributes effectively to the energy loss process. Experimentally, we may determine[19] the effective charge from the measured values of the stopping power $-\left(\frac{\mathrm{d}E}{\mathrm{d}z}\right)_{Z_1}$ for an ion with charge Z_1 from

$$Q_{\mathrm{eff}}^2 = \frac{-\left(\frac{\mathrm{d}E}{\mathrm{d}z}\right)_{Z_1}}{-\left(\frac{\mathrm{d}E}{\mathrm{d}z}\right)_{\mathrm{p}}} \tag{6.2.25}$$

where $-\left(\frac{\mathrm{d}E}{\mathrm{d}z}\right)_{\mathrm{p}}$ is the stopping power for the proton.

Betz *et al.*[20] proposed an empirical formula for the effective charge in the form

$$\zeta = \frac{Q_{\mathrm{eff}}}{Z_1} = 1 - A\exp[-(v/v_{\mathrm{B}}Z_1^{\nu})] \tag{6.2.26}$$

where A and ν are constants, which Brown and Moak[21] determined to be 1·032 and 0·690, respectively. There are slight deviations in experimental results from the scaling law with respect to $v/v_{\mathrm{B}}Z_1^{\nu}$ at low velocities.

Ferrell and Ritchie[22] developed a theory for calculating the heavy-ion stopping power or the effective charge, taking into account the non-local electron distribution $n(\mathbf{r})$ around the ion, which related to the number of orbital electrons Q in the form

$$\int n(\mathbf{r})\,\mathrm{d}\mathbf{r} = Q$$

They derived an expression for the stopping power $\left(-\frac{\mathrm{d}E}{\mathrm{d}z}\right)_{Z_1}$

$$-\left(\frac{\mathrm{d}E}{\mathrm{d}z}\right)_{Z_1}=\frac{2Z_1e^2}{\pi v^2}\int_0^\infty \frac{\mathrm{d}k}{k}|\rho(k)|^2\int_0^{kv}\omega\ln\left[\frac{-1}{\varepsilon(k,\omega)}\right]\mathrm{d}\omega$$

where $\rho(k)$ is the Fourier component of the charge distribution of the projectile ion

$$2\pi\,\rho(k)\,\delta(\omega-\mathbf{k}\cdot\mathbf{v})=\int[Z_1\delta(\mathbf{r}-\mathbf{v}t)-n(\mathbf{r}-\mathbf{v}t)]\exp[i(\mathbf{k}\cdot\mathbf{r}-\omega t)]\,d\mathbf{r}\,\mathrm{d}t$$

Kitagawa[18] evaluated the above integrals, determining $n(\mathbf{r})$ by a variational principle, and obtained the result

$$\zeta=\left(1-\frac{Q}{Z_1}\right)+C(k_\mathrm{F})\frac{Q}{Z_1}\ln[1+(2k_\mathrm{F}\Lambda)^2]$$

where $C(k_\mathrm{F})$ is a constant slightly dependent on k_F, and Λ is the screening length of $n(r)$ as determined by them. Numerical results are compared with experiments of the New York University group in figure 6.5 for $_7$N ions.

The compilation of data for Q_eff was performed by Ziegler[23] and by Yarlagadda *et al.*[24]

6.3. Charge states of channelled ions and the Okorokov effect

The charge state distribution or the equilibrium charge state is greatly modified under channelling conditions, since the capture and loss cross-sections are functions of the position of the ion in the channelling trajectories. However, it is a very difficult problem to calculate them exactly, and we have no exact calculations except for simple ions such as H^+ and He^+. When the ion moves along a crystal axis under channelling conditions we find a (recently discovered) coherent effect, the so called 'Okorokov effect'.

6.3.1. *Equilibrium charge of the channelled ion*

It has been reported experimentally that the charge states or the equilibrium charge is changed by changing the ion's angle of incidence to the channelling axis. For example, Gaillard *et al.*[25] measured the neutral fraction ϕ_0 for H^+ in Ni and Au by changing the ion's energy and channelling condition. Figure 6.6 shows one of their results for Ni foil with a thickness of 70 nm. The neutral fraction is normalized to the neutral fraction of H^+ by the random orientation ϕ_0^{random}. The neutral fraction decreases by ~ 30 per cent from the random value for 2 MeV H^+ in Ni$\langle 110\rangle$. Further, they observed blocking yields for 51 nm Au(110) and Au(111), and detected increased neutral fractions in Au over those in Ni.

Similar problems have also been studied experimentally for lower energy

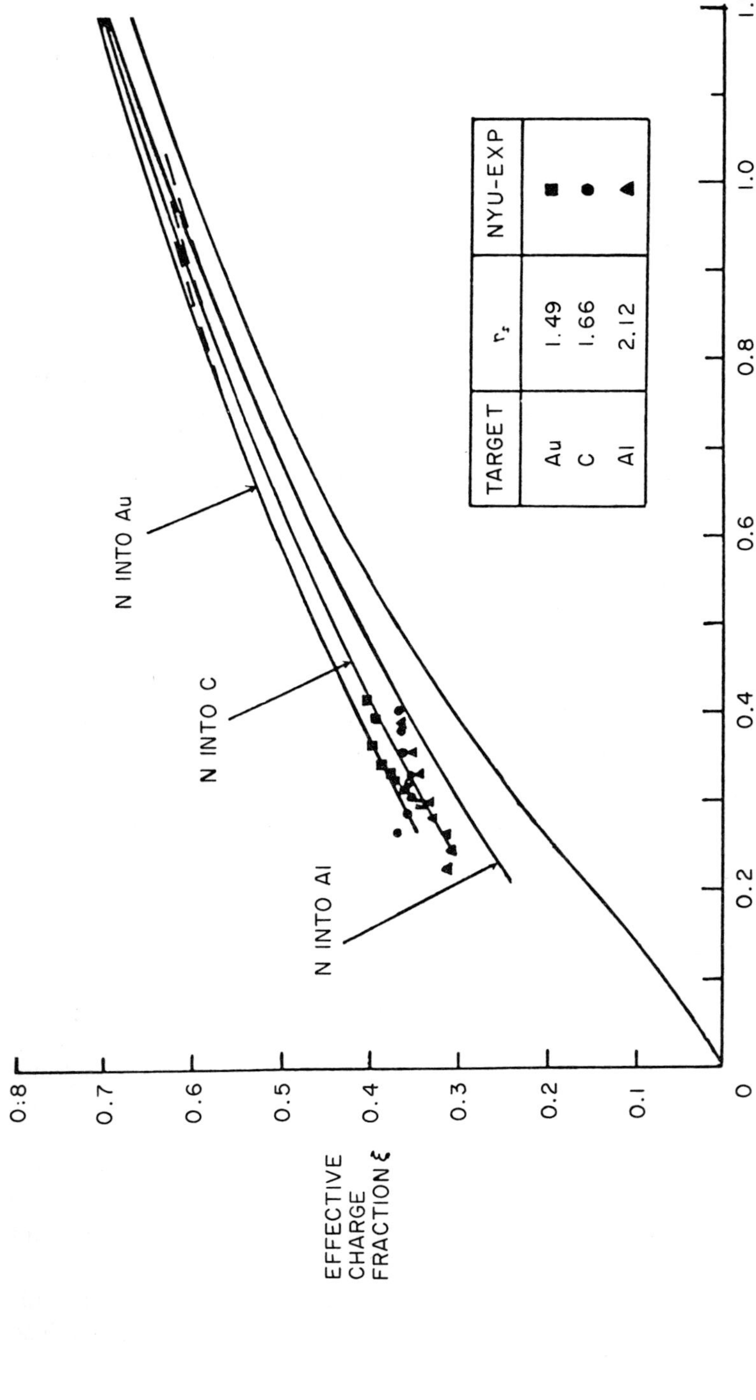

Figure 6.5. Theoretical stopping power according to Kitagawa[18] compared with the experiments of the NYU group.

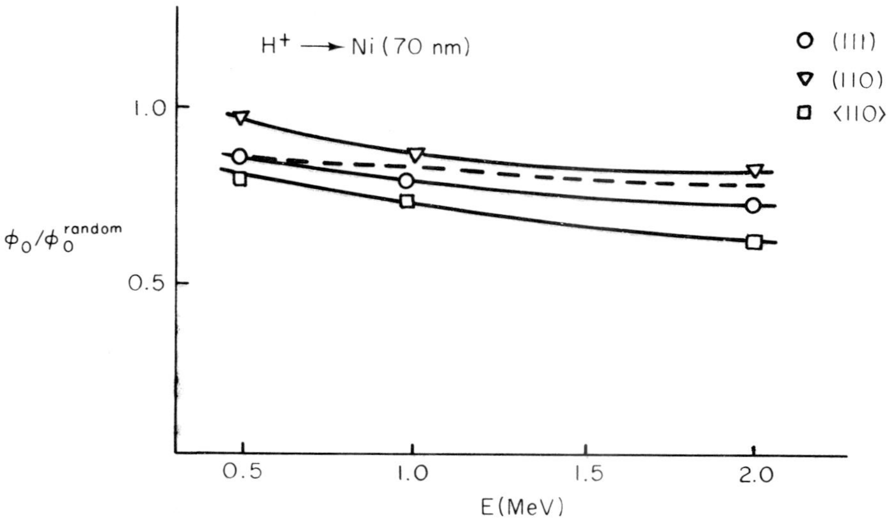

Figure 6.6. Neutral fraction ϕ_0 changing the ion energy and the channelling condition for H^+ in Ni, according to Gaillard *et al.*[25]

H_1^+ and H_2^+ by Matschke.[26] The charged fraction $\phi^+ = 1 - \phi_0$ changed by about 20 per cent for 5 keV H_2^+ into Au(110) on changing the angle of the ion beam relative to the crystal axis. For heavy ions, Kido *et al.*[27] measured the equilibrium charge of 2·28 MeV N^+ ions in the blocking orientations of Au(111).

6.3.2. *The Okorokov effect*

The orbital electrons of the ion moving along the crystal axis interact successively with atoms with a periodicity T. If the frequency $1/T$ coincides with the transition energy divided by h, the orbit electron is excited resonantly. The frequency ν is given by

$$\nu = \frac{1}{T} = \frac{v}{d}$$

for an ion with velocity v along a crystal axis with atom spacing d. Therefore the resonance condition becomes

$$h\nu = E_{n'} - E_n (= \hbar\omega_{n0}) = \frac{hv}{d} \qquad (6.3.1)$$

for the transition $n \rightarrow n'$. This condition was first discovered by Okorokov, and this resonant excitation is called the Okorokov effect.

The theory of the Okorokov effect was developed by Okorokov and his co-workers,[28] starting with the first order perturbation for the excitation. The periodic potential due to the crystal axis takes the form

$$V(\mathbf{R}) = \sum_{\mathbf{g}} V_{\mathbf{g}} \exp(-2\pi i \mathbf{g}\cdot\mathbf{R}) \tag{6.3.2}$$

where $\mathbf{g}$ is the reciprocal lattice vector. $\mathbf{R}$ is the position of the orbital electron of the ion with respect to the centre of the ion $\mathbf{R}_0(t)$

$$\mathbf{R}_0(t) = \mathbf{b} + \mathbf{v}t$$

and the distance $\mathbf{r}$ from it

$$\mathbf{R} = \mathbf{R}_0(t) + \mathbf{r} \tag{6.3.3}$$

In the above, $\mathbf{v}$ and $\mathbf{b}$ are the ion velocity and impact parameter vectors, respectively.

If we take $\mathbf{g}_\perp$ and g_z to the transverse and parallel components, respectively ($g_z = m/d$, $m = 0, 1, 2, \ldots$), we obtain

$$\begin{aligned} V &= \sum_{\mathbf{g}_\perp} \sum_{g_z} V_{\mathbf{g}_\perp, g_z} \exp(-2\pi i \mathbf{g}_\perp \cdot \mathbf{b} - 2\pi i g_z vt - 2\pi i \mathbf{g}\cdot\mathbf{r}) \\ &= i\hbar \sum_m F^{(m)} \exp\left(-i\frac{2\pi mv}{d}t\right) \end{aligned} \tag{6.3.4}$$

where

$$F^{(m)} = \frac{1}{i\hbar} \sum_{\mathbf{g}_\perp} V_{\mathbf{g}_\perp, 2\pi m}/d \exp(-2\pi i \mathbf{g}_\perp \cdot \mathbf{b}) \exp(-2\pi i \mathbf{g}_\perp \cdot \mathbf{r}_\perp - 2\pi i\, mz/d) \tag{6.3.5}$$

Let us calculate the transition probability $P(t)$ for the $0 \rightarrow n$ excitation using first order perturbation. According to Fermi's golden rule, we obtain

$$\begin{aligned} P(t) &= \left| -\frac{i}{\hbar} \int_0^t V_{n0} \exp\left(i\frac{E_n - E_0}{\hbar}t\right) \mathrm{d}t \right|^2 \\ &= \sum_m |F_{n0}^{(m)}|^2 t^2\, \delta(\omega_{n0}, 2\pi mv/d) \end{aligned} \tag{6.3.6}$$

where V_{n0} is the matrix element of equation 6.3.2 and $F_{n0}^{(m)}$ the matrix element of equation 6.3.6. $\delta(\omega_{n0}, 2\pi mv/d)$ is Kronecker's delta.

The transition probability is not zero only when

$$\omega_{n0} = \frac{2\pi mv}{d}$$

or

$$v = v_c[\equiv \omega_{n0} d/(2\pi m)] \tag{6.3.7}$$

due to the factor $\delta(\omega_n, 2\pi mv/d)$ in equation 6.3.6. Equation 6.3.7 is equivalent to the resonance condition 6.3.1. We can take into account the effect of lattice vibrations by the Debye–Waller factor $\exp(-M)$ where (see Chapter 2)

$$M = \frac{1}{6}\left(\frac{2\pi m}{d}\right)^2 \langle \delta R_z^2 \rangle$$

As can be easily seen we cannot find the exact values of the resonant-excitation probability by the above theory (equation 6.3.6) near the velocity $v = \omega_{n0} d/(2\pi m)$, excluding the resonance condition. In order to obtain the exact theory we have to develop the dynamic considerations without using first order perturbation theory. If we assume only two energy levels, these levels are occupied by an electron with different constant probabilities due to the periodic (successive) field from the atomic axis. Our dynamical theory is formulated to obtain this occupation probability. Such theories have been developed by Kondo,[29] and Shindo and Ohtsuki.[30] Here we derive the resonant excitation probability based on these works.

We assume that the Hamiltonian for the orbiting electrons in the ion is H_0, and the interaction Hamiltonian due to the external periodic field is V. The total wave function of the electron corresponding to $H_0 + V$ may be expanded by the eigenfunctions of H_0

$$\psi(t) = \sum_k a_k(t)\psi_k \exp(-iE_k t/\hbar) \tag{6.3.8}$$

where

$$H_0\psi_k = E_k\psi_k \tag{6.3.9}$$

Inserting equation 6.3.8 into the Schrödinger equation

$$i\hbar\frac{\partial\psi}{\partial t} = (H_0 + V)\psi \tag{6.3.10}$$

we obtain

$$\frac{\mathrm{d}a_n}{\mathrm{d}t} = \sum_k \sum_m F_{nk}^{(m)} \exp(i\varepsilon_{nk}^{(m)} t) a_k \tag{6.3.11}$$

Here we defined

$$\varepsilon_{nk}^{(m)} = \frac{E_n - E_k}{\hbar} - 2\pi m v/d \tag{6.3.12}$$

We shall now consider the two state approximation for simplicity. The states are denoted by 0 and n. Equation 6.3.11 is then reduced to

$$\begin{aligned} \frac{\mathrm{d}a_n}{\mathrm{d}t} &= F_{n0}^{(m)} \exp(i\varepsilon_{n0}^{(m)} t) a_0 \\ \frac{\mathrm{d}a_0}{\mathrm{d}t} &= -F_{n0}^{(m)*} \exp(-i\varepsilon_{n0}^{(m)} t) a_n \end{aligned} \tag{6.3.13}$$

Solving the coupled equation with the initial conditions $a_0(0) = 1$ and $a_n(0) = 0$, we get

$$P(t) = |a_n(t)|^2 = \frac{4|F_{n0}^{(m)}|^2}{4|F_{n0}^{(m)}|^2 + \varepsilon_{n0}^{(m)2}} \sin^2 \frac{\sqrt{|\varepsilon_{n0}^{(m)}|^2 + 4|F_{n0}^{(m)}|^2}}{2} t \qquad (6.3.14)$$

The above two-state approximation is good when there is no degeneracy, where the resonance excitation probability is symmetrical with respect to $\varepsilon_{n0}^{(m)}$ or $v - v_c$. Figure 6.7 shows the calculated resonance excitation probability averaged for the channelling condition as a function of ion velocity.

When there are degenerate states, we have to take into account all such states in our coupled equations. For example, if we consider four degenerate p states in the excited level n, equation 6.3.13 is extended to

$$\begin{aligned} \frac{da_0}{dt} &= -\sum_{n=1}^{4} F_{n0}^{(m)*} \exp(-i\varepsilon_{n0}^{(m)} t) a_n + F_{00}^{(0)} a_0 \\ \frac{da_n}{dt} &= F_{n0}^{(m)} \exp(i\varepsilon_{n0}^{(m)} t) a_0 + F_{nn}^{(0)} a_n \end{aligned} \qquad (6.3.15)$$

where $n = 1, 2, 3, 4$. The transition probability $P(t)$ is obtained easily by solving equation 6.3.15 assuming a small splitting of the degenerate levels. This gives

$$\begin{aligned} P(t) &= \sum_{n=1}^{4} |a_n(t)|^2 \\ &= \frac{4\sum_{n=1}^{4} |F_{n0}^{(m)}|^2}{\tilde{\varepsilon}_1^{(m)2} + 4\sum_{n=1}^{4} |F_{n0}^{(m)}|^2} \sin^2 \frac{\sqrt{\tilde{\varepsilon}_1^{(m)2} + 4\sum_{n=1}^{4} |F_{n0}^{(m)}|^2}}{2} t \end{aligned} \qquad (6.3.16)$$

where we assume $\varepsilon_{10}^{(m)} = \varepsilon_{20}^{(m)} = \ldots = \varepsilon_{40}^{(m)}$ and define

$$\tilde{\varepsilon}_1^{(m)} = \varepsilon_{10}^{(m)} + (V_{nn}^{(0)} - V_{00}^{(0)})/\hbar \qquad (6.3.17)$$

It is not so easy to calculate in detail the matrix element $F_{n0}^{(m)}$ and the probability $P(t)$ in this case. Yamashita and Ohtsuki[31] tried to calculate $P(t)$ averaged over the channelling trajectories in Au⟨111⟩ in order to explain the experimental asymmetrical profile of $P(t)$ as a function of $(v - v_c)$ obtained by Datz *et al.*[32] and Moak *et al.*[33,34] The comparison between theory and experiment is made in figure 6.8, from which we can see that the theoretical profile agrees well with the experimental asymmetrical profile.

However, as seen in figure 6.8, the half-width of the experimental profile is broader than the theoretical one. This implies that the experimentally measured coherent length is smaller than the crystal thickness, for as yet unknown reasons. The reasons are probably related to energy loss, lattice

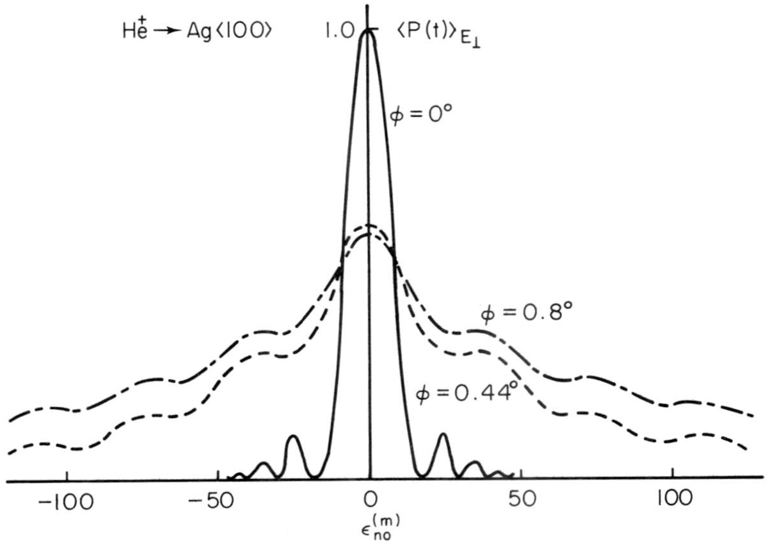

Figure 6.7. Resonance excitation probability as a function of the parameter $\varepsilon_{n0}^{(m)}$, representing the deviation from the Okorokov condition.

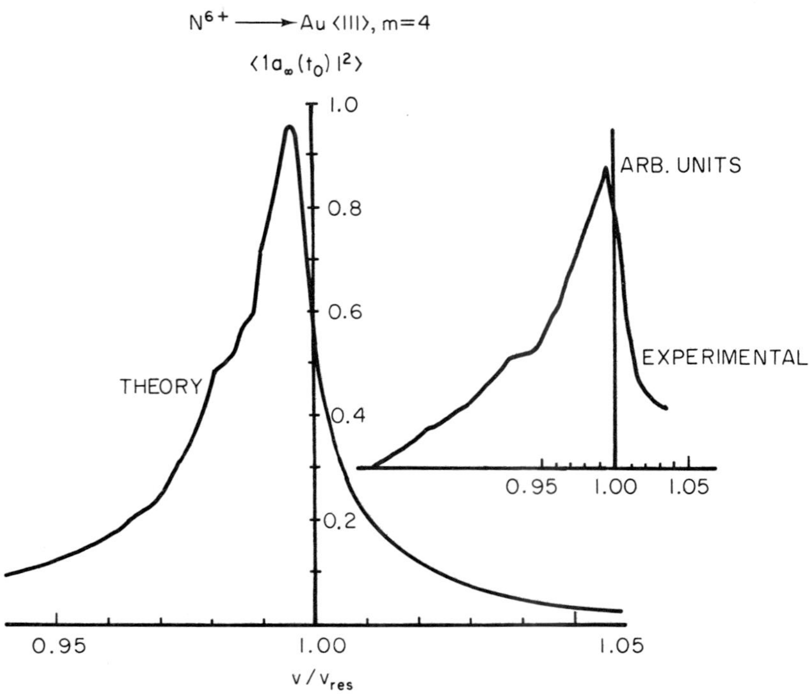

Figure 6.8. Detailed intensity profile of resonant excitation by Yamashita and Ohtsuki's theory,[31] compared with the experimental curve by Datz *et al.*

vibration and charge transfer. Uchida[35] attempted to include the energy loss in the dynamical theory of the Okorokov effect, but, he could not explain the broad profile. However, his theory, which includes the stopping power, is effective in explaining the disappearance of the fine peaks found in the calculation of Yamashita and Ohtuski.[31] Lattice vibration contributes toward decreasing the peak height, but this contribution is insufficient. There are no exact theories to take into account the charge state distribution during resonance in the Okorokov effect.

6.4. Screening and wake

It is well known that the ion potential is screened by the polarization of the medium. For an ion moving with a velocity, there appear retardation effects in space and time for the screening, which we call the wake potential. The screening and wake for an ion moving in a metal are well represented by the virtual excitation of plasmon excitations.

Neufeld and Ritchie[37] first developed the quantitative theory of wake interaction, and many other authors[38–42] have since modified it or developed a different theory, in which they discuss ways to take into account plasmon dispersion, single electron excitation, and the possibility of electron capture in the wake potential. Kitagawa and Ohtsuki,[43] on the other hand, tried to obtain the wake potential between two ions in a self-consistent manner. They employed the canonical transformation of the total Hamiltonian represented in the second quantization form, which is then expanded into the general form. Here we shall start from their theory and discuss and compare their results with other theories and experiments.

6.4.1. *Self-consistent theory*

Our starting Hamiltonian takes the form

$$H = H_0 + H_{\rm pl} + H' \qquad (6.4.1)$$

where H_0 is the Hamiltonian of two ions with mass M_j and charge $Z_j e (j = 1, 2)$

$$H_0 = \sum_{j=1,2} \frac{p_j^2}{2M_j} + \frac{Z_1 Z_2 e^2}{|\mathbf{r}_1 - \mathbf{r}_2|} \qquad (6.4.2)$$

The Hamiltonian $H_{\rm pl}$ is the plasmon-field energy

$$H_{\rm pl} = \sum_{\mathbf{k} < k_c} \left(\frac{\pi_{\mathbf{k}}^+ \pi_{\mathbf{k}}}{2} + \omega_k \frac{q_{\mathbf{k}}^+ q_{\mathbf{k}}}{2} \right) \qquad (6.4.3)$$

where $(q_{\mathbf{k}}, \pi_{\mathbf{k}})$ is the conjugate co-ordinate-momentum operator of a plasmon with energy $\hbar\omega_k$. H' describes the interaction between the ion and the plasmon field

$$H' = -\sum_{\mathbf{k}<k_c} \sqrt{\frac{4\pi V}{k^2}}\,\pi_{\mathbf{k}}^{+}\,\rho(\mathbf{k}) \tag{6.4.4}$$

where

$$\begin{aligned}\rho(\mathbf{k}) &= \frac{1}{V}\int \rho(\mathbf{r})\exp(-i\mathbf{k}\cdot\mathbf{r})\,\mathrm{d}\mathbf{r} \\ &= \frac{1}{V}\sum_{j=1}^{2}(Z_j e)\exp(-i\mathbf{k}\cdot\mathbf{r}_j)\end{aligned} \tag{6.4.5}$$

is the Fourier component of the ion charge-density

$$\rho(\mathbf{r}) = \sum_{j=1}^{2}(Z_j e)\,\delta(\mathbf{r}-\mathbf{r}_j)$$

In equation 6.4.5 V is the arbitrary macroscopic volume of the solid.

Now let us make the canonical transformation in the form

$$H_{\mathrm{T}} = \exp(-iS/\hbar)\,H\exp(iS/\hbar) \tag{6.4.6}$$

using the Hermite operator S, which is in the following form in order to cancel H'

$$S = -\sum_{\mathbf{k}<k_c}\sqrt{\frac{4\pi V}{k^2}}\,q_{\mathbf{k}}\,\rho(\mathbf{k}) \tag{6.4.7}$$

where k_c is the plasmon cut-off wave number.

After the canonical transformation the total Hamiltonian becomes

$$H_{\mathrm{T}} = H_0 + \mathrm{H}_{\mathrm{pl}} + \mathrm{H}_{\mathrm{int}} + \mathrm{H}_{\mathrm{pp}} \tag{6.4.8}$$

$$H_0 = \sum_{j=1}^{2}\frac{p_j^2}{2M_j} + \frac{1}{V}\sum_{\mathbf{k}>k_c}\frac{4\pi Z_1 Z_2 e^2}{k^2}\exp[i\mathbf{k}\cdot(\mathbf{r}_1-\mathbf{r}_2)] \tag{6.4.9}$$

$$H_{\mathrm{pl}} = \sum_{\mathbf{k}<k_c}\left[\frac{\pi_{\mathbf{k}}^{+}\pi_{\mathbf{k}}}{2} + \left(\omega_k^2 + \frac{1}{V}\sum_{j=1}^{2}\frac{4\pi Z_j^2 e^2}{M_j}\right)\frac{q_{\mathbf{k}}^{+}q_{\mathbf{k}}}{2}\right] \tag{6.4.10}$$

$$H_{\mathrm{int}} = -\frac{i}{\sqrt{V}}\left\{\sum_{\mathbf{k}<k_c}\sum_j\left(\frac{Z_j e}{k}\right)\sqrt{\frac{2\pi\hbar}{\omega_k}}\,[t_j^{+}a_{\mathbf{k}}\exp(-i\mathbf{k}\cdot\mathbf{r}_j) - \exp(i\mathbf{k}\cdot\mathbf{r}_j)a_{\mathbf{k}}^{+}t_j^{+}]\right\} \tag{6.4.11}$$

$$H_{\mathrm{pp}} = \sum_{\mathbf{k}\neq\mathbf{k}'<k_c}\sum_j\frac{\pi Z_j^2 e^2\hbar}{VM_j}\frac{(\mathbf{k}\cdot\mathbf{k}')}{kk'\omega_k}(a_{\mathbf{k}}^{+}+a_{-\mathbf{k}})\,(a_{\mathbf{k}}+a_{-\mathbf{k}}^{+})\exp[i(\mathbf{k}-\mathbf{k}')\cdot\mathbf{r}_j] \tag{6.4.12}$$

where $q_{\mathbf{k}} = \sqrt{\hbar/2\omega_k}\,(a_{\mathbf{k}} + a_{-\mathbf{k}}^{+})$ and

$$t_j^{\pm} = [2(\mathbf{p}_j\cdot\mathbf{k}) \pm \hbar k^2]/2M_j \tag{6.4.13}$$

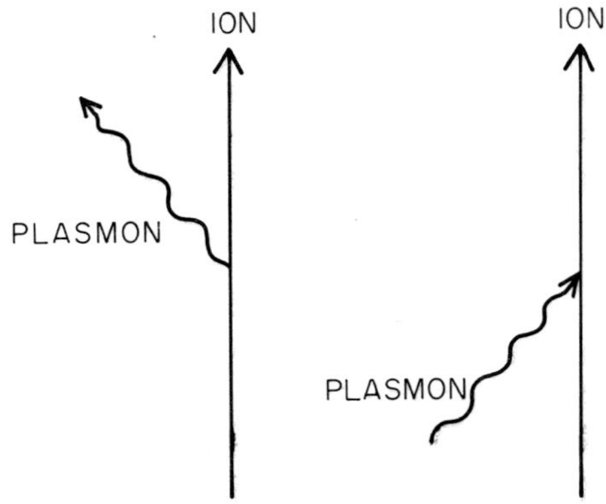

Figure 6.9. Plasmon emission and plasmon absorption.

The second term in H_0 is just the screened potential first derived by Bohm and Pines,[44] which is formed by the virtual excitation of the long wavelength plasmon ($k < k_c$). From H_{pl} we can understand that the plasmon energy $\hbar\omega_k$ is modified by the ion charge, but the modification is very small because the number of ions is very small (only two!). So, equation 6.4.10 is approximated by equation 6.4.3. The term H_{int} represents the interaction Hamiltonian between the ion and the plasmon which is formed by two processes, plasmon emission and plasmon absorption, as shown in figure 6.9. The residual term H_{pp} represents the plasmon–plasmon interaction.

The Hamiltonian that we want to obtain is the effective interaction Hamiltonian between two ions. This is the interaction caused by the exchange of plasmons between ions (see figure 6.10). This process is similar to the nuclear potential between nucleons caused by the exchange of mesons, and the Cooper potential in the BCS theory of super-conductivity caused by the exchange of phonons.

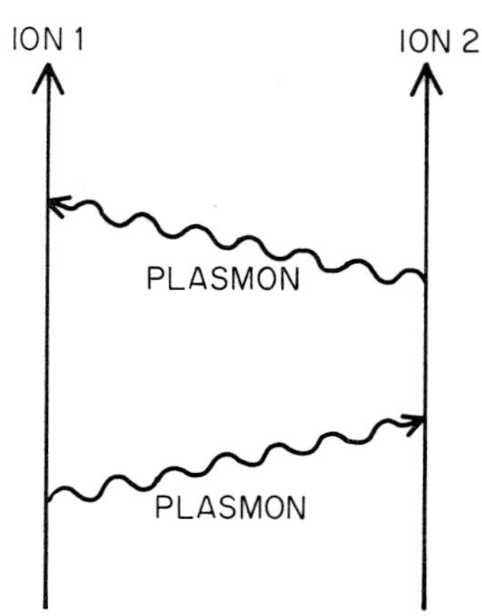

Figure 6.10. Plasmon exchange process.

In order to obtain an effective interaction between ions caused by plasmon exchange we have to make another kind of canonical transformation of H_{T} by the Hermite operator S' to cancel the Hamiltonian H_{int}

$$\mathscr{H} = \exp(iS'/\hbar)\, H_{\mathrm{T}} \exp(-iS'/\hbar) \tag{6.4.14}$$

or

$$H_{\mathrm{T}} = \exp(-iS'/\hbar)\, \mathscr{H} \exp(iS'/\hbar) \tag{6.4.15}$$

Here we represent new operators after the canonical transformation, for example, $A_{\mathbf{k}}^{+}$ and $\mathbf{R}_j$ corresponding to $a_{\mathbf{k}}^{+}$ and $\mathbf{r}_j$, respectively. We can find the Hermite operator S' in the form

$$S' = -\frac{1}{\sqrt{V}} \sum_{\mathbf{k}<k_c} \sum_j \frac{Z_j e}{k} \sqrt{\frac{2\pi\hbar}{\omega_k}} [U_j^- A_{\mathbf{k}} \exp(-i\mathbf{k}\cdot\mathbf{R}_j) + \exp(i\mathbf{k}\cdot\mathbf{R}_j) A_{\mathbf{k}}^{+} U_j^{+}]$$

$$U_j^{\pm} = T_j^{\pm}/(\omega_k + T_j^{\pm}) \tag{6.4.16}$$

The new Hamiltonian $\mathscr{H}$ becomes

$$\mathscr{H} = \mathscr{H}_0 + \mathscr{H}_{\mathrm{pl}} + \mathscr{H}_1 + \mathscr{H}_{\mathrm{int}} \tag{6.4.17}$$

where $\mathscr{H}_0$ and $\mathscr{H}_{\mathrm{pl}}$ are H_0 and H_{pl}, respectively. The Hamiltonian $\mathscr{H}_1$ takes the form

$$\mathscr{H}_1 = \frac{1}{V} \sum_{\substack{\mathbf{k}<k_c \\ j}} \frac{2\pi Z_j^2 e^2}{k^2} \frac{((\mathbf{P}_j\cdot\mathbf{k})/M_j)^2}{(\mathbf{P}_j\cdot\mathbf{k}/M_j)^2 - (\omega_k - \hbar k^2/2M_j)^2} \tag{6.4.18}$$

which should be considered the renormalization term of the kinetic energy part of the ion, $P_j^2/2M$. Since the ion moves with the plasmon cloud, the effective mass or the kinetic energy of the ion should be modified. However, in our case this modification is very small.

$\mathscr{H}_{\mathrm{int}}$ is the term we wanted to obtain

$$\mathscr{H}_{\mathrm{int}} = -\frac{1}{V} \sum_{\mathbf{k}<k} \sum_{j\neq j'} \frac{\pi Z_j Z_{j'} e^2}{k^2 \omega_k} \times \{\exp[i\mathbf{k}\cdot(\mathbf{R}_{j'} - \mathbf{R}_j)] T_j^- U_{j'}^+ + U_{j'}^+ T_j^- \exp[-i\mathbf{k}\cdot(\mathbf{R}_{j'} - \mathbf{R}_j)]\} \tag{6.4.19}$$

If we use the relation

$$U_{j'}^+ T_j^- \exp[-i\mathbf{k}\cdot(\mathbf{R}_{j'}\cdot\mathbf{R}_j)] = \exp[-i\mathbf{k}\cdot(\mathbf{R}_{j'} - \mathbf{R}_j)] T_j^+ U_{j'}^-$$

we can rewrite equation 6.4.19 as

$$\mathscr{H}_{\mathrm{int}} = -\frac{1}{V} \sum_{\mathbf{k}<k_c} \frac{2\pi Z_1 Z_2 e^2}{k^2} \left\{ \exp[i\mathbf{k}\cdot(\mathbf{R}_1 - \mathbf{R}_2)] T_2^- T_1^+ \frac{1}{\omega_k^2 - T_1^{+2}} + \exp[i\mathbf{k}\cdot(\mathbf{R}_2 - \mathbf{R}_1)] T_1^- T_2^+ \frac{1}{\omega_k^2 - T_2^{+2}} \right\} \tag{6.4.20}$$

This is simply the self-consistent wake potential between two ions.

In order to obtain the simplified forms of our wake potential, we have made a few assumptions in equation 6.4.20, i.e., we have neglected the dispersion of the plasmon energy $\hbar\omega_k$ and the recoil of the emission (or absorption) of plasmons

$$\begin{aligned}\omega_k &\simeq \omega_p \\ T_j^{\pm} &= [\mathbf{k}\cdot(\mathbf{P}_j \pm \hbar\mathbf{k})]/M_j \\ &\simeq \mathbf{v}_j\cdot\mathbf{k}\end{aligned}$$

We shall consider the case where two ions move along the z direction with speeds v_1 and v_2. Equation 6.4.20 then becomes

$$\mathscr{H}_{\text{int}} = \frac{Z_1 Z_2 e^2 k_c}{\pi}\left\{\frac{2Si(k_c R)}{k_c R} + \frac{1}{k_c}\int_{-k_c}^{k_c} \mathrm{d}k_z \int_0^{\sqrt{k_c^2 - k_z^2}} \mathrm{d}k_\perp \frac{k_\perp}{k_\perp^2 + k_z^2} J_0(k_\perp R_\perp)\right. \tag{6.4.21}$$

$$\times \cos(k_z Z) \frac{\left[\dfrac{\omega_p^2}{v_1 v_2} + k_z\,\omega_p(v_1 - v_2)\left(\dfrac{1}{v_1 v_2} + \dfrac{k_z^2}{2\omega_p^2}\right)\right]}{k_z^2 - k_z\,\omega_p\left(\dfrac{v_1 - v_2}{v_1 v_2}\right) - \dfrac{\omega_p^2}{v_1 v_2}}$$

where R, $R_\perp$ and Z are the distance between the ions, the distance between the two incident trajectories, and the relative distance of the two ions in the z direction, respectively. The functions $Si(x)$ and $J_0(x)$ are the sine integral function and the Bessel function, respectively.

The first term of equation 6.4.21 is the long-range part of the Coulomb interaction, which should be renormalized with respect to the screened Coulomb interaction. The second term including the integral over the wave number is the wake potential. If we put $v_1 = v_2 (\equiv v)$, we obtain

$$\mathscr{H}_{\text{int}} = \frac{Z_1 Z_2 e^2 k_c}{\pi}\left[\frac{2Si(k_c R)}{k_c R} - \frac{\omega_p}{v k_c}(A^2 + B^2)^{1/2} \sin\left(\frac{\omega_p}{v} Z + \alpha\right)\right] \tag{6.4.22}$$

where

$$A = -\int_0^1 \mathrm{d}t\, F_1(t, R_\perp)[Si(k_c t + \omega_p/v)z + Si(k_c t - \omega_p/v)Z]$$

$$B = -\int_0^1 \mathrm{d}t\, F_1(t, R_\perp)[Ci(k_c t + \omega_p/v)z - Ci(k_c t - \omega_p/v)Z]$$

$$\alpha = \arctan(B/A)$$

$$F_1(t, R_\perp) = \frac{\mathrm{d}}{\mathrm{d}t}\left[\int_0^{\sqrt{1-t^2}} \mathrm{d}t' \frac{t'}{t'^2 + t^2} J_0(k_c R_\perp t')\right]$$

If we consider the case where $v \to \infty$ and $k_c \to \infty$, we can derive the very simple form, which is very close to the expression of Neufeld and Ritchie

$$\phi^{\mathrm{w}} = -\frac{2Z_1 e}{v}\sin\left(\frac{\omega_{\mathrm{p}}}{v}Z\right)K_0\left(R_\perp \frac{\omega_{\mathrm{p}}}{v}\right) \tag{6.4.23}$$

where the function $K_0(x)$ is the modified Bessel function of the second kind. Equation 6.4.23 is derived by considering only a single ion and by making use of classical theory. So, the potential energy of the other ion with charge $Z_2 e$ is $Z_2 e\phi^{\mathrm{w}}$, which differs from expression 6.4.22.

It should be noted that the second term in equation 6.4.21 cancels the first term when $v_1 = v$ and $v_2 = 0$, and the wake potential vanishes. This difference is obvious since the second ion ($Z_2 e$) also contributes some effect to the first ion, and becomes important at low energy. Thus, we may conclude that the Neufeld and Ritchie wake potential can be used only when ions move with the same velocity with high energy (higher than the Fermi energy).

Väger and Gemmell[45] proposed a semi-phenomenological formula for the wake potential

$$\phi^{\mathrm{w}}(R_\perp, z - vt) = -\frac{Ze\omega_{\mathrm{p}}}{v}\int_0^\infty \sin\left(\frac{\omega_{\mathrm{p}}}{v}\zeta\right) \times \left[R_\perp^2 + \left(\frac{\hbar}{mv}\right)^2 + \zeta + z - vt)^2\right]^{-1/2} \mathbf{d}\zeta \tag{6.4.24}$$

and used this formula to analyse the experimental work described below. It should be noted that the term $(\hbar/mv)^2$ appears in the integral, and may be attributed to the uncertainty relation of the closed collision. This term is not included in expression 6.4.22.

Recently, Echenique *et al.*[42] calculated the wake potential in more detail for ions with the same velocity, taking into account plasmon dispersion, single-electron excitation, and plasmon damping. This was done in the form of the complex dielectric constant

$$\varepsilon(k, \omega) = 1 + \frac{\omega_{\mathrm{p}}^2}{\omega_{\mathrm{g}}^2 + v_{\mathrm{AV}}^2 k^2 + \hbar^2 k^4/4m^2 - \omega(\omega + i\gamma)} \tag{6.4.25}$$

Where $\hbar\omega_{\mathrm{g}}$ is the band gap, whence the angular frequency of the collective motion is given by $\Omega_{\mathrm{p}} = (\omega_{\mathrm{p}}^2 + \omega_{\mathrm{g}}^2)^{1/2}$. $v_{\mathrm{AV}} = (3/5)^{1/2} v_{\mathrm{F}}$ is the coefficient of the plasmon dispersion. The terms $\hbar^2 k^4/4m^2$ and γ represent the contribution of single-electron excitation and plasmon damping, respectively.

The wake potential is represented thus in terms of the dielectric constant

$$\phi^{\mathrm{w}} = \frac{Z_1}{\pi v}\int_0^\infty \kappa\, \mathrm{d}\kappa\, J_0(\kappa\rho)\int_{-\infty}^\infty \mathrm{d}\omega \exp\left(\frac{i\omega\bar{z}}{v}\right)\frac{1}{k^2}\frac{1}{\varepsilon(k, \omega)}$$

where $\rho = \sqrt{x^2 + y^2}$, $\bar{z} = z - vt$, and $k^2 = \kappa^2 + \omega^2/v^2$. In this potential, the screening potential is also included.

Echenique *et al.*[42] showed that the electron density due to the wake potential diverges when $v \to v_{AV}$, which will be related to the soliton state described below. They also pointed out that a bow wave appears in front of the ion caused by the knock-on collision in the single-electron excitation.

6.4.2. *Comparison with experiments*

Wake phenomena have been found by many experimenters, especially Gemmell *et al.*,[46] who first predicted the wake interaction between two ions. They measured the energy loss and the angular distribution of two ions, and showed that the calculation agrees well with the experimental results if, and only if, the wake interaction is included in their calculation[47] (their paper also cites almost all important references regarding experiments on cluster ions up to 1980). Another experiment on wake phenomena was that of Laubert and Chen.[48] Figure 6.11 compares calculations with experimental results.

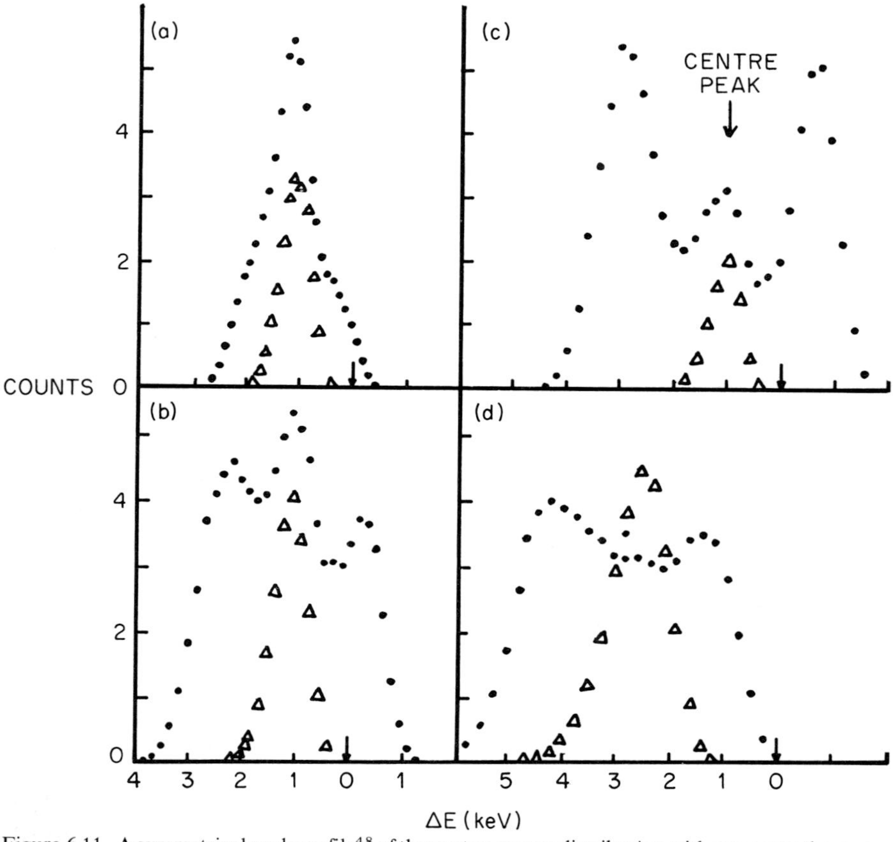

Figure 6.11. A symmetrical peak profile[48] of the proton energy distribution with respect to the centre peak (see (*b*) (*c*) and (*d*)), which is attributed to the alignment effect due to the wake. • = calculations, Δ = experimental results.

The interaction between the conduction electron and the ion due to the wake has been extensively studied, especially the possibility of electron capture in the wake potential ('wake-riding electron'), which was discussed by Neelavathi *et al.*[38] Experimental work on wake-riding electrons was recently reviewed by Brandt.[49]

6.4.3. *Soliton states*

The wake theories studied in this section are not correct for the velocity range $v \lesssim v_F$, when the ion and the plasmon are captured very strongly and move together as a whole, rather like a quasi-particle. Kitagawa and Ohtsuki[50] showed that this strong coupling state becomes a soliton, and calculated the binding energy of the soliton state.

For simplicity we consider the one-dimensional Schrödinger equation for an ion moving along the z axis and the wave equation of the plasma waves

$$i\hbar \frac{\partial \psi_z}{\partial t} = -\frac{\hbar^2}{2M}\frac{\partial^2 \psi_z}{\partial z^2} + G\phi_z^L \psi_z \tag{6.4.26}$$

$$\frac{\partial^2 \phi_z^L}{\partial t^2} = v_{AV}^2 \frac{\partial^2 \phi_z^L}{\partial z^2} + (\omega_p^2 - \Omega^2)\phi_z^L - 2Gb^2 v_z^2 |\psi_z|^2 \tag{6.4.27}$$

It is not too difficult to develop our theory from the three-dimensional Schrödinger equation instead of equation 6.4.26. In the above, G is the coupling constant between the ion and the plasma wave field ($G = Z_1 e$). Equation 6.4.27 has the wave source term which is proportional to the ion density $|\psi_z|^2$, v_{AV} is the mean velocity of the electron cloud and

$$\omega_k^2 = \omega_p^2 + v_{AV}^2 k^2 \tag{6.4.28}$$

Strictly speaking, we should include another term in the wave equation 6.4.27 corresponding to the long-range-type plasmon field. However, we have neglected it since the field is not effective in the range of the ion's wavelength. Further, v_z is the phase velocity of the plasma wave propagating in the z direction, and the parameter Ω defined by

$$b^2 \equiv \frac{\Omega^2}{v_{AV}^2}$$

will be determined later from physical considerations. The exact derivation of equation 6.4.26 is given in a paper by Kitagawa and Ohtsuki.[50]

The important point to consider for the coupled equations 6.4.26 and 6.4.27 is that the ion is affected by the plasma wave, and the plasma wave is affected by the ion. The wake potential was derived in subsection 6.4.1 for $v > v_F$ without the coupling term in the Schrödinger equation 6.4.26.

Now, let us find a special solution of the coupled equation in the form[51]

$$\psi_z = \psi_0(h)\exp[i(v_2 z - \omega_2 t)]$$
$$\phi_z^{L} = \phi_0(h)$$
$$h = v_1 z - \omega_1 t \qquad (6.4.29)$$

Inserting equation 6.4.29 into equation 6.4.26 and 6.4.27, we obtain†

$$(v_1^2 v_{AV} - \omega_1^2)\frac{d^2\phi_0}{dh^2} - \Omega_p\phi_0 = 2Gb^2 v_z^2 \psi_0^2 \qquad (6.4.30)$$

$$\frac{\hbar^2 v_1^2}{2M}\frac{d^2\psi_0}{dh^2} - \left(\frac{\hbar^2 v_2^2}{2M} - \hbar\omega_2\right)\psi_0 = G\phi_0\psi_0 \qquad (6.4.31)$$

$$\left(\frac{\hbar^2 v_1 v_2}{M} - \hbar\omega_1\right)\frac{d\psi_0}{dh} = 0 \qquad (6.4.32)$$

where $\Omega_p = \omega_p^2 - \Omega^2$. According to equation 6.4.32, we obtain

$$\frac{\hbar^2 v_1 v_2}{M} = \hbar\omega_1$$

or, from $\omega_1 = v_z v_1$

$$v_2 = \frac{M\omega_1}{\hbar v_1} = \frac{M v_z}{\hbar} \qquad (6.4.33)$$

Furthermore, we require the solution to be in the form of $\phi_0 = \gamma\psi_0$ (γ is a constant). Inserting $\phi_0 = \gamma\psi_0$ into equation 6.4.30, we can see that this form is possible if

$$\frac{1}{\gamma^2} = \frac{M(v_{AV}^2 - v_z^2)}{\hbar^2 b^2 v_z^2} \qquad (6.4.34)$$

and

$$\hbar\omega_2 = \tfrac{1}{2}M v_z^2 + E_B$$

$$E_B = -\frac{\hbar^2\Omega_p}{2M(v_{AV}^2 - v_z^2)} \qquad (6.4.35)$$

In this case a non-linear equation is obtained

$$\frac{d^2\phi_0}{dh^2} - \frac{\Omega_p}{v_1^2 v_{AV}^2 - \omega_1^2}\phi_0 - \frac{2M}{\hbar^2 v_1^2}G\phi_0^2 = 0 \qquad (6.4.36)$$

It is well known that the solution of equation 6.4.36 for $v_z < v_{AV}$ is the Bell-type soliton.[53]

In equation 6.4.35 the energy E_B should be the binding energy of the soliton state, or the binding energy of the ion in the plasma wave, which becomes

†A modification of equations 6.4.30 and 6.4.31 was performed making use of a density matrix by Warke and Greiner.[52]

infinite for $v_z \to v_{AV}$. If we renormalize the binding energy with respect to the ion mass, we obtain

$$M^* = M - \frac{\hbar^2 \Omega_p}{M v_z^2 (v_{AV}^2 - v_z^2)} \tag{6.4.37}$$

The parameter Ω_p is determined in a complicated manner, details of which will not be given here.

In order to detect the soliton state for a proton whose energy is a few 10 keV, we need an energy resolution of ~ 2 eV, which is not realistic at present. However, the effective mass of positrons in solids may be interpreted well by the wake soliton, as suggested recently by Isii.[54]

6.4.4. *Wake in non-metallic foils*

The band gaps in a non-metallic substance are very small compared with the ion energy. Therefore the wake theory for an electron gas described above does not need to be modified very much. For exciton excitation we can obtain the wake potential by changing $Z_1 e$ in equation 6.4.23 to

$$Z_1 e \sqrt{\frac{\varepsilon_\infty - 1}{\varepsilon_\infty + 1}} = Z_1 e \sqrt{\frac{1 - \dfrac{1}{\varepsilon_\infty}}{1 + \dfrac{1}{\varepsilon_\infty}}}$$

where ε_∞ is the electronic part of the static dielectric constant ε_0. For example, the values of $1/\varepsilon_\infty$ for KCl and CsI are 0·47 and 0·33, respectively. Therefore, the correction to the magnitude of the wake potential is 0·6 for KCl. Thus, the wake potential is large even in non-metallic (ionic) crystals, where the plasmon contribution is small but that of the exciton is large. (Incidentally, for MgO the exciton and the plasmon contributions are comparable.) For oxide foils the plasmon contribution is very large where the wake potential is not modified as much as in metals.

References

1. Bohr, N., *Kgl. Dan. Videns. Sels. Mat-Fys. Medd.* **18**, No. 8 (1923).
2. Remillieux, J., in *Proceedings of Atomic Collision in Solids*, Hamilton, Canada, 1979.
3. Kaneko, T. and Ohtsuki, Y.H., *Phys. Stat. Sol.* **B14**, 491 (1982).
4. Klementi, E. and Roetti, C., *Atom. Data Nucl. Dat. Tables* **14**, 177 (1974).
5. Thomas, L.H., *Proc. Roy. Soc. London* **A114**, 561 (1927).
6. Bohr, N. and Lindhard, J., *Kgl. Dan. Videns. Sels. Mat-Fys. Medd.* **28**, No. 7 (1954).
7. Cross, M.C., *Phys. Rev.* **B15**, 602 (1977).
8. Zener, C., *Proc. Roy. Soc. London* **A137**, 504 (1950).

9. Kido, Y., Kanamori, Y. and Fukuzawa, F., *Nucl. Inst. Methods* **164**, 565 (1979).
10. Knipp, J.K. and Teller, E., *Phys. Rev.* **59**, 659 (1941).
11. Brunings, J.H., Knipp, J.K. and Teller, E., *Phys. Rev.* **60**, 657 (1941).
12. Lamb, W.E., *Phys. Rev.* **58**, 696 (1940).
13. Betz, H., *Rev. Mod. Phys.* **44**, 465 (1972).
14. Nikolaev, V.S. and Dmitriev, I.S., *Phys. Lett.* **28**, 277 (1968).
15. Brandt, W., in *Atomic Collisions in Solids* (Plenum, 1973) p. 263.
16. Kreussler, S., Varelas, C. and Brandt, W., *Phys. Rev.* **B23**, 82(1981).
17. Mann, A. and Brandt, W., *Phys. Rev.* **B24**, 4999 (1981).
18. Kitagawa, M., in *Proceedings of Theoretical Seminar on Atomic Collision in Solids*, Strasbourg, 1981 (Mountain View, USA, World Scientific, 1983)
19. Bloom, S.D. and Sauter, G.D., *Phys. Rev. Lett.* **26**, 607 (1971).
20. Betz, H., Betz, H.D., Hortig, D.G., Leishner, E., Schmelzer, C., Stadler, B. and Weihrauch, J., *Phys. Lett.* **22**, 643 (1966).
21. Brown, M. and Moak, C.D., *Phys. Rev.* **B6**, 90 (1972).
22. Ferrell, T.L. and Ritchie, R.H., *Phys. Rev.* **B16**, 115 (1977).
23. Ziegler, J.F., *Helium Stopping Powers and Ranges in All Elements* (Pergamon, 1977).
24. Yarlagadda, B.S., Robinson, J.E. and Brandt, W., *Phys. Rev.* **B17**, 3473 (1978).
25. Gaillard, M.J., Poizat, J.-C., Remillieux, J., Chatean-Thierry, A., Glandieux, A. and Brandt, W., *Nucl. Inst. Methods* **132**, 547 (1976).
26. Matschke, F.E.P., *Charge State and Energy Distribution of Hydrogen Backscattered from Gold (300 eV–10 KeV)*, Reports from Max-Planck-Institut für Plasmaphysik, Garching (1977).
27. Kido, Y., Yada, T. and Fukuzawa, F., *Nucl. Inst. Methods* **158**, 525 (1979).
28. Okorokov, V.V., Tolchenkov, D.L., Khizhnyakov, I.S., Cheblukov, Yu.N., Lapitski, Y.Y., Iferov, G.A. and Zhukova, Yu.N., *Phys. Lett.* **A43**, 485 (1973).
29. Kondo, J., *J. Phys. Soc. Japan* **36**, 1406 (1974).
30. Shindo, S. and Ohtsuki, Y.H., *Phys. Rev.* **B14**, 3929 (1976).
31. Yamashita, Y. and Ohtsuki, Y.H., *Phys. Rev.* **B22**, 1183 (1980).
32. Datz, S., Moak, C.D., Crawford, O.H., Krause, H.F., Dittner, P.F., Gomez del Campo, J., Biggerstaff, J.A., Miller, P.D., Hvelplund, P. and Knudsen, H., *Phys. Rev. Lett.* **40**, 843 (1978).
33. Moak, C.D., Datz, S., Crawford, O.H., Krause, H.F., Dittner, P.F., Gomez del Campo, J., Biggerstaff, J.A., Miller, P.D., Hvelplund, P. and Knudsen, H., *Phys. Rev.* **A19**, 977 (1979).
34. Moak, C.D., Biggerstaff, J.A., Crawford, O.H., Dittner, P.F., Datz, S., Gomez del Campo, J., Hvelplund, P., Knudsen, H., Krause, H.F., Miller P.D. and Overbury, S.H. in *Atomic Collision in Solids*, ed. by Remillieux, Poizat and Gaillard (Amsterdam: North-Holland, 1981)
35. Uchida, K. (Unpublished Thesis, Waseda Univ., Tokyo, 1980).
36. Crawford, O.H. and Ritchie, R.H., *Phys. Rev.* **A20**, 1848 (1979).
37. Neufeld, J. and Ritchie, R.H., *Phys. Rev.* **98**, 1632 (1955).
38. Neelavathi, V.N., Ritchie, R.H. and Brandt, W., *Phys. Rev. Lett.* **33**, 302 (1974).
39. Day, M.H., *Phys. Rev.* **B12**, (1975).
40. Neelavathi, V.N. and Kner, R.H., *Phys. Rev.* **B14**, 4229 (1976).
41. Ritchie, R.H., Brandt, W. and Echenique, P.M., *Phys. Rev.* **B14**, 4808 (1976).
42. Echenique, P.M., Ritchie, R.H. and Brandt, W. *Phys. Rev.* **B20**, 2567 (1979).
43. Kitagawa, M. and Ohtsuki, Y.H., *Phys. Rev.* **B16**, 5321 (1977).
44. Pines. D., *Elementary Excitations in Solids* (New York: Benjamin, 1964).
45. Väger, Z. and Gemmell, D.S., *Phys. Rev. Lett.* **37**, 1352 (1976).
46. Gemmell, D.S., Remillieux, J., Gaillard, M.J., Holland, R.E. and Väger, Z., *Phys. Rev. Lett.* **34**, 1420 (1975).

47. Gemmell, D.S., in *Proceedings of Atomic Collision in Solids*, 1981, Lyon, edited by Remillieux (1982).
48. Laubert, R. and Chen, F.K., *Phys. Rev. Lett.* **40**, 174 (1978).
49. Brandt, W. in *Proceedings of Atomic Collision in Solids*, 1981, Lyon, edited by Remillieux (1982).
50. Kitagawa, M. and Ohtsuki, Y.H. in *Proceedings of Atomic Collision in Solids*, Hamilton, Canada, 1979.
51. Kubo, S., Namiki, N. and Ohba, I., *Prog. Theor. Phys.* **55**, 860 (1976).
52. Warke, C.S. and Greiner, W., *Phys. Lett.* **A80**, 399 (1980).
53. Scott, A.C., Chu, F.Y.F. and McLaughlin, D.W., *Proc. IEEE* **61**, 1443 (1973).
54. Isii, A., *Phys. Lett.* **A88**, 417 (1982).

CHAPTER 7

particles and solid surfaces

7.1. Introduction

The characteristics of atomic collisions in solids differ from those in gases because in solids there is: (*a*) high atomic density; (*b*) the presence of valence and/or conduction electrons; and (*c*) well defined surfaces. The last two are especially important for atomic collisions in metals. The first point is taken into account as it was done for atomic collisions in gas targets, namely by calculating the effects of increasing atomic density. In the last few chapters we have studied valence- and conduction-electron excitation. In this chapter, we shall discuss interactions between ions and solid surfaces.

Since interaction potentials between atoms at the surface are different from those in the bulk, the structure of atoms at the surface is different. The simplest example of a structural change is the deviation of the lattice constant at the surface from that in the bulk. The extension of the lattice constant (Δa) at the surface in the direction normal to the surface is called 'surface relaxation'. It is worth noting that Δa is negative for some special solid surfaces, which means there is a reduction of the lattice constant.

The structural change at the surface causes abnormal lattice vibrations; the lattice-vibration amplitudes become large in the surface-normal direction, resulting in an increase of the Debye temperature.

Even when the surface structural change is very small, typical band structures of valence electrons appear at the surface, produced by ordinary Bragg reflections due to the periodic potentials, and by surface reflections. In the surface normal direction electrons are confined by Bragg reflection and by reflections due to the work function, which produces surface resonance states (standing waves). This is a cause of the surface band structure. Further, if the surface atomic structure changes very much (Bragg reflection change), the surface band structure changes. However, it should be noted that a band structure change could also change the potentials between atoms at the surface, and determine a new surface structure, which also contributes to changes in the band structure, and so on. For this reason we have to determine the surface structure and the surface band structure simultaneously in a self-consistent manner.

In the following few sections, we shall study the surface structure using low energy electron diffraction (LEED) and ion scattering. We shall discuss surface structure and surface lattice vibrations and introduce ion neutralization (especially proton neutralization) at the surface. The last few sections are devoted to studying the collective motion of valence electrons at the surface (i.e., surface plasmon excitation) in which we present the dynamical surface potential. Making use of the dynamical surface potential we shall discuss special ion motions at the surface, for example, the skipping motion of ions, and surface energy loss.

7.2. LEED

7.2.1. *Kinematical theory*

Electrons with energy less than 100 eV are scattered from the first few atomic planes (mostly from the first) near the surface, since low energy electrons are mainly absorbed in the crystal. In this case there are no periodic atomic planes in the surface-normal direction (z direction), and the kinematical theory or multiple scattering theories may be applied to electron scattering.

Since the lattice periodicity is two-dimensional at the surface, the reciprocal lattice point turns out to be the reciprocal lattice *rod*. The Bragg condition for diffraction is satisfied at the crossing point between the Ewald sphere and the reciprocal lattice rod. Taking the x and y axes parallel to the surface, and $\mathbf{K}_0$ and $\mathbf{K}$ for wave vectors corresponding to incident and scattered waves, respectively, we obtain the Bragg condition

$$\begin{aligned}(\mathbf{K}-\mathbf{K}_0)_x &= 2\pi h_x \\ (\mathbf{K}-\mathbf{K}_0)_y &= 2\pi h_y\end{aligned} \tag{7.2.1}$$

Where $\mathbf{h}=(h_x, h_y)$ is the two-dimensional reciprocal lattice rod vector. The concept of the reciprocal lattice rod may be best understood by thinking of it as an extension of the reciprocal lattice point in the z direction for a crystal of finite thickness. Figure 7.1 shows the Ewald sphere and the reciprocal lattice rod.

The Bragg condition 7.2.1 can always be satisfied by slowly changing the crystal orientation (direction of the reciprocal lattice rod) and/or the electron energy (the radius of the Ewald sphere). This is one characteristic of two-dimensional diffraction. The three-dimensional diffraction pattern should be compared with the two-dimensional diffraction pattern (figure 7.1), noting that the Bragg spots do not disappear by changing the crystal orientation.[1,2]

There are no differences in principle between the Debye–Waller factors for three- and two-dimensional diffraction. However, we have to distinguish the z directed amplitude from the x (or y) directed amplitude of the lattice vibration.

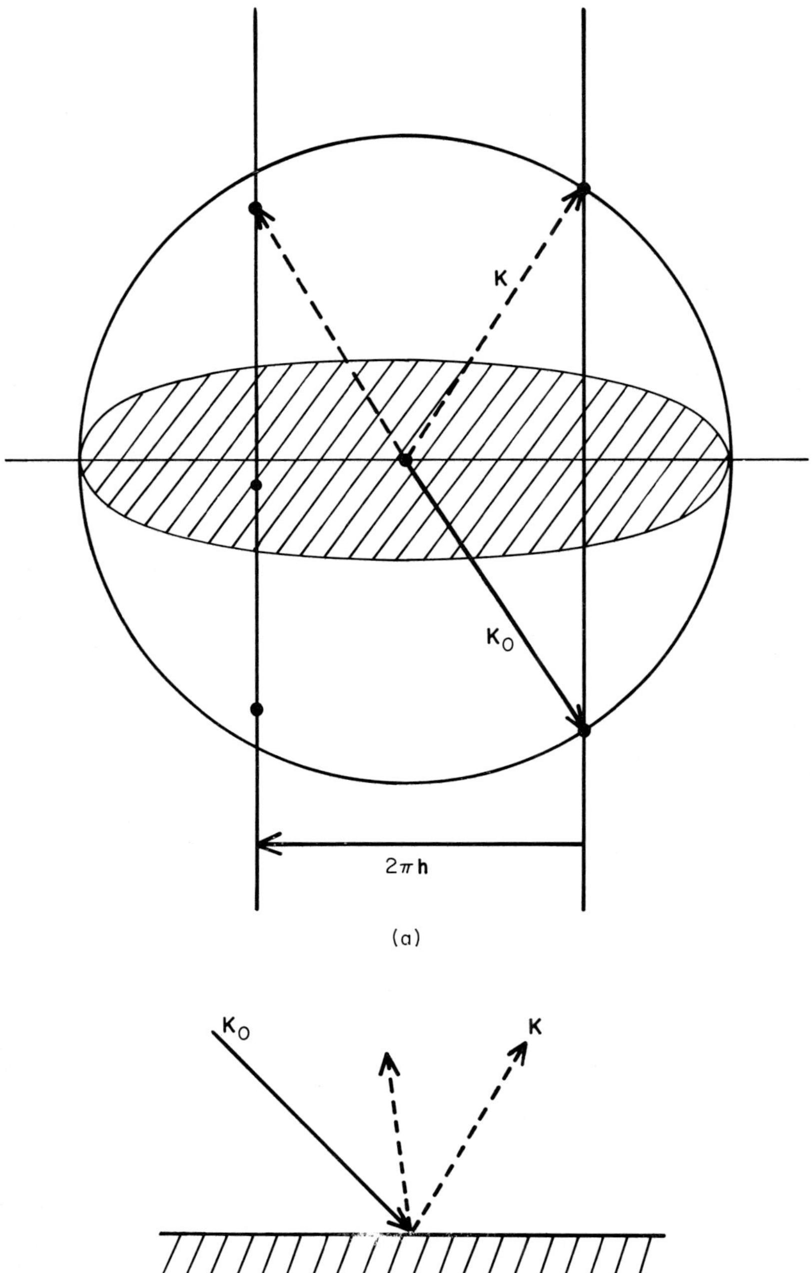

Figure 7.1. The Ewald sphere and the reciprocal lattice rods.

Let us consider the ordinary three-dimensional Debye–Waller factor $\exp(-M)$, where

$$M \propto \langle [(\mathbf{K}-\mathbf{K}_0)\cdot\delta\mathbf{R}_l]^2 \rangle$$

Taking into account the Bragg condition 7.2.1, we obtain

$$M \propto \langle (2\pi\mathbf{h}\cdot\delta\mathbf{R}_{l\parallel})^2 \rangle + \langle (\mathbf{K}-\mathbf{K}_0)_z\cdot\delta\mathbf{R}_{l\perp})^2 \rangle \tag{7.2.2}$$

where $\delta\mathbf{R}_{l\parallel}$ and $\delta\mathbf{R}_{l\perp}$ are the parallel and perpendicular components of the displacement vector of the lattice vibration, respectively. If we consider the specular reflection of the normal incidence, i.e., $\mathbf{h}=0$ (see figure 7.1), the value of M may be represented by $\langle \delta\mathbf{R}_{l\perp}^2 \rangle$ alone, which depends on the number of atomic layers l away from the surface.

Clark, *et al.*[3] studied the layer number dependence of $\langle \delta\mathbf{R}_{l\perp}^2 \rangle$ using computer simulation, which we shall present here in detail. They took a periodic boundary condition for the x and y directions, and expanded it in normal modes

$$\delta R_{mnj,i} = \frac{1}{\sqrt{M_2}} \sum_{\sigma,q_1,q_2} T_\sigma(q_1,q_2)\,\xi_{\sigma,j,i}(q_1,q_2,p)\exp[i\omega_p(q_1,q_2)t]$$

where M_2 is the mass of the atom, $\delta\mathbf{R}_{mnj}$ is the displacement vector of the (m,n)th atom in the jth layer from the surface, the index i stands for the x, y and z components, and the function $\xi_{\sigma,j,i}(q_1,q_2,p)$ is the normal mode whose frequency is $\omega_p(q_1,q_2)$ for the jth layer. Corresponding to phonon branches $\sigma=1$ and 2, $T_\sigma(q_1,q_2)$ is given by

$$T_1(q_1,q_2)=\cos(mq_1+nq_2)$$

$$T_2(q_1,q_2)=\sin(mq_1+nq_2)$$

Since the displacement vector $\delta\mathbf{R}_l$ is represented by its parallel and perpendicular components, the problem of solving the three-dimensional dynamics of the lattice vibration reduces to solving the one-dimensional dynamics problem with interaction between the layers only. So, the dynamical matrix of $3N^3 \times 3N^3$ components reduces to one of $6N \times 6N$ components, N being the number of atoms, where the number $6(=3\times 2)$ comes from the fact that there are two branches for the plane wave. Thus, if we want to perform the calculation for 10 layers, for example, it is sufficient to use a few hundred matrix elements, a problem easy to treat by computer.

Clark *et al.* took 20 layers and calculated the mean square amplitude $\langle \delta R_l^2 \rangle$ for Ni (100), Ni (111), and so on. These results are shown in the table 7.1. This shows that the mean square amplitudes for the parallel and perpendicular directions are the same at the 10th layer, and their values do not change for $j \geq 10$, which corresponds to the bulk values. However, the mean square amplitude in the perpendicular direction for the first layer is twice the value for the parallel direction. This fact may be explained by the weak binding of the atoms in the first layer in the perpendicular direction due to the termination of the interaction in the perpendicular direction at the surface, as was shown directly by Maradudin and Melngailis.[4] They truncated the interaction of the infinite three-dimensional crystal at $z=0$, which is the inter-planar position between two atomic layers. This produces two crystals which have a surface at $z=0$.

Let us take the Hamiltonian H_1 for a crystal with a surface, and H' for the interaction potential which has been cut at $z=0$. Then the total Hamiltonian of the infinite crystal is

$$H = 2H_1 + H'$$

or

$$H_1 = \tfrac{1}{2}H - \tfrac{1}{2}H'$$

We can represent H and H' by the creation and annihilation operators in the harmonic approximation for the Hamiltonian H. Maradudin and Melngailis calculated the Debye–Waller factor $\exp(-M)$ by a perturbation method with the perturbation term $-\tfrac{1}{2}H'$, and also showed that M for the perpendicular direction is twice as large as for the parallel direction.

Experimentally, the Debye–Waller factor for the perpendicular direction is detected by the temperature dependence of the intensity of the normal incidence specular reflection. Figure 7.2 shows the results for the Debye temperature in the Debye–Waller factor first worked out by Jones *et al.*[5] They assumed an absorption coefficient in the form

$$\langle \delta R_l^2 \rangle = \frac{3\hbar^2 T}{mk\Theta_{\mathrm{eff}}^2}$$

defining the effective Debye temperature. High-energy electrons are scattered by atoms that are a great distance from the surface, and the Debye temperature Θ_{eff} is the ordinary bulk Debye temperature introduced in Chapter 3. On the other hand, for very low-energy electrons, electrons are scattered only by the

Table 7.1. The values of $\langle \delta R_l^2 \rangle$ on an arbitrary scale, by Clark *et al.*[3]

	Surface (100)		Surface (111)	
Layer no.	Amplitude parallel to surface	Amplitude perpendicular to surface	Amplitude parallel to surface	Amplitude perpendicular to surface
1	0·624	0·820	0·532	0·821
2	0·458	0·517	0·445	0·518
3	0·426	0·451	0·425	0·457
4	0·414	0·427	0·416	0·434
5	0·407	0·414	0·411	0·422
6	0·403	0·406	0·408	0·416
7	0·400	0·401	0·405	0·412
8	0·398	0·399	0·403	0·409
9	0·397	0·397	0·402	0·407
10	0·396	0·396	0·402	0·407

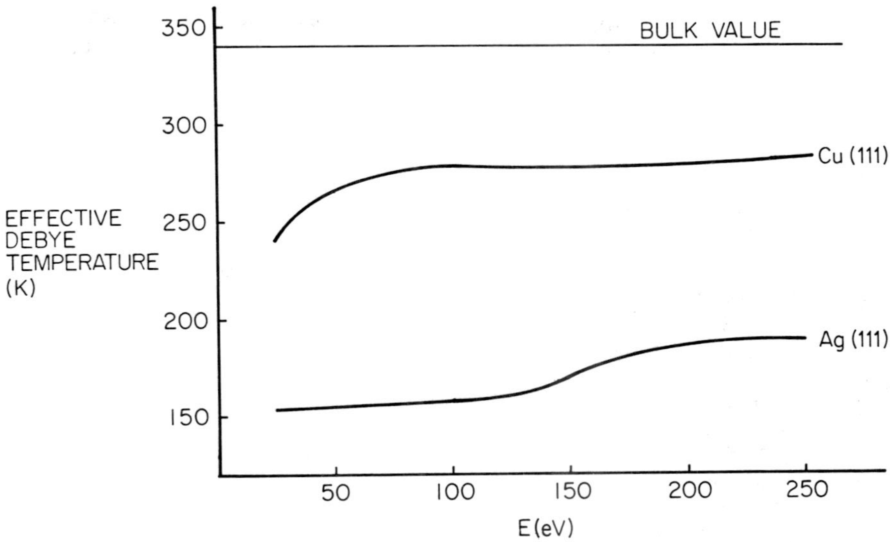

Figure 7.2. Effective Debye temperature measured by Woodruff and Seah[6] for Cu (111) and Jones *et al.*[5] for Ag (111) as a function of the electron energy.

first atomic layer, and $\langle \delta R_l^2 \rangle$ becomes large (Θ_{eff} becomes small). This is confirmed for an Ag(111) surface in figure 7.2. Moreover, Woodruff and Seah[6] measured the energy dependence of the effective Debye temperature for Cu(111), considering not only the absorption but also the temperature dependence of the lattice constant, the mean inner potential, and the Bragg peak profile to be functions of energy. This is shown in the same figure (figure 7.2). It is interesting to note that the Debye temperature does not approach the value Θ by increasing the energy. This means that electrons could not penetrate deep into the crystal, even for an energy of 200 eV.

Table 7.2 shows the minimum values of Θ_{eff} and the ratios of the mean square amplitude in the perpendicular direction to the amplitude in the bulk.

Next, we introduce the fundamental problem of obtaining the scattering

Table 7.2. Effective Debye temperature and mean amplitudes.

			$\sqrt{\langle \delta R_\perp^2 \rangle / \langle \delta R_{bulk} \rangle^2 }$		
Element	Θ(K)	Θ_{eff}	Calculated	Experimental	Reference
Ag(111)	225	155	1·41	1·46	5
Ag(110)	225	152	1·41	1·48	7
Ag(100)	225	104	1·41	2·16	7
Pb(111)	90	55	1·41	1·64	8
Ni(110)	390	220	1·41	1·77	9
Pt(110)	234	110	1·41	2·12	10
Cu(222)	395	244	1·41	1·40	6
Cu(333)	395	278	1·41	1·20	6

Table 7.3. Values of η_l for Si and W (electron energy = 37·5eV).[13]

Element	$\eta_0/2\pi$	$\eta_1/2\pi$	$\eta_2/2\pi$	$\eta_3/2\pi$	$\eta_4/2\pi$
Si	1·63	0·47	0·20	0·072	0·012
W	2·22	1·82	1·30	0·86	0·40

factor for very low energy electrons, where we cannot apply the Born approximation. Under such circumstances it is impossible to represent the atomic scattering factor by the Fourier components of the atomic potential.

As is well known, the general scattering factor $f(\theta)$ may be expanded by the Legendre polynomials $P_l(\cos\theta)$

$$f(\theta)=\frac{1}{2iK}\sum_{l=0}(2l+1)[\exp(2i\eta_l)-1]\,P_l(\cos\theta) \tag{7.2.3}$$

where η_l is the phase shift of the lth partial wave, and is calculated for the atomic potential $V(r)$. When η_l is very small for all l, we obtain[10]

$$\eta_l=-(\tfrac{1}{2}\pi)\frac{8\pi^2 m}{h^2}\int_0^\infty V(r)\,[J_{l+1/2}(Kr)]^2\,r\,\mathrm{d}r \tag{7.2.4}$$

In this case the scattering factor $f(\theta)$ turns out to be the scattering factor for the Born approximation, $f_{\mathrm{Born}}(\theta)$

$$f_{\mathrm{Born}}(\theta)=\frac{1}{K}\sum_l(2l+1)\eta_l\,P_l(\cos\theta) \tag{7.2.5}$$

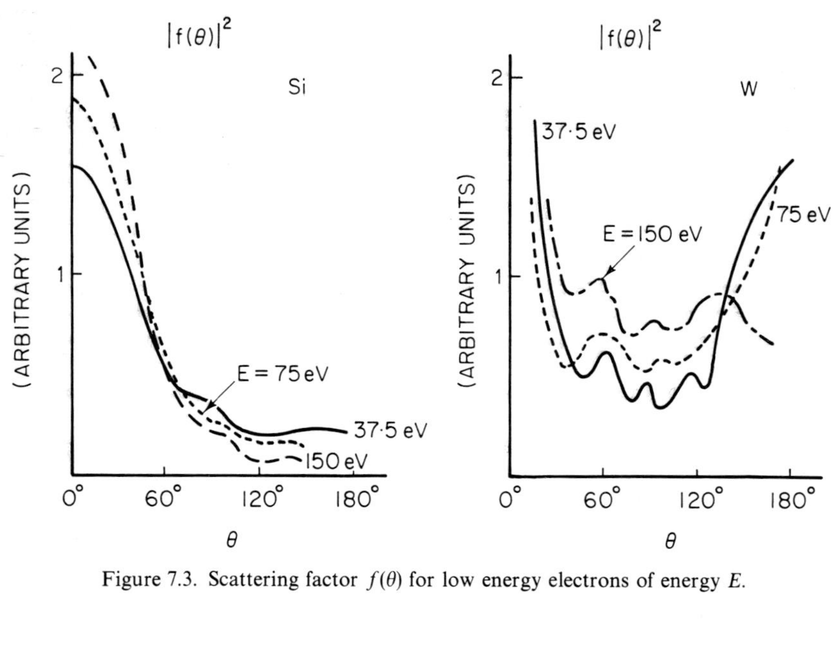

Figure 7.3. Scattering factor $f(\theta)$ for low energy electrons of energy E.

If we insert equation 7.2.4 into equation 7.2.5 the summation over l converges, resulting in the ordinary expression for $f_{\mathrm{Born}}(\theta)$ obtained in Chapter 3. In general it is difficult to calculate η_l in detail. Table 7.3 shows the values of η_l for Si and W at an electron energy of 37·5 eV. The scattering factor $f(\theta)$ is obtained easily by making use of expression 7.2.3, which is plotted[13] in figure 7.3. It is worth noting that η_l decreases rapidly for increasing l, and the number of η_l to be used for the calculation of equation 7.2.3 is not large (see the values in table 7.3). For very low energies we may take η_0 only and neglect η_l (for $l = 1, 2, 3, \ldots$), which is called the s-wave approximation. The s-wave approximation corresponds to considering the potential to be of δ-function form†

$$V(r) \propto \frac{\delta(r)}{r^2} \tag{7.2.6}$$

Under such circumstances the electron wavelength is larger than the atomic potential, and the latter may be considered to be localized like a δ function for the electron waves.

7.2.2. *Dynamical theory of LEED*

In Chapter 3 we said that the extinction distance ξ_G is the minimum distance which electrons penetrate and form standing waves due to multiple scattering by the crystal potential. For low energy electron diffraction (~ 100 eV) the extinction distance ξ_G is of the order of only a few atomic spacings. So, during Bragg reflection at the surface electrons are multiple scattered and the kinematical theory cannot be applied.

For low energy electrons, on the other hand, there is a very high probability of inelastic scattering, which we cannot account for by small perturbations. Experiments also indicate the need for a dynamical theory. As an example, we shall study the energy dependence of the specular reflection intensity, where the incident direction $\mathbf{K}_0$ of the electron beam is almost perpendicular to the surface (z direction). The Bragg condition to be satisfied is

$$(\mathbf{K} - \mathbf{K}_0)_z = 2\pi h_z \tag{7.2.7}$$

corresponding to the points (001), (002), (003), and so on, on the reciprocal lattice rod. If we take the kinematical theory, we obtain Bragg peaks when condition 7.2.7 is satisfied. In our case, $\mathbf{K} = -\mathbf{K}_0$ and $h_z = n/d (n = 1, 2, \ldots)$, d being the inter-plane distance, and the Bragg condition 7.2.7 becomes

$$2K_0 = 2\pi n/d \tag{7.2.8}$$

†Inserting equation 7.2.6 into equation 7.2.4 we obtain

$$\eta_l \propto \int_0^\infty \frac{\delta(r)}{r^2}[J_{l+1/2}(Kr)]^2\, r\, \mathrm{d}r \propto \int_0^\infty [j_l(Kr)]^2\, \delta(r)\, \mathrm{d}r \propto \lim_{\varepsilon \to 0} \varepsilon^l$$

So, η_l becomes 0 except for $l = 0$.

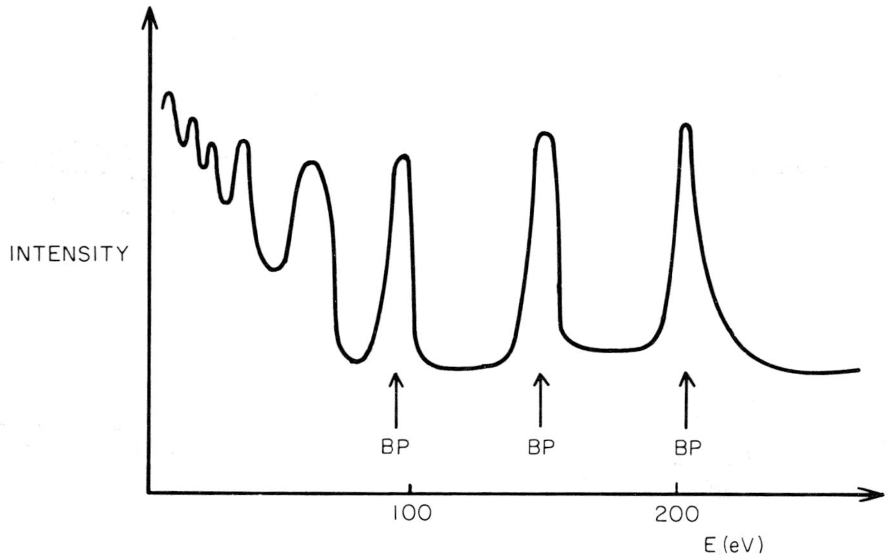

Figure 7.4. Back-scattered intensity profile as a function of the incident energy of electrons for normal incidence. The positions BP show the Bragg condition.

The general Bragg conditions for the z direction, $h \neq 0$, and normal incidence are expressed by

$$K_0 + K_h = 2\pi n/d \tag{7.2.9}$$

where

$$K_h = \sqrt{K_0^2 - (2\pi\mathbf{h})^2} \tag{7.2.10}$$

It is obvious that K_h is the z component of the scattered wave. In figure 7.4 there are a few Bragg peaks (BP) corresponding to equation 7.2.8, especially in the high energy region (> 100 eV). However, in the low energy region (< 100 eV) we can see many peaks not corresponding to BP.[14]

We can now see that we need to develop a dynamical theory for LEED.

Multiple scattering theory

Here let us confine our discussion to the case of strong absorption and/or scattering for very low energy electron diffraction with an electron energy in the range 20–70 eV, where we may consider scattering only by the first layer. The multiple scattering theory by the first layer was developed by McRae[15] making use of the multiple scattering theory of Lax,[16,17] which we shall study first.

We assume that the incident wave is expressed by the plane wave, $\exp(i\mathbf{K}\cdot\mathbf{R})$. The wave function $\psi(\mathbf{R})$ for electrons at the point $\mathbf{R}$ is the sum of the incident plane wave and waves scattered by the sth atom, $\psi^s(\mathbf{R})$. Thus

$$\psi(\mathbf{R}) = \psi_{\text{in}}(\mathbf{R}) + \sum_s \psi^s(\mathbf{R}) \tag{7.2.11}$$

where $\psi_{\text{in}}(\mathbf{R}) = \exp(i\mathbf{K}\cdot\mathbf{R})$. The scattered wave $\psi^s(\mathbf{R})$ is given by

$$\psi^s(\mathbf{R}) = -\int \frac{\exp(iK|\mathbf{R}-\mathbf{R}'|)}{4\pi|\mathbf{R}-\mathbf{R}'|} V(\mathbf{R}', \mathbf{R}_s)\,\psi^s_{\text{in}}(\mathbf{R}')\,d\mathbf{R}' \tag{7.2.12}$$

where $\psi^s_{\text{in}}(\mathbf{R}')$ is the incoming wave at the sth atom, and $V(\mathbf{R}', \mathbf{R}_s)$ is the scattering potential. It is evident that the incoming wave $\psi^s_{\text{in}}(\mathbf{R}')$ is self-consistently obtained from the total wave $\psi(\mathbf{R})$. We define the scattering function U by

$$V(\mathbf{R}', \mathbf{R}_s)\,\psi^s_{\text{in}}(\mathbf{R}') = U(\mathbf{R}', \mathbf{R}_s)\,\psi(\mathbf{R}')$$

It is worth noting that $\psi^s_{\text{in}}(\mathbf{R})$ is the sum of $\psi_{\text{in}}(\mathbf{R})$ and the outgoing waves from the l'th atom where $l' \neq s$

$$\psi^s_{\text{in}}(\mathbf{R}) = \psi_{\text{in}}(\mathbf{R}) + \sum_{l'(\neq s)} \psi^{l'}(\mathbf{R}) \tag{7.2.13}$$

where

$$\psi^{l'}(\mathbf{R}) = -\int \frac{\exp(iK|\mathbf{R}-\mathbf{R}'|)}{4\pi|\mathbf{R}-\mathbf{R}'|} V(\mathbf{R}', \mathbf{R}_{l'})\,\psi^{l'}_{\text{in}}(\mathbf{R}')\,d\mathbf{R}' \tag{7.2.14}$$

Equations 7.2.13 and 7.2.14 give the integral equation for $\psi^s_{\text{in}}(\mathbf{R})$, which should be inserted into equation 7.2.11 and 7.2.12 to determine the total wave $\psi(\mathbf{R})$.

Now, noting that the Fourier component of the potential $V(\mathbf{R}, \mathbf{R}_s)$ is the scattering factor, and $\psi^s_{\text{in}}(\mathbf{R})$ can be expanded in plane waves with wave number $\mathbf{K}$

$$\psi^s(\mathbf{R}) = \int_{|\mathbf{K}|=K} \Gamma_s(\mathbf{K}) \exp[i\mathbf{K}\cdot(\mathbf{R}-\mathbf{R}_s)]\,d\mathbf{K} \tag{7.2.15}$$

we obtain the integral equation[15]

$$\psi(\mathbf{R}) = \exp(i\mathbf{K}\cdot\mathbf{R}) + iK \sum_{s,l} i^l h_l^{(1)}(K|\mathbf{R}-\mathbf{R}_s|)\,P_l(\hat{\mathbf{K}}_s\cdot\hat{\mathbf{K}})\langle f_l^s\rangle\,\psi^s_{\text{in}}(\mathbf{R}_s) \tag{7.2.16}$$

$$\psi^{s'}_{\text{in}}(\mathbf{R}) = \exp(i\mathbf{K}\cdot\mathbf{R}) + iK \sum_{s\neq s'} \sum_l i^l h_l^{(1)}(K|\mathbf{R}-\mathbf{R}_s|)\,P_l(\hat{\mathbf{K}}_s\cdot\hat{\mathbf{K}})\langle f_l^s\rangle\,\psi^s_{\text{in}}(\mathbf{R}_s) \tag{7.2.17}$$

Here we have used the relations

$$\frac{\exp(iK|\mathbf{R}-\mathbf{R}'|)}{|\mathbf{R}-\mathbf{R}'|} = iK \sum_{l=0}^{\infty} (2l+1)\,j_l(KR')\,h_l^{(1)}(KR)\,P_l(\cos\theta_{RR'})$$

and

$$\exp[i\mathbf{K}\cdot(\mathbf{R}-\mathbf{R}_s)] = \sum_{l=0}^{\infty} i^l(2l+1)\sqrt{\frac{\pi}{2(K|\mathbf{R}-\mathbf{R}_s|)}}\,J_{l+1/2}(K|\mathbf{R}-\mathbf{R}_s|)\,P_l(\cos\theta)$$

where $j_l(z)$ and $h_l^{(1)}(z)$ are the spherical Bessell function and the spherical Hankel function of the first kind, respectively. The angle θ_{RR} is the angle between **R** and **R**′, and θ is the angle between **K** and $\mathbf{R} - \mathbf{R}_s$. The mean value of the scattering factor $\langle f_l^s \rangle$ is defined by

$$\langle f_l^s \rangle = \frac{\dfrac{1}{2iK}(2l+1)\left[\exp(2i\eta_l^{(s)}) - 1\right] \displaystyle\int_{\text{angle}} \Gamma_s(\boldsymbol{\kappa})\, P_l\left(\mathbf{K}_s \cdot \frac{\boldsymbol{\kappa}}{\kappa}\right) d\boldsymbol{\kappa}}{P_l(\hat{\mathbf{K}}_s \cdot \hat{\mathbf{K}}) \displaystyle\int_{\text{angle}} \Gamma_s(\boldsymbol{\kappa})\, d\boldsymbol{\kappa}} \qquad (7.2.18)$$

where $\hat{\mathbf{K}}_s$ is the unit vector of the vector $\mathbf{R} - \mathbf{R}_s$, and $\mathbf{K}_s$ is the $\mathbf{R} - \mathbf{R}_s$ component of the vector **K**. The phase $\eta_l^{(s)}$ is that of the *s*th atom.

McRae[15,18] tried to solve the integral equations 7.2.16 and 7.2.17 assuming two-dimensional periodicity parallel to the surface, and 30 atoms per layer in

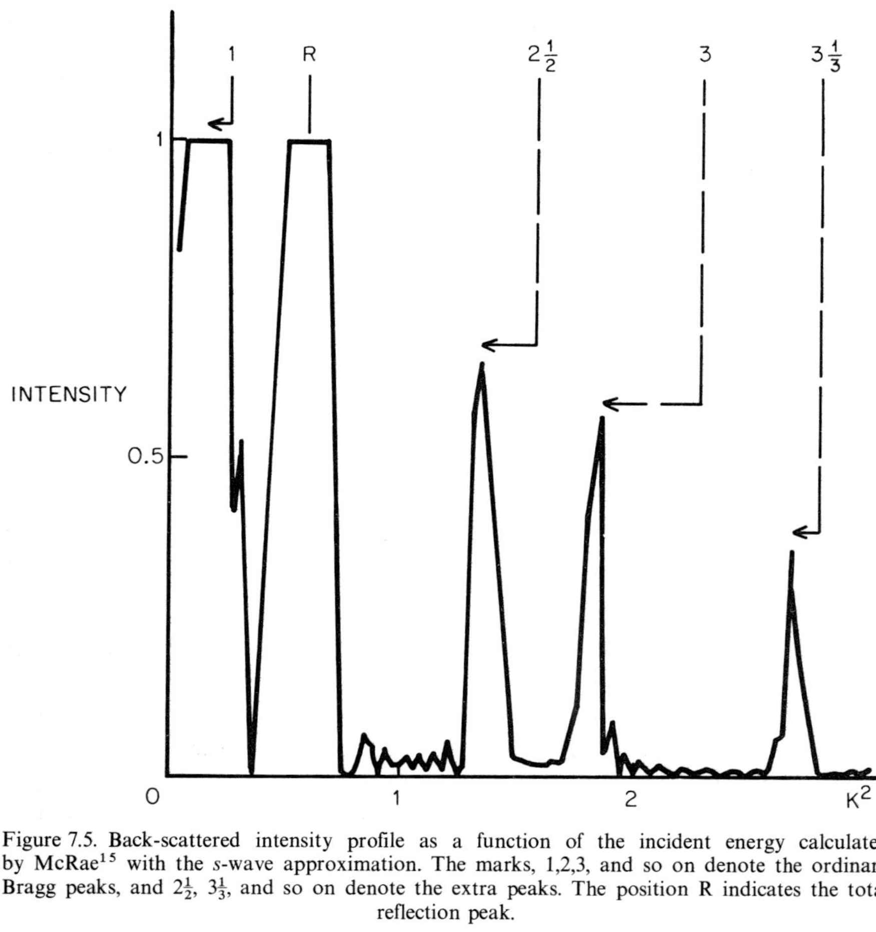

Figure 7.5. Back-scattered intensity profile as a function of the incident energy calculated by McRae[15] with the *s*-wave approximation. The marks, 1,2,3, and so on denote the ordinary Bragg peaks, and $2\frac{1}{2}$, $3\frac{1}{3}$, and so on denote the extra peaks. The position R indicates the total reflection peak.

15 layers. He numerically calculated 30 × 30 coupled linear equations with the s-wave approximation. His results are shown in figure 7.5, where we can see ordinary Bragg peaks marked 1, 2, 3, and so on, and other extra peaks marked by $2\frac{1}{2}, 3\frac{1}{3}$, and so on. Of special interest is the total reflection peak marked R, where the reflection coefficient is unity for low energies. These extra peaks are attributed to the hth reciprocal lattice rod ($\mathbf{h} \neq 0$) wave excitation, which may be explained by the multiple wave effect. Extraordinary peaks $2\frac{1}{2}, 3\frac{1}{3}$, and so on, correspond to the Bragg condition $K_0 + K_h = n'(2\pi/d)$, which may be rewritten as

$$(K_h^2 =)K_0^2 - (2\pi h)^2 = [(2\pi n'/d) - K_0]^2$$

from which we obtain

$$2K_0 = (2\pi/d)[n' + (d^2h^2/n')] \tag{7.2.19}$$

When d^2h^2/n^2 is an integer the above condition reduces to the ordinary Bragg condition, but if it is not an integer the condition gives the extraordinary Bragg peaks such as $2\frac{1}{2}, 3\frac{1}{3}$, and so on, which McRae called 'secondary Bragg peaks'. However, such peaks have been found by Kikuchi and Nakagawa[19] as an intensity anomaly, and have also been discussed by Miyake and Hayakawa.[20]

McRae, on the other hand, called the total reflection peak R the 'resonance peak', which is attributed to $K_h = 0$, where the Ewald sphere touches the hth rod and the hth wave is excited parallel to the surface. This effect will be studied in more detail in subsection 7.2.3.

Other multiple scattering theories have been developed by Kambe[21,22] making use of the theory of Kohn and Rostoker, by Beeby[23] making use of the Green function, and by Marcus and Jepsen[24] making use of the propagation matrix. Absorption effects in LEED theory have been studied by Ohtsuki,[25] and Duke and Tucker.[26]

Band theory in LEED

We shall now study the case where the incident energy of electrons E_0 is smaller than the threshold energy for plasmon excitation E_c. For small reflection coefficients of elastic scattering in a few layers electrons with an energy of $E_0 \lesssim E_c(\sim 50\text{ eV})^\dagger$ penetrate deep into the crystal. It seems to us that the increase of the Debye temperature in the very low energy region in figure 7.2 is a result of the above fact.

For the energy region $E_0 \lesssim E_c$, band theory such as Bethe's dynamical

---

†Note that the value E_c is not the plasmon energy $\hbar\omega_p$. Momentum conservation has to be satisfied at the same time. According to energy momentum conservation we obtain

$$E_c \sim \hbar\omega_p \cdot (K_0/2q_c)$$

q_c being the cut-off wave number.

theory should be applied. However, the band construction is very complicated in this case. The relation between the LEED intensity and the complicated band theory was first pointed out by Boudreaux and Heine,[27] and subsequently by many other authors.[28–32]

The band theory in LEED is not simple. By calculating the bands in crystals, as in solid state physics, we can determine the expansion coefficients of the Bloch waves, and match these Bloch waves with the waves outside the crystal (incident and reflected waves). The matching is the same, in principle, as that explained in Chapter 3. Denoting the perpendicular and parallel components of the wave vector outside the crystal as $\mathbf{K}_\perp$ and $\mathbf{K}_\parallel$, respectively, the wave function outside takes the form

$$\psi_{\text{outside}} = \exp(i\mathbf{K}_\parallel \cdot \mathbf{r} + iK_\perp z) + \sum_{\mathbf{g}} A_{\mathbf{g}} \exp[i(\mathbf{K}_\parallel + 2\pi\mathbf{g})\cdot\mathbf{r} - iK_{\mathbf{g}} z] \qquad (7.2.20)$$

and the Bloch wave function inside the crystal is of the form

$$\psi^{\text{inside}} = \sum_j B_j \sum_{\mathbf{g}} C_{\mathbf{g}}(\mathbf{k}^{(j)}, z) \exp[i(\mathbf{K}_\parallel + 2\pi\mathbf{g})\cdot\mathbf{r}] \qquad (7.2.21)$$

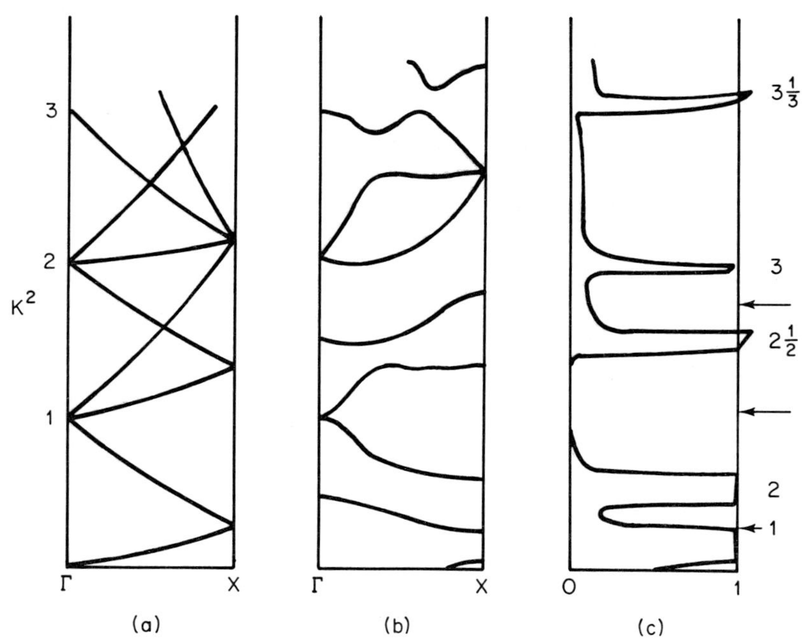

Figure. 7.6. The band structure of electrons in the solid (*a*) and (*b*) corresponding to the back-scattered intensity I_{00} (*c*), according to Capert.[31]

where B_j and $C_{\mathbf{g}}(\mathbf{k}^{(j)},z)$ are coefficients which depend on the band branch j and are determined by band calculations. Here, we show the relation of the band structure in solid state physics to the LEED intensity peaks in figure 7.6 as presented by Capart[31] for cubic crystals. He took the *s*-wave approximation and $\eta_0 = \pi/2$, and calculated the 21 × 21 matrix, obtaining the band structure figure 7.6(*b*). At the same time he determined the $C_{\mathbf{g}}(\mathbf{k}^{(j)},z)$ and $A_{\mathbf{g}}$ coefficients by matching ψ^{outside} and ψ^{inside} at the surface (001). From this he obtained the intensity I_{00} (figure 7.6(*c*)). Comparing figure 7.6(*c*) to figure 7.6(*b*), we can see that the positions of the peaks coincide with the position of the band and the half-widths correspond to the band gap branch. The secondary Bragg peaks and the resonance peak are also seen. Thus, band structure, especially the band gap, has a strong relation to the specular reflection intensity, from which, conversely, we can determine the band structure.

Pendry[33–35] developed a convenient method of calculating the intensity profile of LEED patterns. Notably, he combined[36] the multiple scattering technique with the interaction between atomic layers, and he developed a computer program which is widely used for LEED calculations.

7.2.3. *Surface resonance*

If there is a situation where the reciprocal lattice rods touch at the horizontal position of the Ewald sphere (see figure 7.7), the reflected beam propagates to

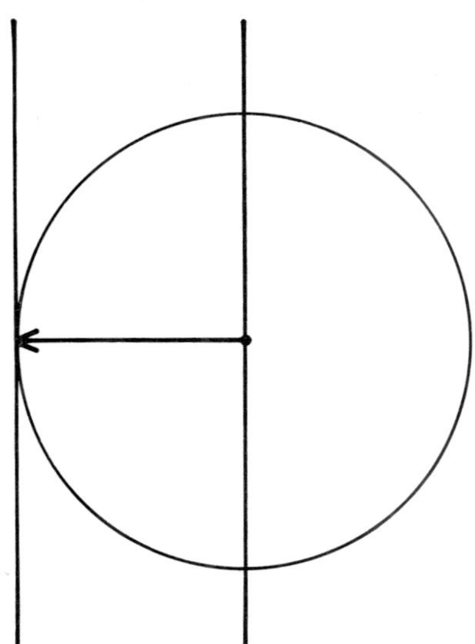

Figure 7.7. An Ewald sphere, touching at a reciprocal lattice rod.

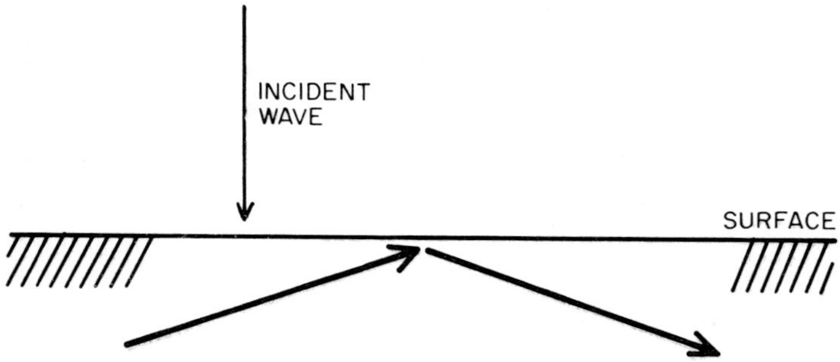

Figure 7.8. Incident wave, and waves reflected by the reciprocal lattice rod and the surface potential.

the surface barrier with a very small angle, and is reflected back by the surface potential V_0. This condition is shown in figure 7.8. In this case the reflected beams are trapped within the surface region and produce a resonance. If the surface resonance takes place, the normally reflected intensity changes in a drastic way, and a peak appears which is attributed to the surface resonance. This effect was first noted by McRae[15] and interpreted in greater detail in his second paper.[18]

7.3 Backscattering, surface scattering and structure analysis

Most ions impinging on a solid penetrate into the solid, but a few are reflected by the surface atomic layers, and this is called 'backscattering'. The backscattering technique had been applied widely to surface analysis. Recently, the method combined with channelling and blocking has become a powerful tool for surface structure analysis.

By decreasing the incident angle (grazing angle) the backscattering intensity increases, finally resulting in total reflection (specular reflection) by the first atomic layer only. This is known as surface channelling or semi-channelling.

7.3.1. *Backscattering*

First we shall consider the case of large-angle single scattering by atoms near the surface, at depth z. The reflection rate (reflection coefficient) η_B is represented by

$$\eta_B = \sigma \Omega N z \tag{7.3.1}$$

where σ is the back scattering cross-section, N is the number density of the atoms, and Ω is the solid angle corresponding to the detector window.

The ions lose energy during large-angle backscattering. As shown in section 4.8, the ion energy E' after scattering by atoms with mass M_2 is obtained by

$$E' = \frac{M_1 \cos\theta_1 + \sqrt{M_2^2 - M_1^2 \sin^2\theta_1}}{M_1 + M_2} E \tag{7.3.2}$$

where θ_1 is the scattering angle of the beam of ions whose mass is M_1. The energy loss due to single-electron excitation is neglected in the case of large-angle scattering. However, we have to take into account energy losses during the ion paths before and after the collision, which we denote by δE_1 and δE_2, respectively. Thus, the total energy loss of the backscattered ions ΔE is given by

$$\Delta E = \left[\frac{M_1 \cos\theta + \sqrt{M_2^2 - M_1^2 \sin^2\theta_1}}{M_1 + M_2} (E_0 - \delta E_1) - \delta E_2 \right] - E_0 \tag{7.3.3}$$

The maximum value of ion energy E_{max} is obtained when $\delta E_1 = \delta E_2 = 0$, corresponding to the case where the target atom is on the surface

$$E_{max} = \frac{M_1 \cos\theta_1 + \sqrt{M_2^2 - M_1^2 \sin^2\theta_1}}{M_1 + M_2} E_0 \tag{7.3.4}$$

Since the scattering cross-section σ becomes large with decreasing ion energy, the energy dependence of η_B is as drawn in figure 7.9.

McCracken and Freeman[38] derived the expression for the reflection coefficient η_B by making use of the LSS theory of the stopping power

$$\eta_B = Z_1^{5/6} (Z_1^{2/3} + Z_2^{2/3})^{3/2} I(E/E_0) 1/E^{3/2} \tag{7.3.5}$$

where $I(E/E_0)$ is a monotonically decreasing function from $I = 4$ (for $E = 0$) to $I = 0$ (for $E = E_0$). They also calculated the 'energy reflection coefficient', defined by the energy transfer carried by reflection.

Recently, unusual enhancement of the yield of low MeV ions backscattered

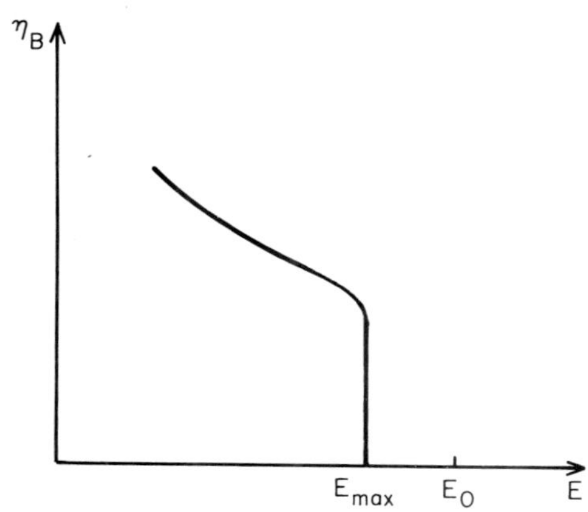

Figure 7.9. Reflection coefficient η_B as a function of the ion energy.

from the near surface regions of solids was reported by Pronko *et al.*[39,40] Barrett *et al.*[41] suggested that this enhancement effect is a consequence of a correlation between the incoming and outgoing paths of the backscattered ion, and computer simulation results by Crawford,[42] Jakes *et al.*[43] and Barrett *et al.*[41] have confirmed this suggestion and the experimental results. Oen[44,45] has developed an analytical theory applying the two-atom scattering model to study the enhancement of ion backscattering.

The backscattering intensity $I(\psi)$ at small exit angle ψ is proportional to the Rutherford backscattering cross-section σ_R. $I(\psi)$ is also proportional to the flux intensity incident on the backscattering plane where scattering atoms are situated, $\frac{b}{l^2 b}\left|\frac{db}{d\theta}\right|$, according to the two atoms model, where b is the impact parameter measured from a second atom lying between the target surface and the backscattering plane (see figure 7.10). The projectile ion makes a glancing collision with this second atom both before and after it is backscattered. θ and l are the scattering angle and the distance between the second atom and the backscattering plane, respectively.

Therefore, we may obtain an expression[44,45] for the backscattering intensity (per steradian)

$$I(\psi) = I_0 \sigma_R \sum_{\Delta l} \iint \frac{b}{l^2 b}\left|\frac{db}{d\theta}\right| f(\theta') \exp(-N\pi b^2 D)\, N l^2 \theta \, d\theta \, d\phi \, \Delta l$$

where $f(\theta')$ is the number of particles scattered per solid angle at the scattering angle θ', and $\exp(-N\pi b^2 D)$ is the probability of a single scattering with an impact parameter b, where N and D are the atomic density and the distance

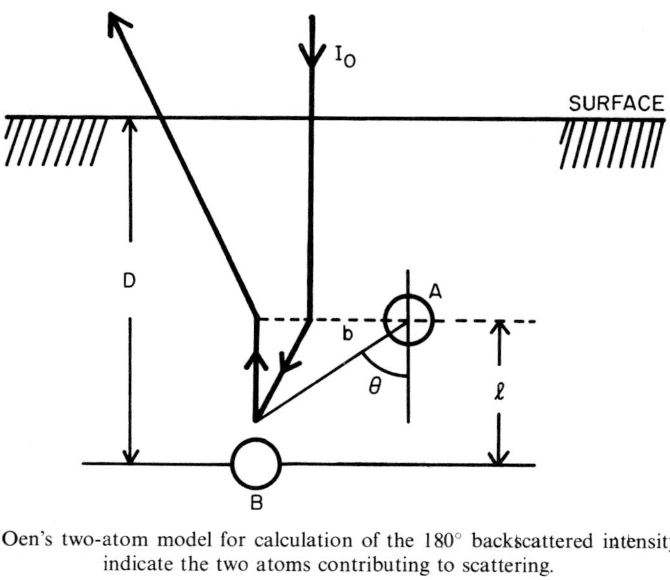

Figure 7.10. Oen's two-atom model for calculation of the 180° backscattered intensity. A and B indicate the two atoms contributing to scattering.

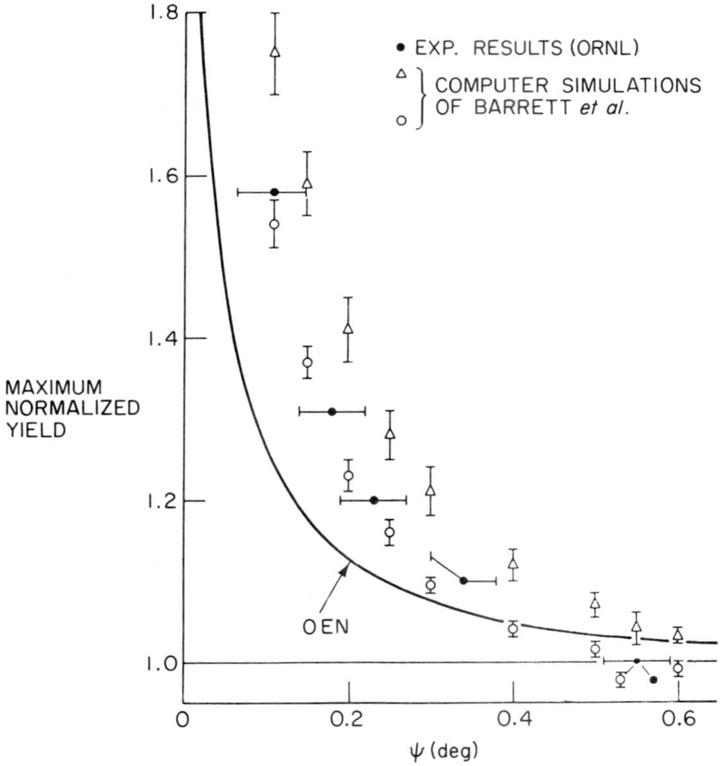

Figure 7.11. Comparison of Oen's theory with computer simulations[41] and experimental results[40] for 1 MeV He ions from amorphous Pt.

between the solid surface and the scattering plane, since the exponential factor gives the probability that there is no atom in a cylindrical volume of radius b and height D. The factor $Nl^2\theta\Delta l\,d\theta\,d\phi$ is the distribution function for finding an atom in a small volume of thickness Δl, where ϕ is the azimuthal scattering angle.

Figure 7.11 compares calculated results by Oen[44,45] and computer simulations[41] for 1 MeV He ions from amorphous Pt with experiments.[40,41]

7.3.2. *Surface scattering and surface channelling*

There is a complicated intensity change corresponding to the various lattice structures at the surface. The effect of the orientation of the atomic axis on the scattering intensity under surface channelling conditions has been studied by Mashkova and Molchanov[46] and many other authors.

The typical effects of the atomic axis are two kinds of ion focusing. The

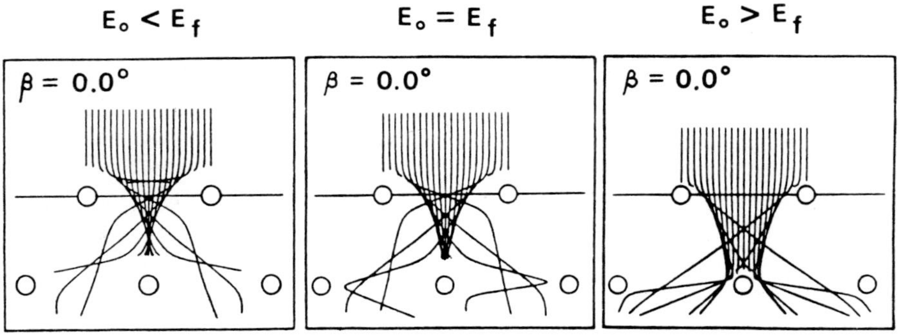

Figure. 7.12. Inward focusing by computer simulation.[49]

first is focusing along the axis, resulting in a change in intensity with respect to the azimuthal angle.[47,48] The second is focusing inward along the surface normal, as shown in figure 7.12, which was obtained by computer simulation.[49] It is interesting that the focal point is in the atomic position of the second layer when the ion energy becomes the critical energy E_f. In this case the scattered intensity increases due to scattering by the second layer.

7.3.3. *Surface structure analysis by ion beam*

When the ion beam direction is along the channelling axis most ions penetrate deep and the reflection coefficient becomes very small (see figure 7.13). However, the intensity curve due to the first atomic layer is not changed by

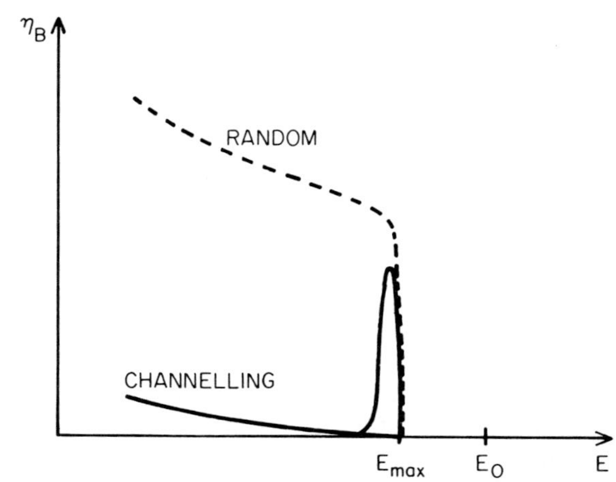

Figure 7.13. Reflection coefficient as a function of ion energy for the channelling condition. The peak near $E \sim E_{max}$ corresponds to the surface scattering.

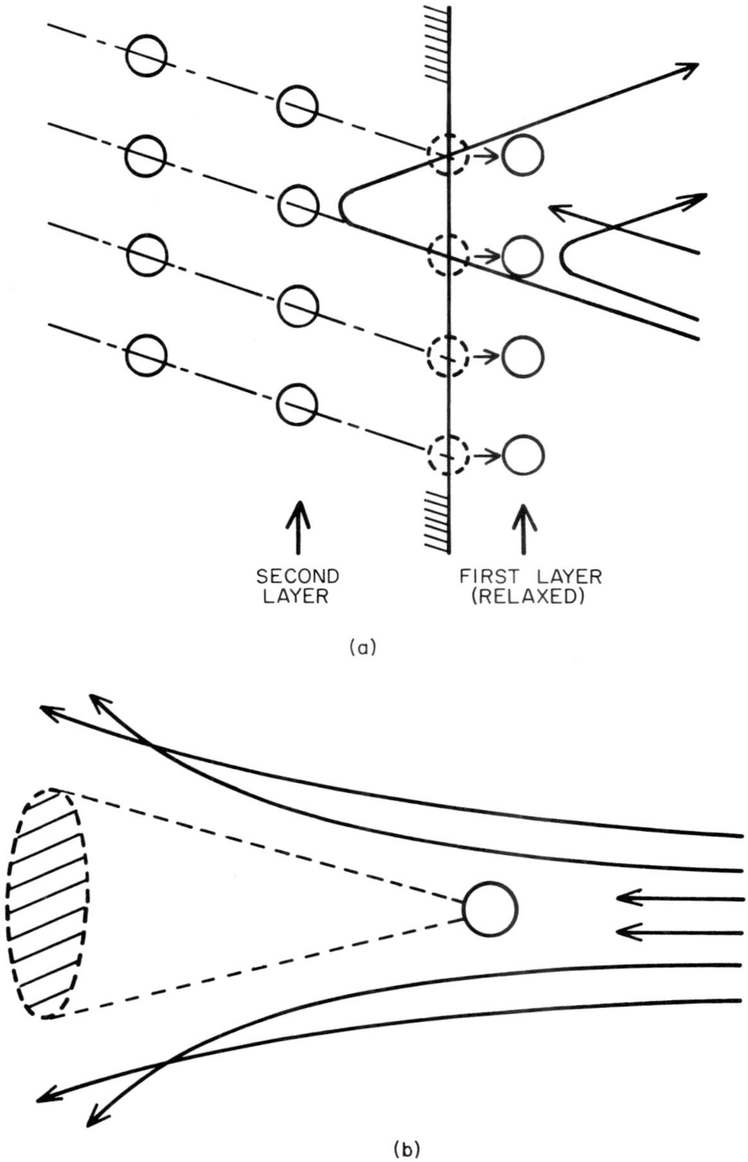

Figure. 7.14. (*a*) Detection of the ion intensity backscattered by the second layer, unmasked by the relaxation of the first layer. (*b*) A Rutherford scattering shadow.

changing the condition from random to channelling incidence, resulting in the surface peak at $E \sim E_{max}$, shown in figure 7.13.

We now consider the case where the position of the atom in the first layer deviates from the normal lattice point. For example, figure 7.14(*a*) shows the relaxation of the atom, where the beam can 'see' the second layer as well as the first layer, and the integrated surface peak intensity is doubled. In other words, in the absence of relaxation the atoms in the second layer are behind the shadow cone of Rutherford scattering (see figure 7.14(*b*)), but when the atoms in the first layer are relaxed the atoms in the second layer are out of the shadow cone, and the ion beam hits both the first and second layers. Changing the ion beam energy, and hence the aperture angle of the shadow cone, we can detect the relaxation length quantitatively.[50] Employing this technique Davies *et al.*[51] measured the relaxation of the first layer of Pt(111) surface to be 0·003 nm.

Saris and his group,[52,53] however, developed another method of surface structure analysis by making use of the blocking effect of the first layer for the scattered ions that are randomly incident on the second or third layers.

7.4. Neutralization of ions at surfaces

It happens that ions passing through the surface capture electrons and are neutralized or decrease their charge state. Since protons whose velocity is smaller than the Fermi velocity ($\sim$ 20 keV) cannot capture electrons inside the crystal (metal) due to the decrease in the Coulomb potential by screening, their neutralization occurs only at the surface. Even for high-energy protons (or other heavy ions), the surface neutralization is significant for the surface channelling condition where the trajectories are confined within the surface region.

Until now, the neutralization of protons has mainly been studied because these processes are very complicated for arbitrary ions. So, in this section, we shall discuss only proton neutralization at the surface, taking into account the Auger process and the resonance transition process.

7.4.1. *Neutralization due to the Auger process*

The Auger process contributes very much to proton neutralization in the very low energy region (<1 keV). In this energy region the experimental neutralization probability is of the order of $10^{16}\,s^{-1}$. The probability of neutralization during an interval of 10^{-15} s, when the ions with an energy of 0·5 keV traverse about 0·1 nm, is therefore $10^{16} \times 10^{-15} = 10$. For this reason we can imagine that the ion will be neutralized completely within a path length of approximately 0·1 nm. We may therefore imagine that the ions do not move during the neutralization process. First we determine the neutralization probability $P(s)$ of the ion at a distance s from the surface. Then we sum $P(s)$ over s corresponding to the ion trajectory, and call this the total neutralization probability for ions moving near surface.

If the function $P(s)$ were already known, the total number of ions N and its decrease $-\dfrac{dN}{dl}$ over the path length dl due to neutralization are related as follows

$$-\frac{dN}{dl} = N\,P(s)/v$$

where v is the ion velocity. Integrating over l along the trajectory, we obtain

$$N = N_0 \exp\left(-\int_{\text{path}} \frac{P(s)}{v}\,dl\right)$$

where N_0 is the number of incident ions. For specular reflection where the perpendicular component of the velocity $v_\perp$ is conserved, N takes the form

$$N = N_0 \exp\left(-\int_{\text{path}} \frac{P(s)}{v_\perp}\,ds\right) \tag{7.4.1}$$

In other words, the total neutralization probability ϕ_0 is expressed by

$$\phi_0 = 1 - (N/N_0) = 1 - \exp(-w) \tag{7.4.2}$$

where

$$w = \int_{\text{path}} \frac{P(s)}{v_\perp}\,ds$$

Now, let us calculate $P(s)$ quantum mechanically considering the Auger process. This is not easy because it is a three body problem. As yet the only theoretical calculations are those by Horiguchi *et al.*[54] They considered the Auger process illustrated in figure 7.15, where the electrons in the crystal are captured by the ion near the surface and the energy difference is transferred to another electron which is excited to a higher level. The three body problem involves electrons being captured, excited, and bound to the ion; a very complicated process.

In general, taking the transition matrix element to be $H(s)$, the transition probability $P(s)$ takes the form

$$P(s) = \left(\frac{4\pi}{\hbar}\right)\frac{V^2}{(2\pi)^6}\iiint |H(s)|^2 \rho_f \,d\Omega''\,d\mathbf{K}\,d\mathbf{K}' \tag{7.4.3}$$

where V is the volume of the solid and the vacuum, and

$$\rho_f = \frac{m\hbar|k''|}{(2\pi\hbar)^3}$$

which is the density of the final states. The variable Ω'' is the solid angle in the direction of the emitted Auger electrons.

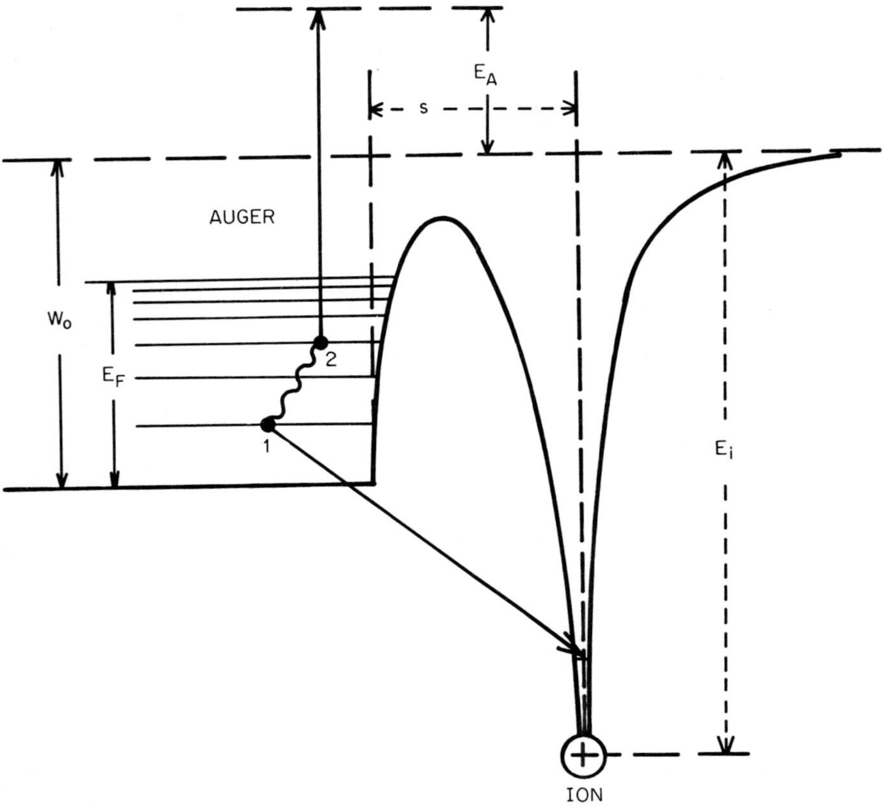

Figure 7.15. Schematic energy relation of the Auger process.

The general expression of the quantum mechanical matrix elements for the Auger process was obtained by Hagstrum.[55] The squared matrix element $|H(s)|^2$ is expressed as

$$|H(s)|^2 = |H_{\mathrm{fi}}(s)|^2 + |J\,H'_{\mathrm{fi}}(s)|^2 - \mathrm{Re}[H^*_{\mathrm{fi}}(s)\,H'_{\mathrm{fi}}(s)] \qquad (7.4.4)$$

where

$$H_{\mathrm{fi}}(s) = \iint u^*_{\mathrm{A}}(\mathbf{r}_2)\,u^*_{\mathrm{G}}(\mathbf{r}_1 - \mathbf{s})\,(e^2/r_{12})\,u_k(\mathbf{r}_1)\,u_{k'}(\mathbf{r}_2)\,\mathrm{d}\mathbf{r}_1\,\mathrm{d}\mathbf{r}_2$$

and

$$H'_{\mathrm{fi}}(s) = \iint u^*_{\mathrm{A}}(\mathbf{r}_1)\,u^*_{\mathrm{G}}(\mathbf{r}_2 - \mathbf{s})\,(e^2/r_{12})\,u_{k'}(\mathbf{r}_2)\,u_k(\mathbf{r}_1)\,\mathrm{d}\mathbf{r}_1\,\mathrm{d}\mathbf{r}_2$$

In the above, $\mathbf{r}_1$ and $\mathbf{r}_2$ are the position vectors for electrons 1 and 2, respectively, and $r_{12} = |\mathbf{r}_1 - \mathbf{r}_2|$. J is the exchange operator. $u^*_{\mathrm{A}}(\mathbf{r}_2)$ is the wave function for the Auger electron, which may be expressed by the plane wave function

$$u_{\mathrm{A}}^{*}(\mathbf{r}_2) = \exp(-i\mathbf{k}''\cdot\mathbf{r}_2) \tag{7.4.5}$$

On the other hand, the bound electron (ground state) within the ion is represented by

$$u_{\mathrm{G}}(\mathbf{R}) = (\pi a_{\mathrm{B}}^3)^{-1/2}\exp(-R/a_{\mathrm{B}}) \tag{7.4.6}$$

The function u_k is the wave function of a free electron near the surface and is represented by

$$u_k(\mathbf{r}) = \sqrt{\frac{2}{V}}\,u_{kz}(z)\exp(i\mathbf{k}_{\parallel}\cdot\mathbf{r}_{\parallel})$$

$$u_{kz}(z) = k_z(k_z^2+\kappa^2)^{-1/2}\exp(-\kappa z)$$

for $z>0$ and

$$u_{kz}(z) = \tfrac{1}{2}\{\exp[i(k_z z+\gamma)] + \exp[-i(k_z z+\gamma)]\} \tag{7.4.7}$$

for $z<0$. We assume a vacuum for $z>0$ and a solid for $z<0$, separated by a step potential whose value is W_0. We also assume that

$$\begin{aligned} k_z &= [2m(E_z+W_0)/\hbar^2]^{1/2} \\ \kappa &= (-2mE_z/\hbar^2)^{1/2} \\ E_z &= E-\hbar^2|\mathbf{k}_{\parallel}|^2/2m \\ \tan\gamma &= \kappa/k_z \end{aligned} \tag{7.4.8}$$

The wave function $u_k(\mathbf{r})$ is divided into two parts: a plane wave as the parallel component and an exponentially damped wave as the perpendicular component. This means that the electron exudes into the vacuum from the surface of the solid and travels a small distance κ^{-1}. We shall employ the same type of wave function in sections 7.5 and 7.8. It should be noted that the exchange effect between positions 1 and 2 for free electrons is included in the matrix element H'_{fi}.

Inserting equations 7.4.5, 7.4.6 and 7.4.7 into equation 7.4.4 and integrating, the matrix element takes the form

$$H_{\mathrm{fi}}(s) = \frac{M_{\mathrm{fi}}(s)}{q_{\parallel}} + N_{\mathrm{fi}}(s) \tag{7.4.9}$$

where $q_{\parallel} = |\mathbf{k}'_{\parallel} - \mathbf{k}''_{\parallel}|$. Since the parallel component of the wave vector $\mathbf{k}'_{\parallel}$ of the free electron in the surface is nearly equal to $\mathbf{k}''_{\parallel}$ of the Auger electron because of the symmetry, the first term of equation 7.4.9 is significant. The matrix element $M_{\mathrm{fi}}(s)$ is expressed by

$$M_{\mathrm{fi}} = \sum_{j=1}^{4} M_{\mathrm{fi}}^{(j)}(s) \tag{7.4.10}$$

The important point here is that $H_{\mathrm{fi}}(s)$ takes the form of linear combinations of terms such as $(s+A)\exp(-us)$ and $\exp[-(\kappa+q_{\parallel}+u)s]$.

It is very complicated to perform the integration in order to obtain $P(s)$ by inserting equations 7.4.9 and 7.4.10 into equation 7.4.3. Here we assume that the electrons with wave vectors perpendicular to the surface in the Fermi sphere are captured because they go far into the vacuum from the solid.[56] For this reason we take $\mathbf{k} = \mathbf{k}' \doteqdot (0, 0, k_F)$, whence $H_{fi}(s) \doteqdot H'_{fi}(s)$. Taking $|M_{fi}(s)|^2$ from the integral we arrive at the form

$$P(s) = \frac{4\pi}{\hbar} \frac{V^2}{(2\pi)^6} [|M_{fi}(s)|^2 \rho_f]_F \int \frac{1}{|k''_{\parallel}|^2} \mathrm{d}\mathbf{k}\, \mathrm{d}\mathbf{k}'\, \mathrm{d}\Omega'' \qquad (7.4.11)$$

where $[\ldots]_F$ is the value at $\mathbf{k} = \mathbf{k}' = (0, 0, k_F)$, and $q_{\parallel} \doteqdot 0$. The integral in equation 7.4.11 diverges at $\mathbf{k}''_{\parallel} = 0$, as a result of the bare Coulomb interaction. We may avoid this divergence by assuming the screened Coulomb interaction with the screening length λ. Further, it is obvious that

$$\frac{4}{(2\pi)^6} \int\int \mathrm{d}\mathbf{k}\, \mathrm{d}\mathbf{k}' = n^2 \qquad (7.4.12)$$

n being the free-electron density. Finally we obtain

$$P(s) = A\{\exp(-2\kappa_0 s) + (Bs + C)\exp[-(\kappa_0 + a_B^{-1})s] + (Ds^2 + Es + F)\exp(-2a_B^{-1}s)\} \qquad (7.4.13)$$

Where A, B, C, D, E and F are constants of very complicated analytical forms (we have no space here to give these expressions). The constant κ_0 is the value of κ at the Fermi surface

$$\kappa_0 = (-2mE_F/\hbar^2)^{1/2}$$

Now, it is worth noting that the simple expression for $P(s)$, which Hagstrum derived in 1954 and which has been used widely since then, was derived phenomenologically

$$P(s) = A' \exp(-\eta s) \qquad (7.4.14)$$

where $\eta \doteqdot 2a_B^{-1}$, and A', shown in table 7.4, was determined by experiment. Figure 7.16 shows a comparison between equation 7.4.13 and equation 7.4.14 as a function of distance s for a proton of energy 10·3 keV at an Au surface ($E_F = 5{\cdot}5$ eV, $n = 5{\cdot}90 \times 10^{22}\ \mathrm{cm}^{-3}$). It should be noted that the neutralization probability $P(s)$ deviates from that of Hagstrum by one order of magnitude for small values of s.

Table 7.4. The value of A' in Hagstrum's theory ($P(s) = A' \exp(-\eta s)$).

Ion	For $\eta = 2 \times 10^8\ \mathrm{cm}^{-1}$	For $\eta = 5 \times 10^8\ \mathrm{cm}^{-1}$
He^+	$7{\cdot}1 \times 10^{16}$	$1{\cdot}5 \times 10^{20}$
Ne^+	$4{\cdot}8 \times 10^{16}$	$1{\cdot}6 \times 10^{20}$

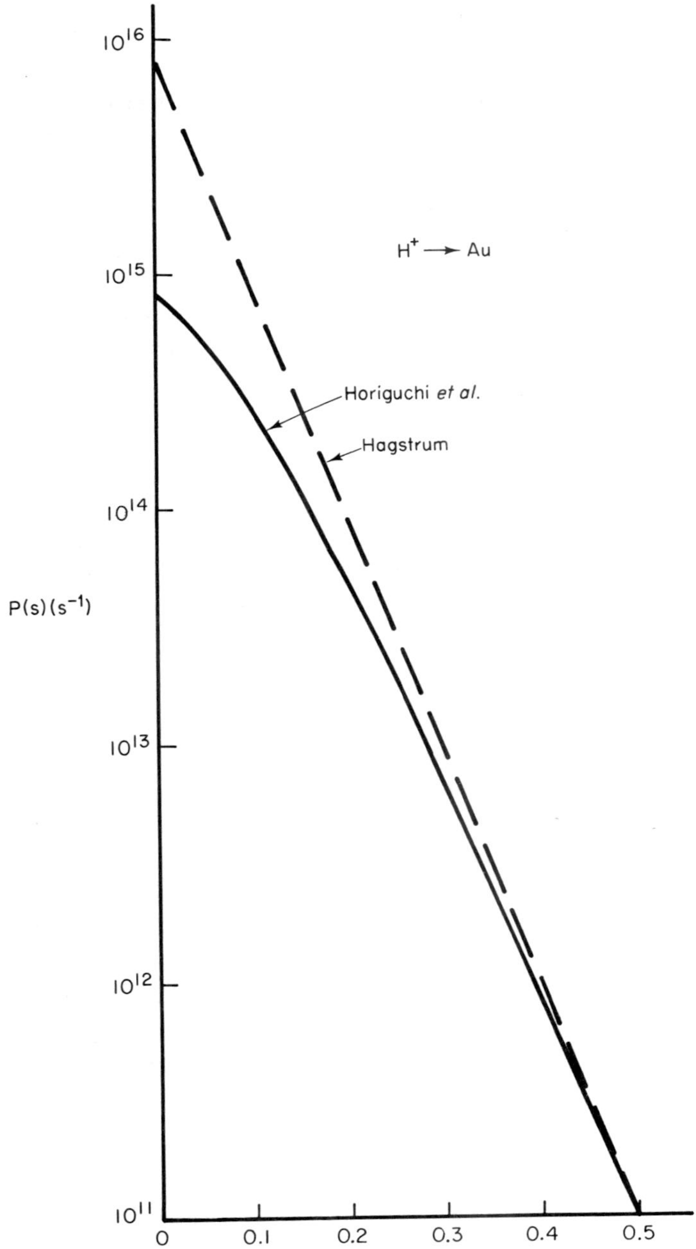

Figure 7.16. Comparison of the theory of Horiguchi *et al.*[54] with Hagstrum's[55] expression for neutralization probability $P(s)$ as a function of the distance from the surface s in the case of 10·3 keV H^+ at Au surface.

It should also be noted that the above calculation of Auger neutralization includes a few rough approximations for the matrix elements $H_{fi}(s)$ and integration over **k**. It seems to be very difficult to perform more exact calculations.

7.4.2. *Resonance capture*

When the energy of the incident electron becomes approximately 50 keV, the ion velocity is 10 times larger than the $10^{14}\,\mathrm{nm\,s^{-1}}$ mentioned at the beginning of subsection 7.4.1. Therefore we cannot use the static ion model which was adopted there for the calculation of the total neutralization probability, since the neutralization probability per ångström is less than unity. In other words, we cannot model neutralization by first obtaining the function $P(s)$ quantum mechanically.

The neutralization probability for the resonance capture was calculated by Trubnikov and Yablinskii[57] making use of time-dependent perturbation theory, showing that the resonance capture process becomes important at energies of a few 10 keV. The resonance capture process was also taken into account for very low energies in the static ion models by Cobas and Lamb[58] and Hagstrum.[55,59] However, there are a few parameters not determined from first principles in the above calculations, and unfortunately their results are not consistent with the experimental data. Kitagawa and Ohtsuki[60] developed a theory based on the Bates method[61] to solve from first principles the time-dependent Schrödinger equation and gave an expression for the neutralization of protons of arbitrary velocity. Here we shall study this theory in detail.

Let us denote the ion velocity and the position of the electron by **v** and **r**, respectively. We take the origin of the time scale to be the time when the ion passes through the surface plane. The Coulomb potential $V(\mathbf{r}, t)$ of interaction between the ion and the electron takes the form

$$V(\mathbf{r}, t) = \frac{-e^2}{|\mathbf{r} - \mathbf{v}t|} \tag{7.4.15}$$

The electrons in the crystal are bound within the crystal by the step potential whose value is W_0, as in subsection 7.4.1. Then, our initial Schrödinger equation for electrons is given by

$$i\hbar\frac{\partial\Phi}{\partial t} = H\Phi$$

$$H = -\frac{\hbar^2}{2m}\Delta + V(\mathbf{r}, t) + (-W_0)\theta(-z) \tag{7.4.16}$$

Let us consider the physical interpretation of $\Phi(\mathbf{r}, t)$. The electron may be partly in a free-electron state of the crystal, and partly in the bound state of a hydrogen atom (where we neglect the capture of electrons into excited states of the hydrogen atom). Thus the wave function $\Phi(\mathbf{r},t)$ may be represented by

$$\Phi(\mathbf{r},t) = C_1(t)\phi_1(\mathbf{r},t) + C_2(t)\phi_2(\mathbf{r},t) \tag{7.4.17}$$

where $\phi_1(\mathbf{r},t)$ is the wave function of the free electron

$$\phi_1(\mathbf{r},t) = u_k(\mathbf{r})\exp\left[-\frac{i}{\hbar}\left(-W_0 + \frac{\hbar^2k^2}{2m}\right)t\right] \tag{7.4.18}$$

$u_k(\mathbf{r})$ being the wave function given by equation 7.4.7. $\phi_2(\mathbf{r},t)$ is a wave function of the bound state of the hydrogen atom $\phi_2(|\mathbf{r}-\mathbf{v}t|)$

$$\phi_2(\mathbf{r},t) = \phi_2(|\mathbf{r}-\mathbf{v}t|)\exp\left[\frac{i}{\hbar}(m\mathbf{v}\cdot\mathbf{r} - \tfrac{1}{2}mv^2t - W_0t)\right] \tag{7.4.19}$$

The second factor, exp[...], in the above expression originates from the Galilei transformation of the moving atom into the static laboratory system.

Now, inserting equation 7.4.17 into equation 7.4.16, and defining $C_2'(t)$ by

$$C_2'(t) = C_2(t)\exp\left(\frac{i}{\hbar}\int^t \frac{h_{22} - S_{21}h_{12}}{1-|S|^2}\,\mathrm{d}t\right) \tag{7.4.20}$$

where

$$h_{22} = \langle\phi_2|(-W_0)\theta(-z)|\phi_2\rangle \tag{7.4.21}$$
$$h_{12} = \langle\phi_1|(-W_0)\theta(-z)|\phi_2\rangle \tag{7.4.22}$$
$$S_{12} = \langle\phi_1|\phi_2\rangle = S_{21}^*, |S|^2 = S_{12}S_{21} \tag{7.4.23}$$

we obtain a second-order differential equation for $C_2'(t)$, as in the Bates method

$$\ddot{C}_2' + \left\{\frac{i}{\hbar}[\varepsilon_1(t) + E_1 - \varepsilon_2(t) - E_2] - \frac{\dot{\varepsilon}_{21}'(t)}{\varepsilon_{21}'(t)}\right\}\dot{C}_2' + \frac{\varepsilon_{12}(t)\varepsilon_{21}(t)}{\hbar^2}C_2' = 0 \tag{7.4.24}$$

where $\varepsilon_{12}(t)$, $\varepsilon_{21}(t)$, $\varepsilon_{12}'(t)$, $\varepsilon_{21}'(t)$ and $\varepsilon_2(t) - \varepsilon_1(t)$ are defined by the formulae

$$\varepsilon_{12}(t) = \frac{1}{1-|S|^2}(h_{12} - S_{12}h_{22}) \tag{7.4.25}$$

$$\varepsilon_{21}(t) = \frac{1}{1-|S|^2}(d_{21} - S_{21}d_{11}) \tag{7.4.26}$$

$$\varepsilon_{12}'(t) = \varepsilon_{12}(t)\exp[i(E_2 - E)t/\hbar] \tag{7.4.27}$$
$$\varepsilon_{21}'(t) = \varepsilon_{21}(t)\exp[i(E_2 - E)t/\hbar] \tag{7.4.28}$$

$$\varepsilon_2(t) - \varepsilon_1(t) = \frac{1}{1-|S|^2}(h_{22} - d_{11} - S_{21}h_{12} + S_{12}d_{21}) \tag{7.4.29}$$

where

$$d_{ij} = \langle\phi_i|V|\phi_j\rangle \tag{7.4.30}$$

$$E = -W_0 - \frac{\hbar^2k^2}{2m} \tag{7.4.31}$$

$$E_2 = E_n + \tfrac{1}{2}mv^2 \tag{7.4.32}$$

It should be noted that we arrive at the theory of Trubnikov and Yablinskii if we assume a time independent $C_1(t)$, which leads to a first-order differential equation for $C_2'(t)$ instead of equation 7.4.24.

It is impossible to solve the differential equation 7.4.24 analytically without some simplifying assumptions. Kitagawa and Ohtsuki[60] assumed an 'instantaneous' interaction of the resonance capture, which implies that the interaction time t is very short. Thus

$$\Delta E t \ll \hbar \tag{7.4.33}$$

where ΔE is the energy transfer of the interaction. The velocity of a proton with an energy of 50 keV is of the order of $10^{15}\,\mathrm{nm\,s^{-1}}$, and the interaction time of a proton within 0·01 nm of the surface is of the order of 10^{-17} s. The energy transfer ΔE, on the other hand, is the order of a few electron volts. We obtain

$$\frac{\Delta E\, t}{\hbar} \sim \frac{10\dfrac{10^{-18}}{6} \times 10^{-17}}{10^{-34}} \sim 0{\cdot}16$$

So, we may conclude that the instantaneous interaction condition 7.4.33 is satisfied for a proton with energy higher than 50 keV.

Under the instantaneous interaction assumption, we may expand the time-evolving function $\varepsilon_1(t) - \varepsilon_2(t)$ in powers of t

$$\varepsilon_1(t) - \varepsilon_2(t) \simeq \{\varepsilon_1(0) - \varepsilon_2(0)\} + A_2 t \tag{7.4.34}$$

$$A_2 = \left.\frac{\partial[\varepsilon_1(t) - \varepsilon_2(t)]}{\partial t}\right|_{t=0} \tag{7.4.35}$$

On the other hand, $|S_{21}|$ and $|S_{12}|$ are very small compared to unity as the ion is well separated from the surface. We can therefore neglect S_{12}, S_{21} and $|S|^2$ in $\varepsilon_{12}(t)$, $\varepsilon_{21}(t)$ and $\varepsilon_2(t) - \varepsilon_1(t)$. So, as seen from equations 7.4.27 and 7.4.28, the t dependence of $\varepsilon_{12}'(t)$ and $\varepsilon_{21}'(t)$ is neglected. Then we may take $\dot{\varepsilon}_{21}'(t) = 0$ in equation 7.4.24. Further, we may write

$$\varepsilon_{12}(t)\varepsilon_{21}(t) \approx \varepsilon_{12}(0)\varepsilon_{21}(0) \tag{7.4.36}$$

Thus we arrive at the following form for equation 7.4.24

$$\ddot{C}_2' + \left\{\frac{i}{\hbar}[E_1 - E_2 + \varepsilon_1(0) - \varepsilon_2(0) + A_2 t]\right\}\dot{C}_2' + \frac{\varepsilon_{12}(0)\varepsilon_{21}(0)}{\hbar^2} C_2' = 0 \tag{7.4.37}$$

Defining

$$\tau = [E_1 - E_2 + \varepsilon_1(0) - \varepsilon_2(0)]/A_2 + t$$

we obtain

$$\frac{\mathrm{d}^2 C_2'}{\mathrm{d}\tau^2} + \frac{i}{\hbar} A_2 \tau \frac{\mathrm{d}C_2'}{\mathrm{d}\tau} + \frac{\varepsilon_{12}(0)\varepsilon_{21}(0)}{\hbar^2} C_2' = 0 \tag{7.4.38}$$

We may calculate $C'_2(\infty)$ by solving this equation for the initial conditions

$$|C'_1(-\infty)| = 1 \qquad (7.4.39)$$

$$|C'_2(-\infty)| = 0 \qquad (7.4.40)$$

which show that the electron is in state 1 initially (at $t = -\infty$). Then, making use of the Landau–Zener method,[62] we can obtain the neutralization probability ϕ_0

$$\phi_0 = |C'_2(\infty)|^2 = 1 - \exp(-w) \qquad (7.4.41)$$

$$w = \frac{2\pi}{\hbar}\frac{|\varepsilon_{12}(0)\varepsilon_{21}(0)|}{A_2} \qquad (7.4.42)$$

Here we present a more realistic expression for equation 7.4.41. We assume that the ion velocity v takes the form

$$\mathbf{v} = (0, v\sin\theta, v\cos\theta) \qquad (7.4.43)$$

where θ is defined as the angle between the ion beam direction and the surface normal. The calculations of the quantity A_2 yield

$$A_2 = \frac{\partial}{\partial t}[\langle\phi_2|(-W_0)\theta(-z)|\phi_2\rangle - \langle\phi_1|V|\phi_1\rangle|_{t=0}$$
$$\simeq -\left[\frac{W_0}{2a_B} + \frac{2}{V}\frac{k_z^2}{k_z^2+\kappa^2}\cdot\frac{\pi e^2}{\kappa}\right]v\cos\theta \qquad (7.4.44)$$

Further, we obtain

$$|\varepsilon_{12}(0)\varepsilon_{21}(0)| = |h_{12}(0)d_{21}(0)|$$
$$= \frac{2\pi e^2 W_0}{Va_B^4}\frac{1}{\alpha^3}\left|\left\{\frac{1}{\alpha + i\left(\dfrac{mv\cos\theta}{\hbar} - k_z\right)}\left[\frac{1}{\alpha} + \frac{1}{\alpha + i\left(\dfrac{mv\cos\theta}{\hbar} - k_z\right)}\right]\right.\right.$$
$$\left. + \frac{\exp(2i\gamma)}{\alpha + \left(\dfrac{mv}{\hbar}\cos\theta + k_z\right)}\left[\frac{1}{\alpha} + \frac{1}{\alpha + i\left(\dfrac{mv}{\hbar}\cos\theta + k_z\right)}\right]\right\}$$
$$\left.\times\left\{\frac{1}{\alpha - i\left(\dfrac{mv\cos\theta}{\hbar} - k_z\right)} + \frac{\exp(-2i\gamma)}{\alpha - i\left(\dfrac{mv\cos\theta}{\hbar} + k_z\right)} + \frac{\left[\dfrac{2k_z\exp(i\gamma)}{\sqrt{k_z^2+\kappa^2}}\right]}{\alpha + \kappa + i\dfrac{mv}{\hbar}\cos\theta}\right\}\right| \qquad (7.4.45)$$

where $\alpha = \left[a_B^{-2} + k_x^2 + \left(k_y - \dfrac{mv}{\hbar}\sin\theta)^2\right.\right]^{1/2}$

Since equations 7.4.44 and 7.4.45 are too complicated to employ, we have to obtain a much more simplified form. If we take into account the Bohr velocity criterion for the value of the wave number **k**

$$\hbar\mathbf{k} = m\mathbf{v}$$

we can obtain the simpler formula

$$|\varepsilon_{12}(0)\varepsilon_{21}(0)|^2 = 4\pi n W_0 \left[4 + \frac{1}{F} + \frac{3}{F^2} + \frac{8}{W_0 F}\left\{ v^2\cos^2\theta\,\kappa + \frac{1}{F}[(1+2\kappa)v^2\cos^2\theta - W_0] \right\} \right] \left[1 + \frac{1}{F} + \frac{4}{GW_0}v^2\cos^2\theta + \frac{2}{W_0 F} \times [(1+2\kappa)v^2\cos^2\theta - W_0] + \frac{2v^2\cos^2\theta}{W_0^2 FG}\{ [v^2\cos^2\theta(1+2\kappa) - W_0] + [v^2\cos^2\theta + \kappa(1+\kappa)][\kappa - 2(v^2\cos^2\theta - W_0)] \} \right] \tag{7.4.46}$$

where (in atomic units)

$$F = 1 + 4v^2\cos^2\theta$$
$$G = 1 + 2\kappa + 2W_0$$

In the above, we take $(1/V) = n$ to be the electron density. This relation is easily interpreted in a different manner from first principles.[63]

For the determination of n, Kitagawa and Ohtsuki performed the average over **k** in equation 7.4.42. However, it would be more reasonable† to take into account the enhancement factor G for the electron density n in the region of the electron orbit of the ion. The size of electron orbit in the neutralized ion is $\sim a_B^3$. Therefore the electron density n is given by

$$n = \frac{1}{\frac{4}{3}\pi k_F^3}\left(\frac{1}{a_B^3}\right)\cdot\frac{1}{G}n_0 \tag{7.4.47}$$

where n_0 is the average electron density.

Unfortunately, we have no reasonable enhancement factor G near the surface at present. Hardly any predictions of G for low velocities depend on ion velocity. However, if we take a very simple relation (in atomic units) like

$$G = a\exp(-E/8{\cdot}15) \tag{7.4.48}$$

we can obtain a good results which agrees well with experiments. More detailed discussions are given below.

---

†The author is indebted to Y. Matsuda and R. Kawai for valuable discussion about this.

7.4.4. *Other theories and discussions*

The fraction of protons and light ions that are neutralized at the surface has been measured by many authors over a very large range of energies. For H^+, H_2^+, H_3^+ and He^+ in an energy range from several hundred eV to several tens of MeV the neutralization fraction has been studied in detail, especially for metal surfaces. From such measurements the following facts have become clear:

(a) in reflection experiments neutralization is caused at the surface, and the bulk effect is small in comparison;

(b) in transmission experiments the fraction of neutralized protons with energy smaller than a few tens of MeVs is also greatly affected by the surface.

Figure 7.17 shows the experimental results[64] for the fraction of neutralized protons reflected at Be and Mo surfaces† in the energy range of 10–150 keV along with the theoretical results of Kitagawa and Ohtsuki (equation 7.4.46) for an Au surface (30° emergence angle). The experimental points for V, Cu, Nb and Ta are between the points for Be and Mo (or Au). The predicted results of Yablinskii and Trubnikov[57] are also given in the same figure. Zaidins[65] and Brandt and Sizmann[66] also proposed formulae for the neutralized fraction of protons; unfortunately, we cannot find any theoretical derivation of these formulae in their paper at present and therefore cannot give any comments on these formulae.

Kleber and Zwiegel[67] proposed an electron-density enhancement model for proton neutralization. They assumed that the ratio between ionization

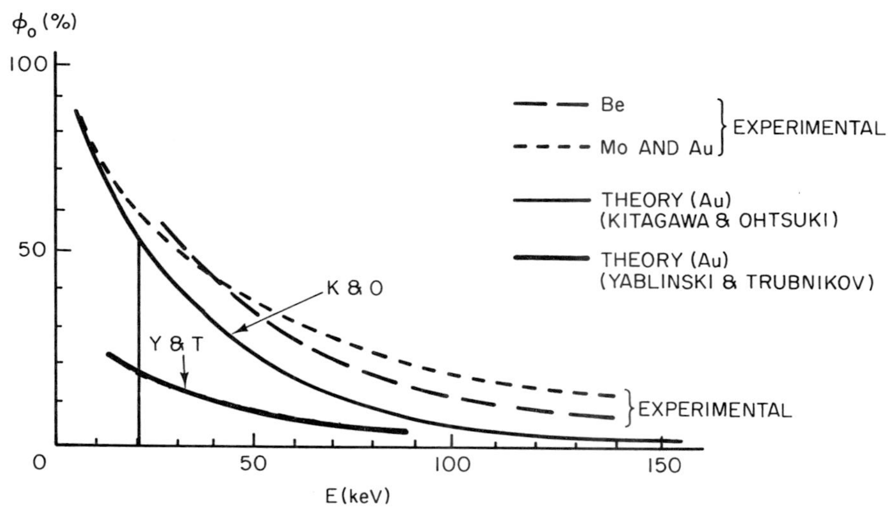

Figure 7.17. Neutral function ϕ_0 due to the resonant process. The theoretical results by Kitagawa and Ohtsuki (K & O) and Yablinskii and Trubnikov (Y & T) are compared with experiments.

† There were no differences between the Mo and Au experimental data.

probability P_{ion} and total excitation $P_{\text{ion}} + P_{\text{ex}}$ equals the ratio between the ion fraction ϕ^+ and the excited state fraction $1 - \phi_0$

$$\frac{P_{\text{ion}}}{P_{\text{ion}} + P_{\text{ex}}} = \frac{\phi^+}{1 - \phi_0} (\simeq f) \tag{7.4.49}$$

In the above, $P_{\text{ion}} + P_{\text{ex}}$ is expressed in terms of the non-ionization probability P_0^{1s}

$$P_{\text{ion}} + P_{\text{ex}} = 1 - P_0^{1s} \tag{7.4.50}$$

where P_0^{1s} is defined by

$$P_0^{1s} = |\langle 1s| \exp[i(m\mathbf{v}\cdot\mathbf{r}/\hbar)] |1s\rangle|^2 \tag{7.4.51}$$

and $\langle 1s|$ is the hydrogen ground state. For small v, P_0^{1s} is calculated simply

$$P_0^{1s} = 1 - \left(\frac{mv}{\hbar} a_{\text{B}}\right)^2 \tag{7.4.52}$$

The ionization probability P_{ion} is obtained easily for small v

$$P_{\text{ion}} \simeq 0{\cdot}28 (mv\, a_{\text{B}}/\hbar)^2 \tag{7.4.53}$$

Zwiegel,[67] however, assumed that the neutralization is governed by the electron cloud which contributes to the screening of the moving ion potential. This enhancement of the electron density is proportional to $1-2(mv/\hbar k_{\text{TF}})^2$ for small v, or

$$\phi_0 = 1 - 2(mv/\hbar k_{\text{TF}})^2 \tag{7.4.54}$$

Inserting equations 7.4.50, 7.4.52 and 7.4.53 into equation 7.4.49, we obtain

$$f = \frac{P_{\text{ion}}}{1 - P_0^{1s}} = 0{\cdot}28$$

From this and equation 7.4.49, we can get

$$\begin{aligned} \phi^+ &= 0{\cdot}28(1 - \phi_0) \\ &= 0{\cdot}28 \times 2 \times \left(\frac{mv}{\hbar k_{\text{TF}}}\right)^2 \end{aligned}$$

Thus, we may conclude that the positive fraction is proportional to the ion energy in Kleber and Zwiegel's model.

As is seen in figure 7.17, the theory of Kitagawa and Ohtsuki agrees well with experiments for protons with energies less than 20 keV. The disagreement between theory and experiment at the high energies comes from the bulk effect, as studied in section 6.2. We should note that the emergence angle was 30° in the experiments, which is large enough for bulk effects to contribute at the high energies. For this energy region the theory of Cross, which was introduced in section 6.2, could be applied, taking into account the surface effect.

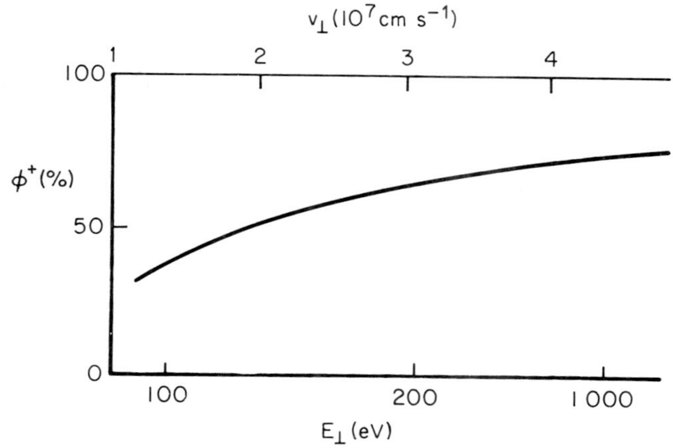

(a)

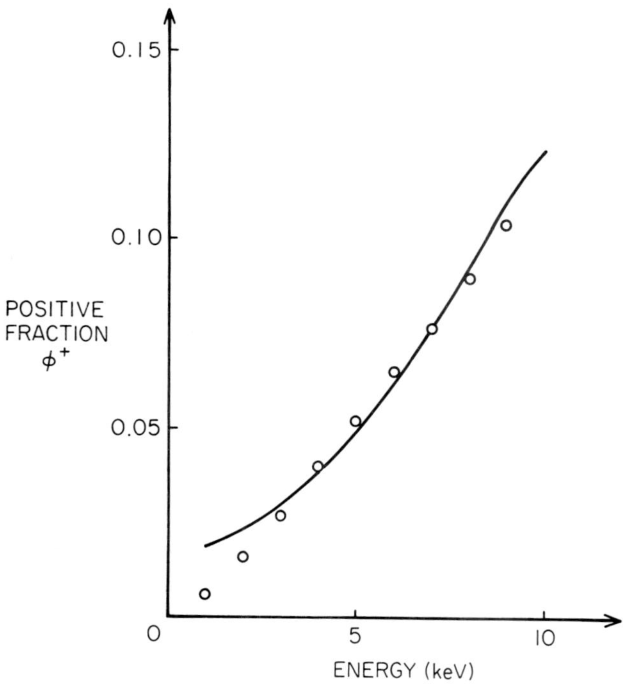

(b)

Figure 7.18. Charged fraction ϕ^+ due to (*a*) the Auger process (Horiguchi *et al.*) in Au, and (*b*) the resonant process (K & O) in Au compared with the experiments of Matschke.[69]

Rausch and Thomas[68] measured the neutral fraction (especially neutralization to excited states) for H^+ impinging on to a Mo surface at a range of energies from a few keV to 200 keV. They normalized the formulae of Kitagawa and Ohtsuki and of Cross to their experimental data at 30 keV, and used the former for $E < 30$ keV and the latter for $E > 30$ keV. Thus, Rausch and Thomas obtained good agreement over a large energy range between their experimental data and the theoretical work of Kitagawa and Ohtsuki, and of Cross.

The Auger process becomes important for proton energies smaller than 1 keV. Figure 7.18 shows the charged fraction (non-neutralized protons) $\phi^+ = 1 - \phi_0$ for Au at energies of a few keV. The reflected charged fraction due to the Auger process, calculated by assuming simple straight line trajectories at the surface, is shown in figure 7.18(*a*).

The total reflected fraction of non-neutralized protons due to Auger and resonant processes ϕ_{tot}^+ is given by

$$\phi_{\text{tot}}^+ = \exp(-w)\exp\left(-\int_{\text{path}} P(s)/v_\perp \, \mathrm{d}s\right)$$

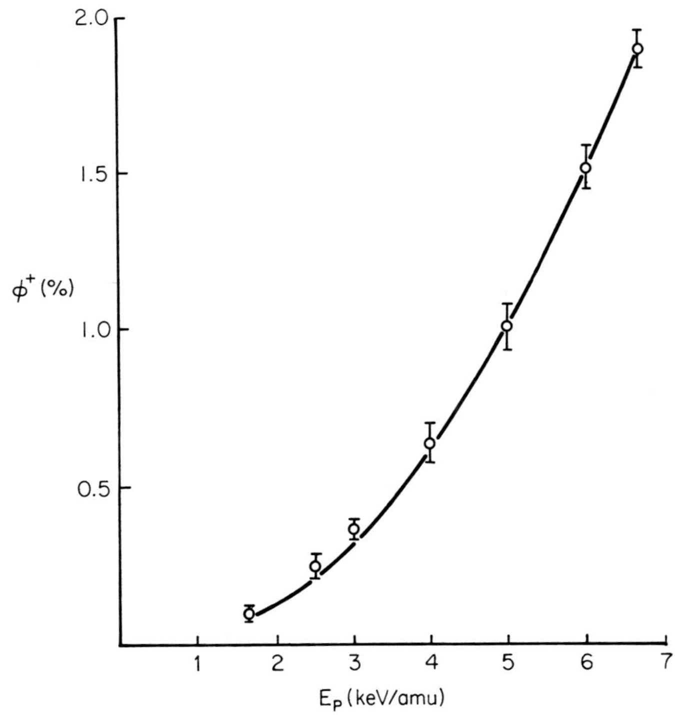

Figure 7.19. Charged fraction ϕ^+ under grazing conditions by Gemmingen and Sizmann.[70] The solid curve was obtained by making use of the uncertainty principle combined with Kitagawa and Ohtsuki's matrix elements for resonant neutralization.

where the factor $\exp(-w)$ may be calculated from the formula of Kitagawa and Ohtsuki (equation 7.7.46), and the second factor, $\exp\left(-\int P(s)/v_\perp \, ds\right)$, by the formula at Horiguchi *et al.* (or figure 7.18(*a*)).

Theoretical results for ϕ^+_{tot} and the experimental ones for $H^+ \rightarrow$ Au and Cu by Matschke[69] are compared in figure 7.18(*b*), where the transverse velocity $v_\perp$ is taken to be $v/\sqrt{2}$ in order to be consistent with the experiments. The calculated values of $\exp(-w)$ from Kitagawa and Ohtsuki (simply from equations 7.4.46 and 7.4.47) are plotted in the same figure, which is derived directly by changing $v_\perp$ from $v \cos 30°$ to $v \cos 45°$ in figure 7.17 and by assuming a fitting value for the enhancement factor G in equation 7.4.47. The calculated values for ϕ^+_{tot} agree well with the experimental data. It should be noted that the Auger process is not so important in this energy region because the contribution of the resonant process $\exp(-w)$ is sufficient to fit the experimental data.

Recently, Gemmingen and Sizmann[70,71] performed very precise measurements under grazing conditions for metals which were prepared very carefully. Figure 7.19 shows the experimental results as a function of the beam energy. The solid line in the figure is the theoretical prediction by Gemmingen and Sizmann based on the uncertainty principle, employing Kitagawa and Ohtsuki's matrix element 7.4.46, and assuming the electron-density enhancement factor to be $G = 5{\cdot}3$ for Ni and Cu.

7.5. Dynamical surface potential

The dynamical surface potential is induced by the surface polarization caused by an ion moving near the surface. For very slow ions the dynamical surface potential is reduced to the static image potential (or the work function), corresponding to the bulk screened potential (see section 6.3). For high velocity ions the dynamical surface potential is represented in a very complex manner i.e., in the form of the dynamical image potential, the surface wake potential and the 'wavelet' potential. It should be noted that the motion of the ion is affected by the dynamical surface potential induced by the ion itself.

The dynamical surface potential can be well understood by the reflection of the bulk plasmon, surface plasmon excitation, surface exciton excitation, and so on. We shall begin with the theory of surface plasmon excitation.

7.5.1. *Surface plasmons*

Consider the closed rectangular surface Γ including the metal surface of the electron gas where there are no free charges (see figure 7.20). The summation

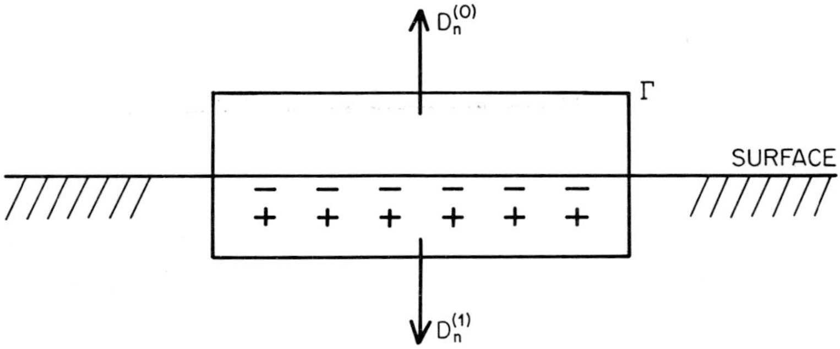

Figure 7.20. Conservation of the electrical displacement at the surface.

of the electrical displacement **D** over the closed surface vanishes

$$\int_{\Gamma} D_n \, dS = 0 \tag{7.5.1}$$

Assuming $D_n = D_n^{(0)}$ in the vacuum and $D_n = D_n^{(1)}$ in the metal, we obtain

$$D_n^{(0)} + D_n^{(1)} = 0$$

Inserting $\mathbf{D}^{(\nu)} = \varepsilon_\nu \mathbf{E}$ we get

$$\varepsilon_0 E_n + \varepsilon_1 E_n = 0$$

where $\varepsilon_\nu (\nu = 0, 1)$ is the dielectric constant of the vacuum ($\varepsilon_0 = 1$) or the metal (ε_1). Accordingly, we obtain

$$\varepsilon_0 + \varepsilon_1 = 0$$

or

$$1 + (1 - \omega_p^2/\omega^2) = 0$$

Therefore

$$\omega = \frac{\omega_p}{\sqrt{2}} \tag{7.5.2}$$

which is the angular frequency of the surface plasmon, first derived by Ritchie.[72,73]

We shall now study the properties of surface plasmons in more detail. Considering the electric field near the surface, we take the surface-normal component (z component, for $z > 0$ representing the solid region) of the electric field to be in the form

$$E_z = \exp[i(K_x x - K_z z) - i\omega t] \tag{7.5.3}$$

where the x axis is parallel to the surface. Inserting equation 7.5.3 into the wave equation we get the relation

$$K_x^2 + K_z^2 = \frac{\varepsilon}{c^2}\omega^2$$

or

$$K_z = [\varepsilon(\omega/c)^2 - K_x^2]^{1/2} \qquad (7.5.4)$$

corresponding to $\varepsilon = \varepsilon_1$ or ε_0, $K_z = K_{z1}$ or $K_z = K_{z0}$.

Now, for $\varepsilon_0(\omega/c) < K_x$, or $\omega/K_x = v_{\text{phase}} < c, K_{z0}$ becomes imaginary, say $i\kappa$, and therefore the plane wave 7.5.3 takes the form

$$E_z = \exp[i(K_x x - \omega t)]\exp(\kappa z)$$

which dampens out for large $|z|, z \to -\infty$. Therefore, we can conclude that the surface plasmon field is localized within the surface region of κ^{-1}. However, for $v_{\text{phase}} > c, K_{z1}$ is real, and the electric field extends to infinity. We call the localized surface plasmon 'non-radiative', and the non-localized surface plasmon 'radiative'. The non-radiative surface plasmon corresponds to the surface plasmon mode $\omega_p/\sqrt{2}$ introduced by Ritchie. The radiative surface plasmon was studied in detail by Kliewer and Fuchs[74] and others (e.g., reference 75).

Inserting equation 7.5.3 into Maxwell's equations, and taking into account the continuity condition for the field **D** at the surface (equation 7.5.1) of the metal slab of thickness d, we obtain†

$$(\varepsilon_1 K_{z0} + \varepsilon_0 K_{z1}) + (\varepsilon_1 K_{z0} - \varepsilon_0 K_{z1})\exp(iK_{z1}d) = 0$$

or

$$\varepsilon_1 K_{z0} - iK_{z1}\tan(K_{z1}d/2) = 0 \qquad (7.5.5)$$

If we consider the case of $K_x \ll \omega_c$, we get the dispersion relation from equation 7.5.5

$$\omega = \omega_p\left[1 + \frac{1}{8}(K_p d)^2\frac{K_x^2}{K_p^2}\right] \qquad \text{for } K_p d/2 \ll 1$$

$$\omega = \omega_p\left(1 + \frac{K_x^2}{2K_p^2}\right) \qquad \text{for } K_p d/2 \gg 1$$

where $K_p = \omega_p/c$. We note that $\omega = \omega_p$ for $K_x \to 0$.

The detailed profile of the dispersion relation of the surface plasmon for the semi-infinite solid is shown in figure 7.21. For small K_x the radiative surface plasmon is clearly divided into the plasmon branch and the photon branch. However, for large K_x the photon branch and the plasmon branch couple strongly and produce a new radiative plasmon. As stated before, the non-radiative surface plasmon is in the region $\omega < cK_x$ (the shadowed region in

†This is for p polarization. The s-polarization wave is not so important.

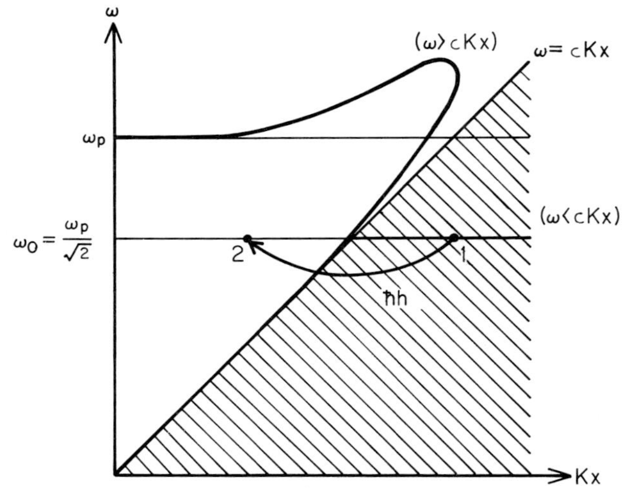

Figure 7.21. Detailed profile of the dispersion relation of the surface plasmon for the semi-infinite solid. The shadowed region corresponding to $\omega < cK_x$ includes the dispersion for the non-radiative surface plasmon. The dispersion for the radiative surface plasmon is in the region $\omega > cK_x$.

figure 7.21), and the radiative surface plasmon is in the region $\omega > c\,K_x$ (the white region in figure 7.21). There are a few cases where a new momentum transfer appears in the collision between the photon and the plasmon. For example, the roughness of the surface and the periodic atomic structure produce an extra momentum transfer $\hbar \mathbf{h}$ in the photon surface–plasmon collision, which is shown in figure 7.21 by an arrow $1 \rightarrow 2$. By such a momentum transfer, position 1 in the region $\omega < c\,K_x$ could be changed to position 2 in the region $\omega > c\,K_x$. Accordingly, the non-radiative surface plasmon is changed to the radiative. The non-radiative surface plasmon excited by the ion beam does not decay directly into the photon, but for surfaces with a periodic roughness the excited non-radiative surface plasmon emits radiation of frequency $\omega_p/\sqrt{2}$.

The quantization or the second quantization is performed in the same way as for the bulk plasmon. The interaction Hamiltonian between the ion beam and the surface plasmon will be studied in a later section.

7.5.2. *Derivation of the dynamical surface potential*

The general formula for the dynamical surface potential due to the surface plasmon was derived by Suzuki *et al.*[76] for an ion moving with arbitrary velocity relative to the surface. Prior to their theory the dynamical image potential was derived by many authors[77–85] for special cases. Ritchie,[79] especially, showed that the image potential could be attributed to the excitation of surface plasmons, and Muscat and Newns[86] (see also reference 87) calculated the dynamical image potential for an electron moving parallel to the surface. Here we shall study the method of Suzuki *et al.*[76] and derive the general form of the dynamical surface potential, including the dynamical image potential, the surface wake potential and the 'wavelet' potential.

The potential $\varphi_{\text{eff}}(\mathbf{r}_1, t_1)$ is induced at position $\mathbf{r}_1$ and time t_1 by the external charge distribution $\rho_{\text{ext}}(\mathbf{r}_2, t_2)$ at $\mathbf{r}_2$ and t_2. According to the linear response theory the induced potential φ_{eff} is represented in the form of a linear combination of ρ_{ext}

$$\varphi_{\text{eff}}(\mathbf{r}_1, t_1) = \int \mathrm{d}\mathbf{r}_2 \int \mathrm{d}t_2\, W(\mathbf{r}_1, \mathbf{r}_2; t_1 - t_2)\, \rho_{\text{ext}}(\mathbf{r}_2, t_2) \tag{7.5.6}$$

where $W(\mathbf{r}_1, \mathbf{r}_2; t_1 - t_2)$ is the response function.

In order to find the response function W we take into account multiple polarization. It is helpful to consider the process with a diagram (see figure 7.22). The single polarization can be represented by the single loop in the diagram. With no polarization the potential is simply the Coulomb interaction v, which is represented by the wavy line. The response function W due to multiple polarization corresponds to the double wavy line. Therefore the function W becomes the sum of the single, double and multiple polarizations, as shown in figure 7.22(*a*), which is reduced to the form of figure 7.22(*b*). The marks 1, 2, and so on, in the figure refer to position $\mathbf{r}_1$ at time t_1, $\mathbf{r}_2$ at time t_2, and so on.

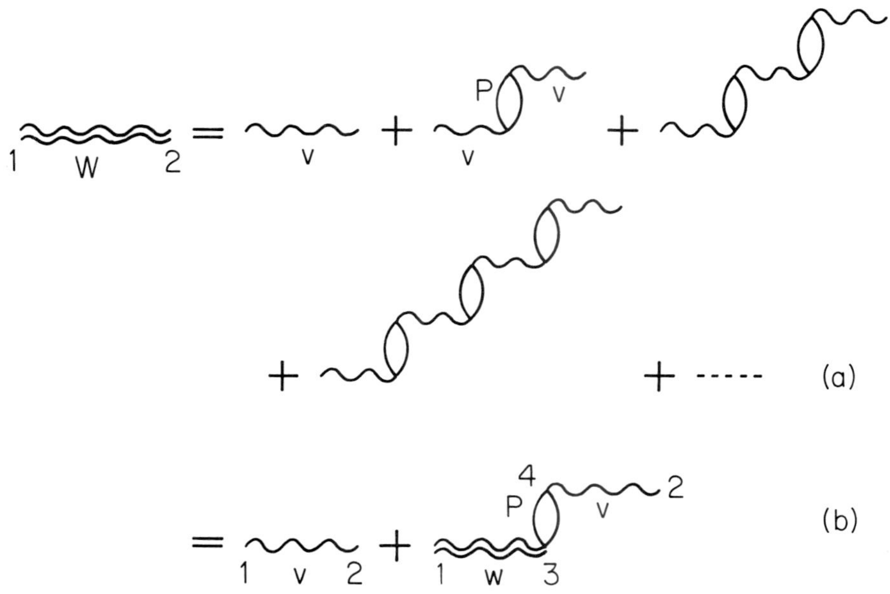

Figure 7.22. Diagram for calculating the response function W. A single loop corresponds to a single polarization, a single wavy line to the Coulomb potential, and a double wavy line to the response function.

We may represent diagram (*b*) by the integral equation

$$W(1,2) = v(1,2) + \int W(1,3)\,P(3,4)\,v(4,2)\,d(3,4) \tag{7.5.7}$$

where the Coulomb potential $v(1,2)$ takes the form

$$v(1,2) = \frac{-e}{|\mathbf{r}_1 - \mathbf{r}_2|}\delta(t_1 - t_2) \tag{7.5.8}$$

The function $P(1,2)$ is the so called polarization function,[88] whose Fourier transform in momentum and angular frequency is represented by

$$P(\mathbf{q}_1, \mathbf{q}_2; \omega) = -\frac{\mathbf{q}_1\mathbf{q}_2}{m\omega^2} n(q_1 + q_2) \tag{7.5.9}$$

where $n(q)$ is the Fourier component of the electron density.

In order to take into account the surface effect we assume a step function for the change in electron density, i.e.

$$n(\mathbf{r}) = n_0\,\theta(z)$$

where

$$\theta(z) = 1 \qquad \text{inside the solid, } z > 0$$

$$\theta(z) = 0 \qquad \text{outside the solid, } z < 0$$

It is not so difficult to solve the integral equation 7.5.7. If we insert equations 7.5.8 and 7.5.9 into equation 7.5.7, we obtain[89]

$$\begin{aligned} W(\mathbf{r}_1, \mathbf{r}_2; t_1 - t_2) &= \frac{1}{[(\boldsymbol{\rho}_1 - \boldsymbol{\rho}_2)^2 + (z_1 - z_2)^2]^{1/2}}\delta(t_1 - t_2) \\ &- \frac{1}{[(\boldsymbol{\rho}_1 - \boldsymbol{\rho}_2)^2 + (z_1 - z_2)^2]^{1/2}}\omega_p \sin[\omega_p(t_1 - t_2)]\theta(t_1 - t_2)\theta(z_1) \\ &+ \frac{1}{[(\boldsymbol{\rho}_1 - \boldsymbol{\rho}_2)^2 + (|z_1| + |z_2|)^2]^{1/2}}\omega_p \sin[\omega_p(t_1 - t_2)]\theta(t_1 - t_2)\theta(z_1) \\ &- \frac{1}{[(\boldsymbol{\rho}_1 - \boldsymbol{\rho}_2)^2 + (|z_1| + |z_2|)^2]^{1/2}}\omega_s \sin[\omega_s(t_1 - t_2)]\theta(t_1 - \theta_2) \end{aligned} \tag{7.5.10}$$

where we have assumed that $\mathbf{r} = (\boldsymbol{\rho}, z)$.

In the above equation, the first term is the ordinary Coulomb interaction, the second is the interaction due to bulk plasmon excitation, and the third and fourth are attributed to surface interactions. The third is the interaction due to bulk plasmon reflection at the surface, and the fourth is the interaction due to surface plasmon excitation with energy $\hbar\omega_s$. Inserting equation 7.5.10 into equation 7.5.6 we can obtain† the dynamical surface potential φ_{eff}.

† Hereafter we neglect the dispersion relation of the plasmon. If we take into account the relation, the potential $u_s^{\pm}$ is extended outwards from the surface, but this shifting is small compared with the 'surface region'.

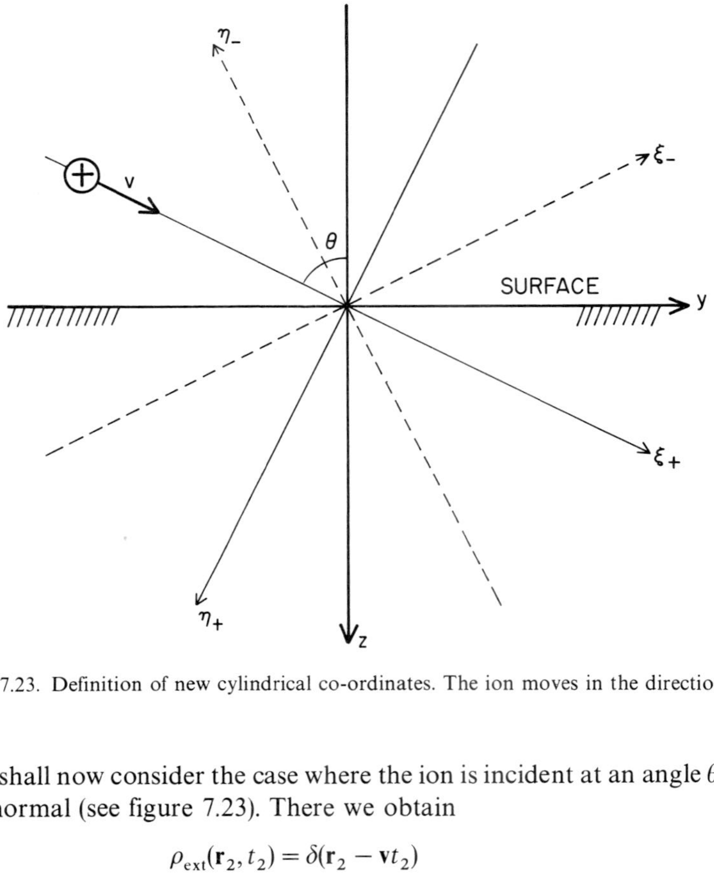

Figure 7.23. Definition of new cylindrical co-ordinates. The ion moves in the direction ξ_+.

We shall now consider the case where the ion is incident at an angle θ to the surface normal (see figure 7.23). There we obtain

$$\rho_{\text{ext}}(\mathbf{r}_2, t_2) = \delta(\mathbf{r}_2 - \mathbf{v}t_2)$$
$$\mathbf{v} = (0, v \sin\theta, v\cos\theta)$$

In this case it is convenient to use cylindrical co-ordinates, (ξ_+, ρ_+) and (ξ_-, ρ_-), defined by

$$\begin{aligned} \xi_+ &= -z\cos\theta + y\sin\theta \\ \rho_+^2 &= x^2 + \eta_+^2 \end{aligned} \tag{7.5.11}$$

where $\eta_+ = z\sin\theta - y\cos\theta$, and

$$\begin{aligned} \xi_- &= -z\cos\theta + y\sin\theta \\ \rho_-^2 &= x^2 + \eta_-^2 \end{aligned} \tag{7.5.12}$$

where $\eta_- = -z\sin\theta - y\cos\theta$. By making use of these co-ordinates we may obtain the dynamical surface potential φ_{eff} as a function of the $(\xi_\pm, \rho_\pm)$. Since our expression is very complicated we divide the solution into two parts. One is the surface potential induced by the ion before impinging ($t_2 < 0$) on the surface, the other after impinging ($t_2 > 0$).

Surface potential induced before impinging

In this case the dynamical surface potential may be represented by

$$\varphi_{\text{eff}} = v - u_s^- - w_s^- \theta(vt - \xi_-) \tag{7.5.13}$$

for $z < \theta$ (outside), and

$$\varphi_{\text{eff}} = v - u_s^+ - w_s^+ \theta(vt - \xi_+) \tag{7.5.14}$$

for $z > 0$ (inside).
We drop the subscript 1 in the co-ordinate $(\xi_\pm, t)$ for simplicity and define $u_s^\pm$ and $w_s^\pm$ as

$$u_s^\pm = \int_0^{k_c} \mathrm{d}k \frac{(\omega_s/v)^2}{k^2 + (\omega_s/v)^2} J_0(k\rho_\pm) \exp(-k|\xi_\pm - vt|) \tag{7.5.15}$$

$$w_s^\pm = 2\left(\frac{\omega_s}{v}\right) \sin \omega_s \left(t - \frac{\xi_\pm}{v}\right) \int_0^{k_c} \mathrm{d}k \frac{k}{k^2 + (\omega_s/v)^2} J_0(k\rho_\pm) \tag{7.5.16}$$

It is worth noting that only the surface plasmons contribute to φ_{eff}, since the ion is outside the solid and cannot excite bulk plasmons.

The term $u_s^\pm$ is the screened potential due to the surface plasmon. u_s^- is simply the ordinary dynamical image potential. Recently, Schmeits and Lucas[90,91] derived the 'self-image potential', which is identical to u_s^- if $\theta = 90°$. The term $w_s^\pm$ is the wake potential caused by the surface plasmon, so we call this the 'surface wake potential'. The factor $\theta(vt - \xi_\pm)$ means that the surface wake potential is defined only for the region behind the ion position (or its image position) vt.

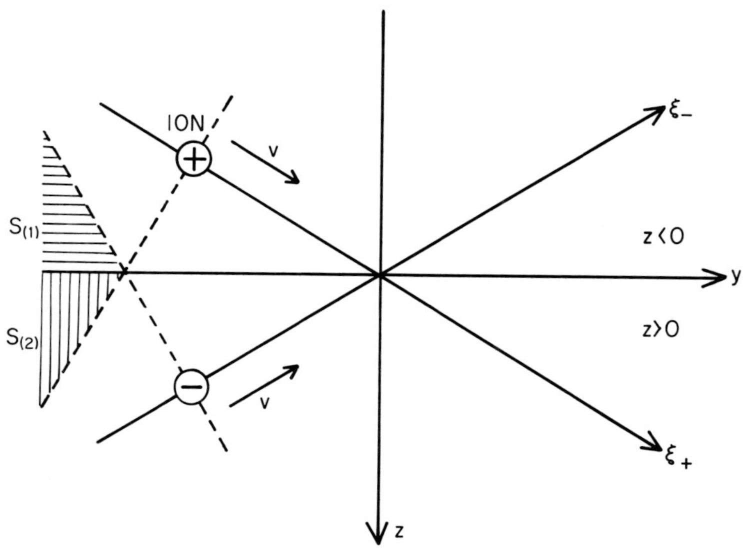

Figure 7.24. Region where the wake potential is defined.

For example, the wake potential in equation 7.5.13 is defined by the shadowed region $S_{(1)}$ of figure 7.24, which is $z<0$, $\xi_- - vt<0$. Therefore, excluding parallel motion, we may conclude that the ion motion could not be affected by the surface wake potential induced by the ion itself since the ion is not in the shadowed region. According to equation 7.5.14, for $z>0$ the surface wake potential is created in the shadowed region $S_{(2)}$ in figure 7.24. As a result we obtain the region $S_{(1)}+S_{(2)}$ where the surface wake potential is induced.

What is the dynamical image potential u_s^+? This is interpreted as follows. The free electrons are attracted to the surface by the external ion in $z<0$, from which the number of free electrons at a random point in $z>0$ are reduced. The test charge in $z>0$ is attracted toward the surface, as if the negative (image of image) ion is in the position of the external ion. This is simply u_s^+.

The self-potential arises from the fact that the ion is affected by the dynamical surface potential induced by the ion itself. The self-potential is represented by

$$\varphi_{\text{eff}}^{\text{self}} = v - u_s^- \tag{7.5.17}$$

for $\theta \neq 90°$, and

$$\varphi_{\text{eff}}^{\text{self}} = v - u_s^- - w_s^- \theta(vt - \xi_-) \tag{7.5.18}$$

for $\theta = 90°$.

We must, however, be careful: the dynamical image potential u_s^- is not the ordinary image potential ϕ_s introduced in electrodynamics. ϕ_s is defined as the work done to separate the ion a distance Δz from the surface, against the force F_s

$$\phi_s = F_s \Delta z$$

However, if we separate the ion a distance Δz from the surface, then the image charge is also separated Δz from the surface. This fact is included automatically in our expression of φ_{eff}. The work done during the separation of $2\Delta z$ is u_s^-

$$eu_s^- = F_s \cdot 2\Delta z$$

Therefore we obtain

$$\phi_s = \tfrac{1}{2} u_s^-$$

and

$$F_s = -\frac{\partial \phi_s}{\partial z} = -\frac{1}{2}\frac{\partial u_s^-}{\partial z} \tag{7.5.19}$$

Surface potential induced after impinging

Next we have to obtain the dynamical surface potential when the ion is moving inside the solid after impinging on the surface. In this case, besides the image part u_s and the wake part w_s, we get additional terms

$$\varphi_{\text{eff}} = v - u_s^+ - w_s^+ \theta(vt - \xi_+)\theta(\xi_+) - w_s^- \theta(-\xi_-) - x_s^- + x_s^+ \tag{7.5.20}$$

for $z<0$ (outside) and

$$\varphi_{\text{eff}} = (v - u_p^+) - (u_s^- - u_p^-) - w_p^+\theta(vt-\xi_+)\theta(\xi_+) - (w_s^- - w_p^-)\theta(vt-\xi_-)\theta(\xi_-) - w_s^+\theta(-\xi_+) + x_p^+ - x_p^- - x_s^+ + x_s^- \quad (7.5.21)$$

$z>0$ (inside) where u_p^+ and w_p^+ are expressions obtained from u_s^+ and w_s^+ by replacing ω_s with ω_p. The new potential term χ is defined by†

$$\chi_{s,p}^{\pm} = \cos(\omega_{s,p}t)\int_0^{k_c}\frac{(\omega_{s,p}/v)^2}{k^2+(\omega_{s,p}/v)^2}J_0(k\rho_\pm)\exp(-k|\xi_\pm|)\,dk + \operatorname{sgn}(\xi_\pm)\left(\frac{\omega_{s,p}}{v}\right)\sin(\omega_{s,p}t)\int_0^{k_c}\frac{k\,dk}{k^2+(\omega_{s,p}/v)^2}J_0(k\rho_\pm)\exp(-k|\xi_\pm|) \quad (7.5.22)$$

For expression 7.5.20, when $z<0$ the dynamical image potential u_s^+ is the same as 7.5.14, and formed by the image of the image (negative image charge at the external ion position). In other words, from the viewpoint of the test charge in $z<0$ the image charge moves in the direction ξ_-, is reflected at the surface, and then moves in the direction ξ_+ with the external ion. Further, we get the surface wake w_s^+ behind the ion. The region of w_s^+ is the shadowed region $S_{(1)}$ in figure 7.25. However, we should note that the wake ω_s^- formed in the shadowed region $S_{(2)}$ of $\xi_-<0$ before impinging (i.e., $t<0$) does not

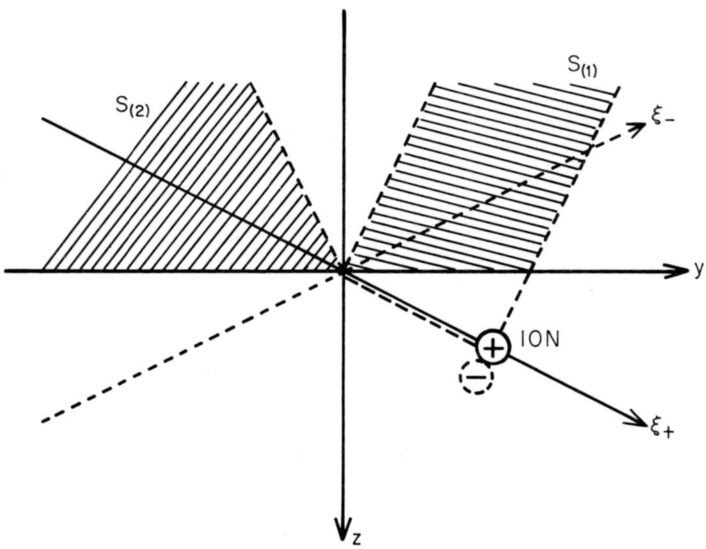

Figure 7.25. Region where the wake potential is defined.

†The expressions given in equation 7.5.15 are the same as those obtained by Suzuki *et al.*,[76] but the expression of the first term in equation 2.13(*b*) of their paper, $J_0(k\rho_\pm)\exp(-k|vt\mp\xi_\pm|)$ should be corrected to $J_0(k\rho_\pm)\exp(-k|\xi_\pm - vt|)$, and the second term, $\theta(\xi)$, should be $\theta(\pm\xi)$.

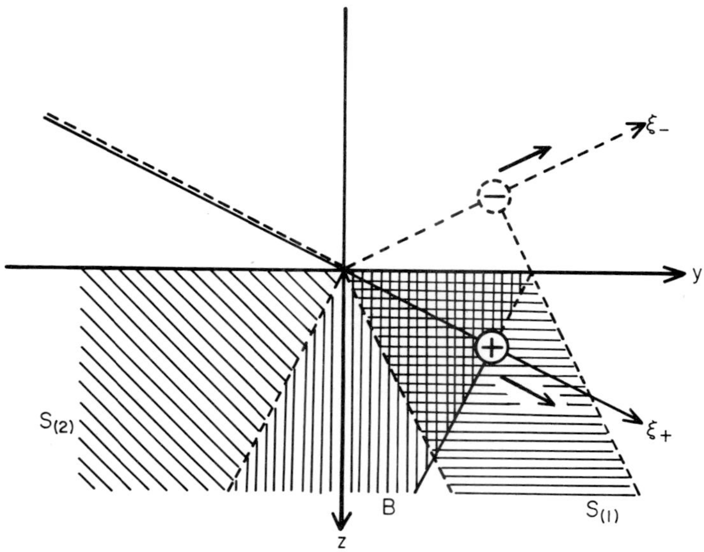

Figure 7.26. Region where the wake potential is defined.

disappear for a while because of $\theta(-\xi_-)$. Another point to be noted is that the term $\chi_s^{\pm}$ depends on the time t after impinging, and does not depend on the ion's position. This means that the potential is excited in the surface only for $t=0$. So we can call this the 'wavelet potential'[†].

Next, let us consider equation 7.5.21 for $z>0$. The term $v-u_p^+$ represents the decrease in the Coulomb potential by the screening of the bulk plasmons, and $(u_s^- - u_p^-)$ is the decrease in the image potential due to the reflected bulk plasmons. Further, we have the wake potential due to the bulk plasmon w_p^+, the reflected bulk plasmon w_p^- and the surface plasmon w_s^-. Figure 7.26 shows the potential in the region $z>0$. The image ion is reflected at the surface and moves in the ξ_- direction. The wake potential due to the image is in region $S_{(1)}$, corresponding to $-(w_s^- - w_p^-)\theta(vt-\xi_-)\theta(\xi_-)$ in equation 7.5.21. The region $S_{(2)}$ is the potential region $-w_s^+\theta(-\xi_+)$ formed for $t<0$. In the same figure we show the ordinary wake potential, region B, due to the bulk plasmon. The external ion is in the regions B and $S_{(1)}$. So, the ion is decelerated, and is forced out by the weakened image force caused by the reflected plasmon.

Next, we consider the case when the ion moving in the solid is ejected from the surface. We should note that the situation is not related directly to above cases by changing $v \to -v$.

[†]The wavelet introduced here does not propagate since we neglected the dispersion of the surface plasmon.

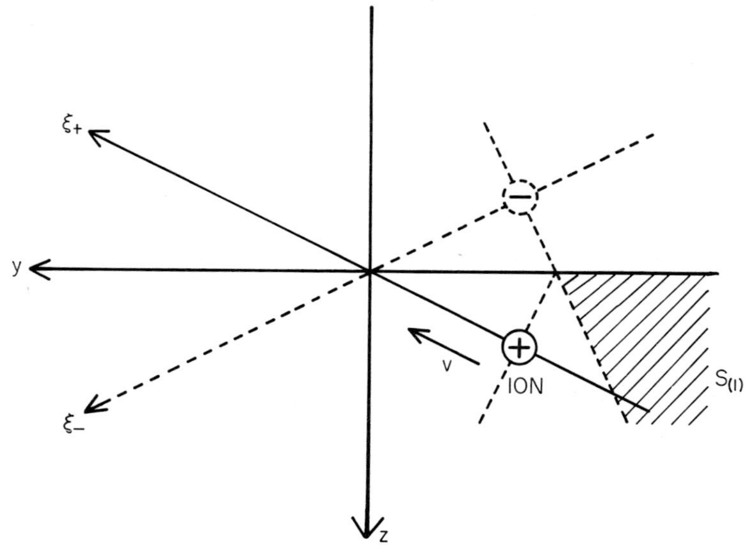

Figure 7.27. Region where the wake potential is defined.

Surface potential induced before ejection (t < 0)

Taking the axis ξ_+, ξ_- as shown in figure 7.27, for $t < 0$ we obtain

$$\varphi_{\text{eff}} = (v - u_s^+) - w_s^+ \,\theta(vt - \xi_+) \tag{7.5.23}$$

for $z < 0$, and

$$\varphi_{\text{eff}} = (v - u_p^+) - (u_s^- - u_p^-) - w_p^+ \,\theta(vt - \xi_+) - (w_s^- - w_p^-)\,\theta(vt - \xi_-) \tag{7.5.24}$$

for $z > 0$.

Surface potential induced after ejection (t > 0)

Taking the same axis in figure 7.27 again, for $t > 0$ we obtain

$$\varphi_{\text{eff}} = v - u_s^- - w_s^- \,\theta(vt - \xi_-)\theta(\xi_-) - w_s^+ \,\theta(-\xi_+) - \chi_s^+ + \chi_s^- \tag{7.5.25}$$

for $z < 0$, and

$$\varphi_{\text{eff}} = v - u_s^+ - w_s^+ \,\theta(vt - \xi_+)\,\theta(\xi_+) - w_p^+(-\xi_+) - (w_s^- - w_p^-)\,\theta(-\xi_-)$$
$$- \chi_p^+ + \chi_p^- - \chi_s^- + \chi_s^+ \tag{7.5.26}$$

for $z > 0$.

Here we compare equation 7.5.13 with equation 7.5.24. The image potential parts are exactly the same. There is no residual wake potential in equation 7.5.13, which is replaced by equation 7.5.25. Therefore the wake potential region is changed to $S_{(1)}$ indicated in figure 7.27. So, the decelerating force 'after ejection' differs from that 'after impinging'.

Finally, we shall summarize the special case when the ion moves parallel to the surface at a distance s from the surface. We obtain

$$\begin{aligned}\varphi_{\text{eff}} &= v - u_s^- - w_s^-\,\theta(vt-y) - [u_p^+ + w_p^+\,\theta(vt-y) \\ &\quad - u_p^- - w_p^-\,\theta(vt-y)]\,\theta(z)\,\theta(s)\end{aligned} \tag{7.5.27}$$

where

$$u_p^+ = \int \mathrm{d}k \frac{(\omega_p/v)^2}{k^2+(\omega_p/v)^2} J_0\left[k\sqrt{x^2+(z-b)^2}\right] \exp(-k|y-vt|)$$

$$w_p^+ = 2\frac{\omega_p}{v}\sin\left[\omega_p\left(t-\frac{y}{v}\right)\right] \int \mathrm{d}k \frac{k}{k^2+(\omega_p/v)^2} J_0\left[k\sqrt{x^2+(z-b)^2}\right]$$

$$u_{s,p}^- = \int \mathrm{d}k \frac{(\omega_{s,p}/v)^2}{k^2+(\omega_{s,p}/v)^2} J_0\left[k\sqrt{x^2+(|z|+|b|)^2}\right] \exp(-k|y-vt|)$$

$$w_{s,p}^- = 2\frac{\omega_{s,p}}{v}\left[\sin\omega_{s,p}\left(t-\frac{y}{v}\right)\right] \int \mathrm{d}k \frac{k}{k^2+(\omega_{s,p}/v)^2} J_0\left[k\sqrt{x^2+(|z|+|b|)^2}\right] \tag{7.5.28}$$

In the next section we shall study the special effect of dynamical surface potential on ion motion near the surface.

7.6. Skipping motion, surface energy loss and photon emission

7.6.1. *Skipping motion*

The possibility of a skipping motion of ions incident at a grazing angle to the surface has been proposed by Ohtsuki,[92] and confirmed by computer simulation by Ohtsuki *et al.*[93] The ions scattered by the first and second layers

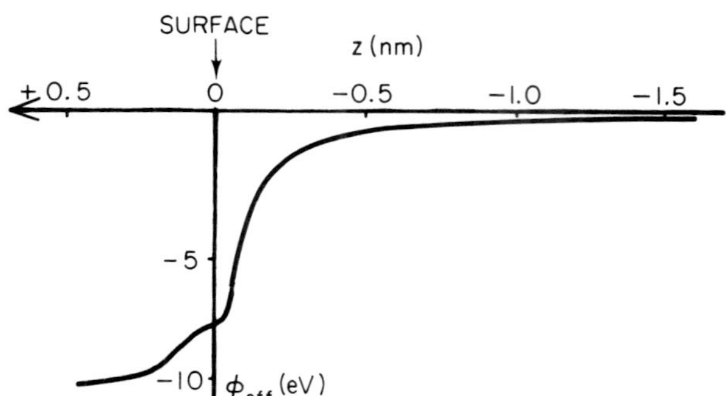

Figure 7.28. Dynamical surface potential as a function of the distance from the surface for 30 keV H^+ at Ni with a grazing angle $\sim 1°$.

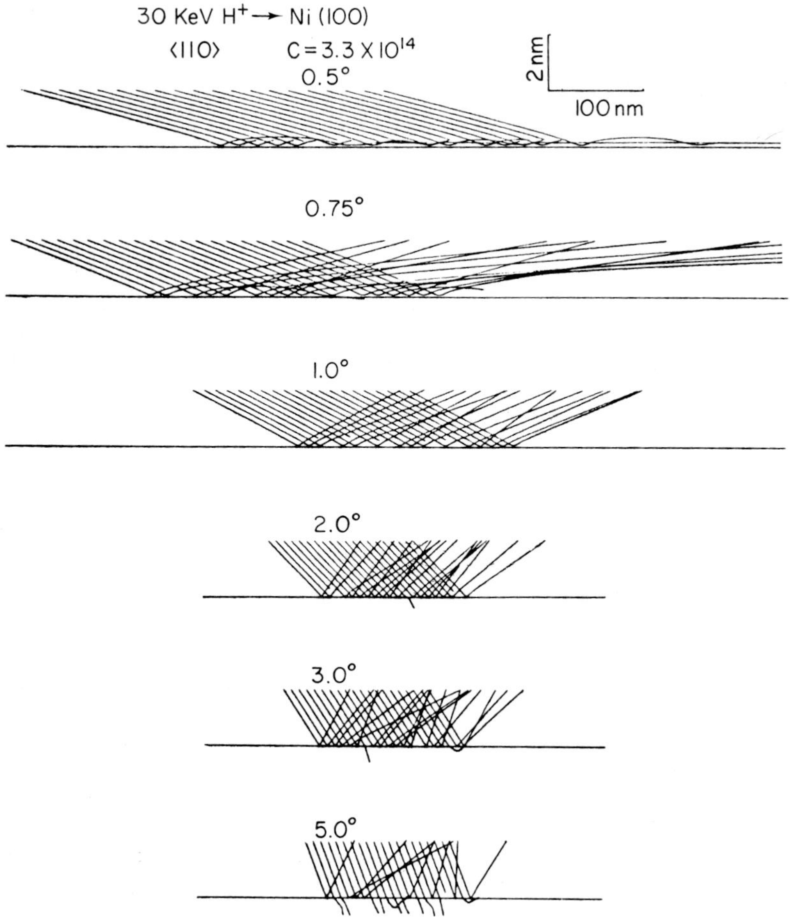

Figure 7.29. Computer simulated trajectories of ions, taking into account the dynamical surface potential and neutralization for 30 keV H^+ near a Ni (100) surface. Skipping motions can be seen for small incident angles.

are forcibly drawn further inward by the dynamical surface potential. For example, figure 7.28 shows the dynamical surface potential for 30 keV H^+ near a Ni surface with an emergence angle θ_i equal to 1°. The potential agrees with the ordinary static image for large $|z|$ for $z < 0$.

Figure 7.29 shows the results of the computer simulated trajectories of ions, taking account of the dynamical surface potential and surface neutralization for 30 keV ions near the Ni (100) surface as the incident angle θ_i changes from 0·5° to 5·0°. We can see one, two, three, and so on, skips ranging in length from 10 nm to 100 nm for ions with $\theta_i = 0{\cdot}5°$. The trajectories sometimes disappear due to neutralization. For an ion with $\theta_i = 0{\cdot}75°$, we find a very long skipping motion whose period is greater than 100 nm. However, we cannot see the skipping motion for ions with large incident angles. For those ions under the skipping motion, the specular reflection angle or the surface channelling angle deviate greatly, and the emergence angle α (the angle between the ejected ion direction and the surface) is not θ_i but $\sim\frac{1}{2}\theta_i$.

7.6.2. *Surface energy loss*

It should be noted that the force corresponding to the dynamical surface potential is conservative, because it is caused by the *virtual* excitation of plasmons. But we have to note that there are other forces due to *real* (bulk and surface) plasmon excitation, which cause energy loss. The energy loss of the ion in a solid is mainly attributed to single-electron excitation in the atom (Chapter 4), where the plasmon excitation makes a very small contribution. However, the situation is very much changed for the energy loss of ions near a surface. The problem was first studied by Ritchie,[94] and more recently by Lucas.[91] Here we start with the function $W(\mathbf{r}_1, \mathbf{r}_2; t_1 - t_2)$ introduced in section 7.5, since it is well known[88] that the energy loss can be obtained from the imaginary part of the propagation function W. The imaginary part of W is derived easily from equation 7.5.10 to be in the form[95]

$$W^{\mathrm{Im}}(\mathbf{r}_1, \mathbf{r}_2; t_1 - t_2)$$
$$= -\left[\frac{\omega_p \sin \omega_p (t_1 - t_2)}{\sqrt{(\boldsymbol{\rho}_1 - \boldsymbol{\rho}_2)^2 + (z_1 + z_2)^2}} - \frac{\omega_p \sin \omega_p (t_1 - t_2)}{\sqrt{(\boldsymbol{\rho}_1 - \boldsymbol{\rho}_2)^2 + (|z_1| + |z_2|)^2}}\right]\theta(z_1)$$
$$- \frac{\omega_s \sin \omega_s (t_1 - t_2)}{\sqrt{(\boldsymbol{\rho}_1 - \boldsymbol{\rho}_2)^2 + (|z_1| + |z_2|)^2}} \tag{7.6.1}$$

Inserting equation 7.6.1 into equation 7.5.7 we obtain the potential $\varphi_{\mathrm{eff}}(\mathbf{r}_1, t_1)$. However, we cannot get the force (a friction force) from the derivative of $\varphi_{\mathrm{eff}}(\mathbf{r}_1, t_1)$, since the force is not conservative. First we have to derive the electric field E from W^{Im}, and then we can obtain the friction force F_{eff} by

$$F_{\mathrm{eff}} = Z_1 e E$$

which gives the stopping power

$$\frac{\mathrm{d}W}{\mathrm{d}\xi_+} = F_{\mathrm{eff}} \tag{7.6.2}$$

From equation 7.6.2 we can find a simple† expression for E for an ion moving

† We neglect the dispersion of the plasmon, which does not give effects important in the calculation of the stopping power.

in the y direction parallel to the surface and at a distance s from it

$$-\frac{dE}{dy}=(Z_1e)^2\frac{\omega_p^2}{v^2}\log\left[\frac{k_c+q_0}{(\omega_p/v)}\right]\theta(s)$$

$$-\frac{(Z_1e)^2\,\omega_p^2}{v^2}\theta(s)\int_0^{k_c(p)}\frac{1}{\sqrt{k^2+(\omega_p/v)^2}}\exp\left[-2|s|\sqrt{k^2+(\omega_p/v)^2}\right]dk$$

$$+\frac{(Z_1e)^2\,\omega_s^2}{v^2}\int_0^{k_c(s)}\frac{1}{\sqrt{k^2+(\omega_s/v)^2}}\exp\left[-2|s|\sqrt{k^2+(\omega_s/v)^2}\right]dk \qquad (7.6.3)$$

where $k_c(\text{p or s})=\sqrt{q_0^2-(\omega_{p,s}/v)^2}$, q_0 being the cut-off wave number. The first term is the well-known stopping power due to the bulk plasmon excitation. The second term is the extra term coming from the surface effect which has an effect on $\frac{dW}{dy}$ opposite to that of first term. The third term is the stopping power due to surface plasmon excitation. Outside the solid ($s<0$), only the third term contributes to the energy loss. The third term is an even function of s and localized within a narrow region of s. If we make $q_0\to\infty$, the third term of equation 7.6.3 can be integrated easily, $-\frac{dW}{dy}=(Z_1e)^2(\omega_s/v)^2K_0(2|s|\omega_s/v)$, which coincides with the expression derived by Núñez *et al.*[87] For example, figure 7.30 shows the surface plasmon stopping power (third term of equation 7.6.3) for

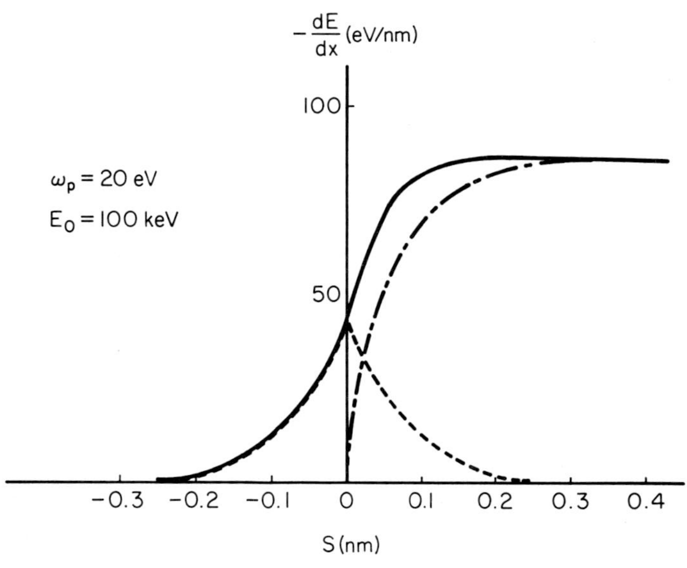

Figure 7.30. Surface stopping power as a function of the distance from the surface for 100 keV H^+ near a Ni (100) surface.

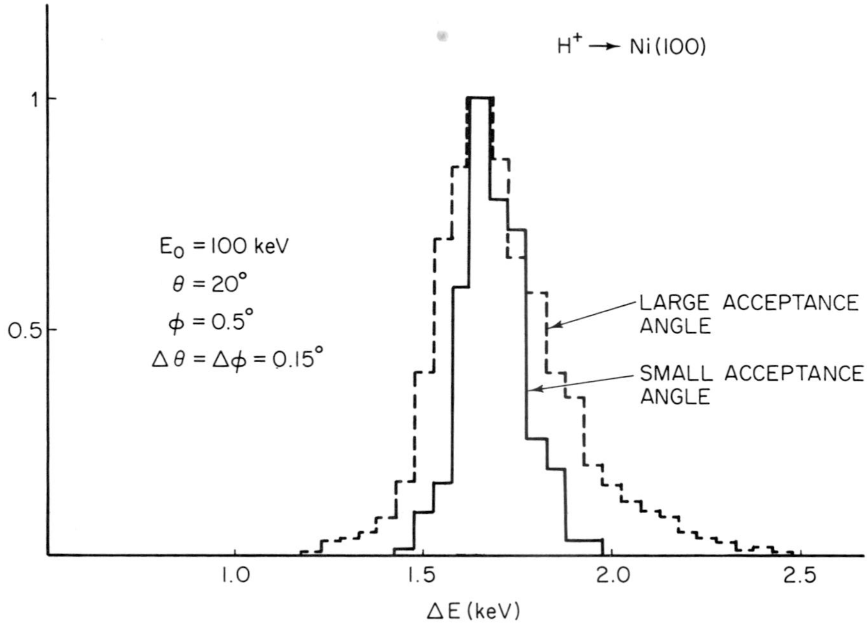

Figure 7.31. Computer simulation of the energy loss spectrum of 100 keV H^+ near a Ni (100) surface.

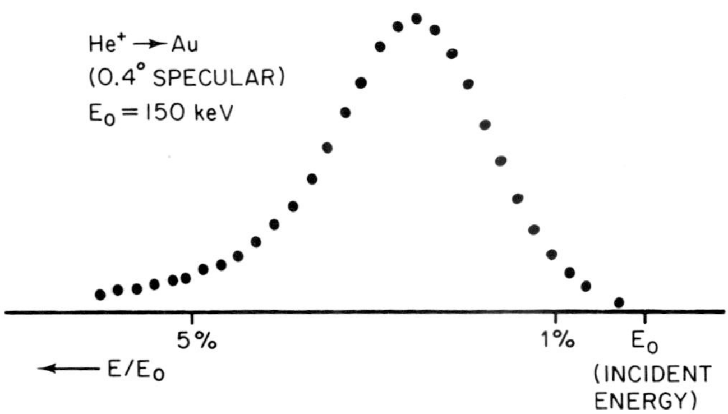

Figure 7.32. Experimental energy loss spectrum for 150 keV He^+ near an Au surface according to Andrä *et al.*[48]

100 keV H^+ near a Ni surface. From this we can calculate the mean energy loss value for surface channelling with a channelling length (the half-length of a single channelling trajectory) of approximately 40 nm to be about 1·2 keV, since the stopping power for $s = 0$ is 30 eV nm^{-1}. On the other hand, using the channelling stopping power due to the single-electron excitation obtained in Chapter 4, the mean energy loss is approximately 400 eV. Therefore, we may

conclude that the surface stopping power is more important than the single electron stopping power for certain surface scattering conditions.

Kawai *et al.*[95] made a computer simulation of the energy loss spectrum for 100 keV ions scattered from the Ni(100) surface with a grazing angle $\theta_i = 0{\cdot}5°$, the results of which are shown in figure 7.31. Note that the mean energy loss value is $1{\cdot}7$ keV, which is much larger than the value (~ 600 eV) expected by Firsov's stopping power. Many experiments have been done to measure the energy loss spectrum for the grazing incidence. Figure 7.32 shows the experimental energy loss spectrum for 150 keV $He^+ \rightarrow Au$ surface by Andrä *et al.*,[48] where the mean energy loss value is about 3 keV, which is very large, as is expected for surface plasmon excitation.

7.6.3. *Photon emission during surface scattering*

After transmission of an ion through a solid foil, the orbital electrons in the ion can be in an excited state, and immediately after ejection the excited ion emits a photon, which can be analysed to study the atomic state. This study is called 'beam foil spectroscopy'. Recently, it was found that by making the incident angle smaller and smaller, the emitted photon becomes elliptically or circularly polarized, both for transmission and surface scattering.[96–99]

This phenomenon was first studied theoretically by Eck,[100] Hermann,[101] and Schröder and Kupfer.[102] Eck, and Schröder and Kupfer considered the unbalanced capture of electrons from the surface into the various magnetic-quantum-number levels to be due to the action of the surface potential, but Hermann attributed the phenomenon to the reconstruction of the electron states

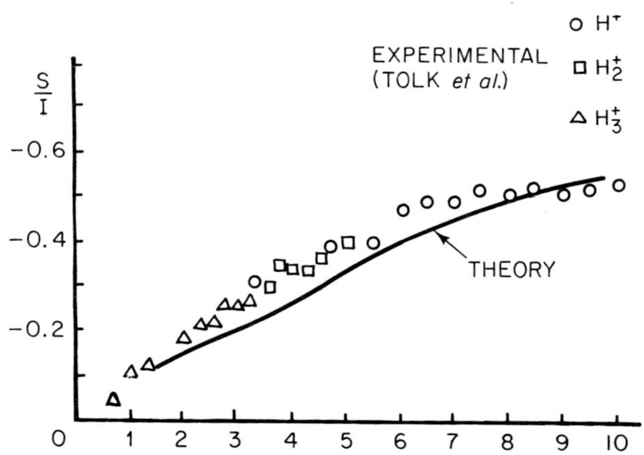

Figure 7.33. Degree of polarization (*S*/*I* in Stokes parameter) of photons emitted from H^+ and H_2^+ scattered by an Au surface, comparing the calculation with experiments of Tolk *et al.*[104]

near the surface. These two models might be correct for individual specialized cases, as stated in section 7.4. For low energy H^+, H_2^+, the electron capture at the surface is the main process, corresponding to the Eck, and Schröder and Kupfer cases. However, for heavy ions Hermann's consideration becomes more important.

Unfortunately these theories cannot provide any quantitative conclusion for comparison with detailed experimental results. Recently, Ohtsuki *et al.*[103] made a calculation of the degree of polarization (S/I in Stokes parameter) as a function of the ion energy for H^+ and $H_2^+ \rightarrow$ Au, making use of the method of Kitagawa and Ohtsuki for electron capture studied in section 7.4. These results are shown in figure 7.33 and compared with the experimental results of Tolk *et al.*[104,105]

There is another process which results in photon emission at the surface due to surface excitation. It is well known that an energetic electron beam transmitted through a solid foil emits ultraviolet radiation, which we call 'transition radiation'. There is a peak near the plasmon energy, which may be attributed to the radiative surface plasmon (Section 7.5).

The theory of transition radiation was developed by Ginzburg and Frank,[106] and refined by many other workers.[107–110] By solving Maxwell's equation for the energetic electron beam, we obtain the number of photons emitted in the direction θ from the incident-electron direction in the energy range $\hbar\omega$ to $\hbar\omega + \mathrm{d}(\hbar\omega)$. Thus

$$\frac{\mathrm{d}^2 N}{\mathrm{d}(\hbar\omega)\,\mathrm{d}\Omega} = \frac{1}{\pi^2}\cdot\frac{e^2}{\hbar c}\,\frac{1}{\hbar\omega}|\varepsilon - 1|^2\left(\frac{v}{c}\right)^2 M \sin^2\theta\cos^2\theta$$

$$M = \left|\frac{\sin(\omega d/2v)}{i(c/\omega)L^+} \pm \frac{i\cos(\omega d/2v)}{i(c/\omega)L^-}\right|$$

$$L^+ = \varepsilon_1 K_{z0} - iK_{z1}\tan(K_{z1}d/2)$$

$$L^- = \varepsilon_1 K_{z0} + iK_{z1}\cot(K_{z1}d/2)$$

where $K_{z1} = [(\omega/c)^2\varepsilon_1 - K_x^2]^{1/2}$ (see subsection 7.5.1). The plus sign in the equation for M refers to $\theta < \pi/2$ and the minus to $\pi > \theta > \pi/2$.

In the above L^+ and L^- vanish at $\omega = \omega_\mathrm{p}$ and M becomes large. This peak in the intensity is found in experiments, for example, by Hattendorf,[111] for 80 keV electrons through 60 nm Ag foil ($\theta = 160°$), as shown in figure 7.34.

7.7. Interaction between two ions near the surface

As shown in Chapter 6, an interaction (attractive force) between two ions in a solid occurs because of the virtual exchange of bulk plasmons, from which the alignment effect arises. What kinds of interaction occur between two ions near the surface due to the exchange of bulk, reflected and surface plasmons?

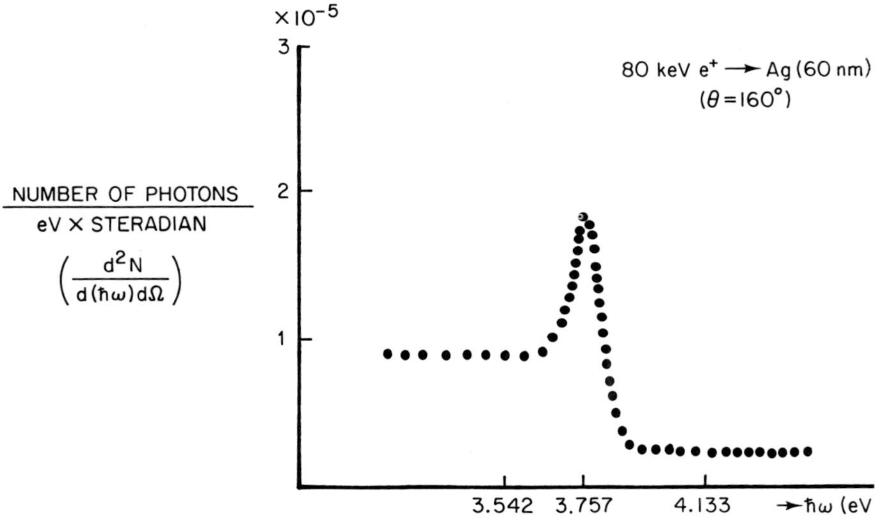

Figure 7.34. Experimental intensity profile of the transition radiation for 80 keV electrons into Ag foil, according to Hattendorf.[111]

Here the dynamical image terms play an important role. One of the interesting consequences of this interaction is a new type of force which gives a moment in the direction of an axis of rotation perpendicular to the velocity of the ion pair and parallel to the surface. The theory of the interaction between the two ions was developed by O'hori and Ohtsuki,[112] and here we give a brief description of the process and their results.

For simplicity we only take into account the surface plasmon exchange. The total Hamiltonian takes the form[113]

$$H = H_0 + H_{\mathrm{spl}} + H_{\mathrm{int}} \tag{7.7.1}$$

where H_0 is the Hamiltonian of the two ions as in section 6.4, and H_{spl} is the Hamiltonian of the surface plasmon

$$H_{\mathrm{spl}} = \sum_{\mathbf{k}<\mathbf{k}_c} \hbar\omega_s (a_{\mathbf{k}}^+ a_{\mathbf{k}} + \tfrac{1}{2}) \tag{7.7.2}$$

The last term H_{int} in equation 7.7.1 is the interaction Hamiltonian between the ion and the surface plasmon

$$H_{\mathrm{int}} = \sum_{\mathbf{k}<\mathbf{k}_c} \sum_j \sqrt{\frac{\pi\hbar e^2 \omega_s}{kA}} Z_j \exp(-k|z_j| + i\mathbf{k}\cdot\rho_j)(a_{\mathbf{k}} + a_{-\mathbf{k}}^+) \tag{7.7.3}$$

In the above $\mathbf{r}_j = (\boldsymbol{\rho}_j, z_j)$ (see figure 7.35), and a_k^+ is the creation operator of the surface plasmon. The other symbols have the same meaning as section 6.4. The quantity A is the macro-area of the surface.

In order to describe the interaction between two ions it is useful to employ the canonical transformation, as shown in section 6.4. The canonical transforma-

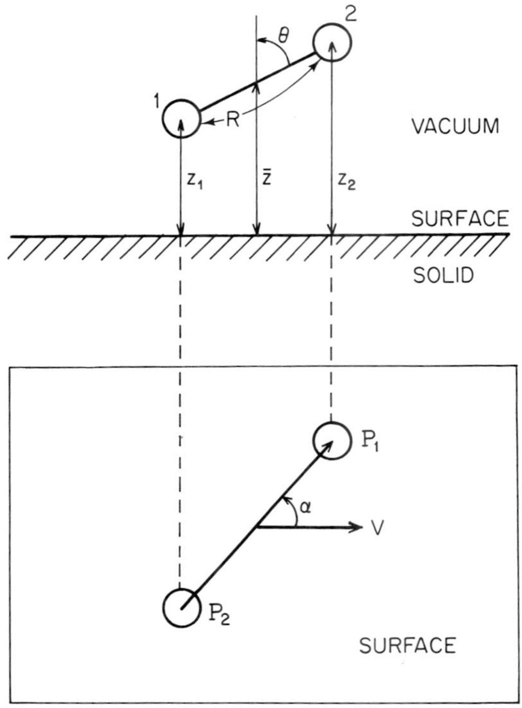

Figure 7.35. Sketch of two ions near the surface.

tion to cancel the term H_{int} is performed by

$$S_1 = -i \sum_{\mathbf{k}<k_c} \sum_j \frac{Z_j}{\omega_s}\left(\frac{\pi\hbar e^2\omega_s}{kA}\right)\exp(-k|z_j| + i\mathbf{k}\cdot\boldsymbol{\rho}_j)(a_{\mathbf{k}} - a^+_{-\mathbf{k}}) \qquad (7.7.4)$$

The total Hamiltonian becomes

$$H' = H_0 + H_{\text{spl}} + V_{\text{si}} + H_{\text{di}} + \ldots \qquad (7.7.5)$$

where

$$H_0 = \sum_j \frac{P_j^2}{2M_j} + \sum_k \frac{2\pi Z_1 Z_2 e^2}{kA}\exp(-k|z_1 - z_2|)\exp[i\mathbf{k}\cdot(\boldsymbol{\rho}_1 - \boldsymbol{\rho}_2)] \qquad (7.7.6)$$

$$H_{\text{spl}} = \sum_{\mathbf{k}<k_c} \hbar\left[\omega_s + \sum_{j=1}^{2} \frac{2\pi Z_j^2 e^2}{\omega_s^2 M_j A} k\exp(-2k|z_j|)\right](a_{\mathbf{k}}^+ a_{\mathbf{k}} + \tfrac{1}{2}) \qquad (7.7.7)$$

$$V_{\text{si}} = -\sum_{ij}\sum_{\mathbf{k}<k_c} \frac{\pi e^2}{kA} Z_i Z_j \exp[-k(|z_i| + |z_j|)]\exp[i\mathbf{k}\cdot(\boldsymbol{\rho}_i - \boldsymbol{\rho}_j)] \qquad (7.7.8)$$

$$H_{\text{di}} = \sum_{\substack{\mathbf{k}<k_c \\ j}} \frac{Z_j}{\omega_s}\left(\frac{\pi\hbar e^2\omega_s}{kA}\right)^{1/2}(a_{\mathbf{k}} - a^+_{-\mathbf{k}})\exp(-k|z_j|) \quad \exp(i\mathbf{k}\cdot\boldsymbol{\rho}_j)\frac{\mathbf{k}\cdot\mathbf{P}_{j\parallel} + i\sigma(z_j)^k P_{jz}}{M_j} \qquad (7.7.9)$$

In the above we define

$$\sigma(z) = 1 \quad \text{for } z > 0$$
$$\sigma(z) = -1 \quad \text{for } z < 0 \tag{7.7.10}$$

$\mathbf{P}_{j\parallel}$ and P_{jz} are the parallel and perpendicular components of $\mathbf{P}_j$, the momentum of the ion, respectively.

Next, we perform the second canonical transformation of S_2 in order to cancel H_{di}

$$S_2 = -i \sum_{\mathbf{k}<k_c} \sum_j \frac{Z_j}{\omega_s}\left(\frac{\pi\hbar e^2\omega_s}{kA}\right)^{1/2}$$
$$\times\left[\exp(-k|z_j| + i\mathbf{k}\cdot\boldsymbol{\rho}_j)\frac{T(\mathbf{k},\mathbf{P}_j)}{\omega_s - T(\mathbf{k},\mathbf{P}_j)}a_k + \text{h.c.}\right] \tag{7.7.11}$$

$$T(\mathbf{k},\mathbf{P}_j) = \frac{\mathbf{k}\cdot\mathbf{P}_{j\parallel} + i\sigma(z_j)P_{jz}}{M_j} \tag{7.7.12}$$

The total Hamiltonian after the canonical transformations becomes

$$H = H_0 + H_{\text{spl}} + \sum_{i(=j)} V_{i\bar{i}} + \sum_{i\neq j} V_{i\bar{j}} \tag{7.7.13}$$

where H_0 and H_{spl} are given by equations 7.7.6 and 7.7.7. The Hamiltonian V_{ij} represents the interaction between the ith ion and the image of the jth ion. Therefore $V_{i\bar{i}}$ is the interaction with its self-image

$$V_{i\bar{j}} = -\sum_{\mathbf{k}<k_c} \frac{\pi e^2 Z_j Z_i}{kA}\exp[-k(|z_i| + |z_j|)]\exp[(i\mathbf{k}\cdot(\boldsymbol{\rho}_1 - \boldsymbol{\rho}_2)]\left(1 + \frac{T_i^+ T_j}{\omega_s^2 - T_j^2}\right) \tag{7.7.14}$$

Let us discuss a simple special case. When the ions move parallel to the surface with the same velocity v, we obtain

$$V_{i\bar{i}} = -\frac{Z_i^2 e^2 \omega_s}{2v}\int_0^{u(v)} \mathrm{d}x \frac{\exp(-2\omega_s|z_i|x/v)}{\sqrt{1-x^2}} \tag{7.7.15}$$

where we define

$$u(v) = v/v_F \quad \text{for } v < v_F$$
$$u(v) = 1 \quad \text{for } v > v_F$$
$$v_F = \omega_s/k_c \tag{7.7.16}$$

Further, we obtain

$$\sum_{\substack{i,j\\(i\neq j)}} V_{i\bar{j}} = -\frac{2Z_1Z_2e^2\omega_s}{v}\int_0^{u(v)} \frac{\mathrm{d}x}{\sqrt{1-x^2}}\left[J_0\left(\frac{\omega_s\rho x}{v}\right) + 2\sum_{\substack{n=2\\(\text{even})}}^{\infty} J_n\left(\frac{\omega_s\rho x}{v}\right)\right.$$

$$\times \cos\left(\frac{n\pi}{2}\right)\cos(n\alpha)\left(\frac{x}{1+\sqrt{1-x^2}}\right)^{2n}\Bigg]\exp[-\omega_s(|z_1|+|z_2|)x/v]$$

$$-\theta(v-v_F)\frac{4Z_1Z_2e^2\omega_s}{v}\int_0^{v_F}\frac{dx}{\sqrt{x^2-1}}\sum_{\substack{n=2\\(\text{even})}}^{\infty}J_n\left(\frac{\omega_s\rho x}{v}\right)$$

$$\times \cos(n\alpha)\sin\left(n\tan^{-1}\frac{1}{\sqrt{x^2-1}}\right)\exp[-\omega_s(|z_1|+|z_2|)x/v]$$

(7.7.17)

In the above we take α to be the angle between the ion velocity $\mathbf{v}$ and the line connecting the two ions (see figure 7.35). When we change the variable $x=\sin\varphi$ in equation 7.7.15, and $x=\sec\varphi$ in equation 7.7.17, we obtain the effective interaction between the two ions

$$H_{\text{eff}}=\frac{Z_1Z_2e^2}{|\mathbf{r}_1-\mathbf{r}_2|}+\sum_i V_{i\bar{i}}+2V_{1\bar{2}} \tag{7.7.18}$$

$$V_{i\bar{i}}=-\frac{Z_i^2e^2\omega_s}{v}\int_0^{\varphi_0(v)}d\varphi\exp[-2\omega_s|z_i|\sin(\varphi/v)] \tag{7.7.19}$$

$$V_{1\bar{2}}=-\frac{Z_1Z_2e^2\omega_s}{2v}\int_0^{\varphi_0(v)}d\varphi\Bigg[J_0\left(\frac{\omega_s\rho\sin\varphi}{v}\right)+2\sum_{\substack{n=2\\(\text{even})}}^{\infty}J_n\left(\frac{\omega_s\varphi\sin\varphi}{v}\right)$$

$$\times\cos\left(\frac{n\pi}{2}\right)\cos(n\alpha)\tan^n\left(\frac{\varphi}{2}\right)\exp[-\omega_s(|z_1|+|z_2|)\sin\varphi/v]$$

$$+\theta(v-v_F)\frac{Z_1Z_2e^2\omega_s}{v}\sum_{\substack{n=2\\(\text{even})}}^{\infty}\int_0^{\sec^{-1}(v/\bar{v}_F)}d\varphi\sec\varphi\,J_n\left(\frac{\omega_s\rho\sec\varphi}{v}\right)$$

$$\times\cos\left(\frac{n\pi}{2}\right)\cos(n\alpha)\sin(n\varphi)\exp[-\omega_s(|z_1|+|z_2|)\sec\varphi/v]\Bigg] \tag{7.7.20}$$

In the above the upper limit of the integration $\varphi(v)$ is defined as

$$\sin\varphi(v)=u(v)$$

Figure 7.36 shows the effective potential V_{eff} as a function of the angle θ between the surface normal and the line connecting the two ions moving with the same velocity $2v_F$ parallel to the surface in the condition $\bar{z}=0{\cdot}1$ nm and $\alpha=0$ (see figure 7.35). It is worth noting that the effective potential decreases with $\theta\to 0$, which implies that a new kind of the rotating force appears. Therefore, the two ions 'stand up' to the surface.

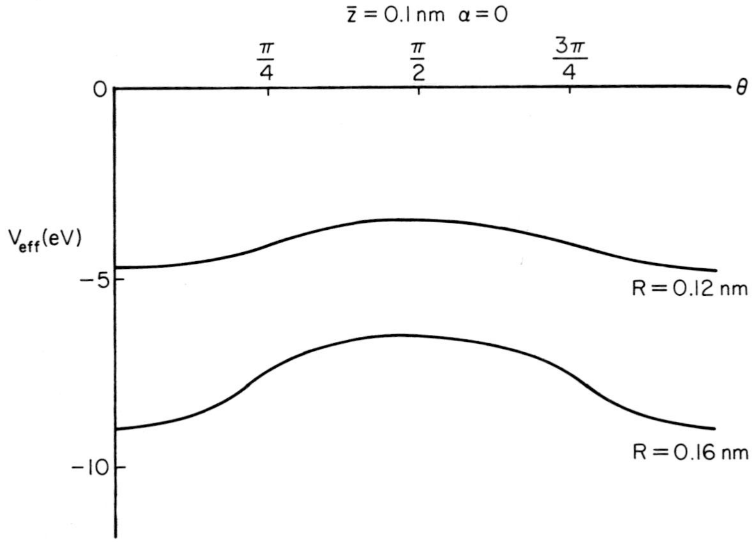

Figure 7.36 Effective potential between two ions as a function of the angle θ (see Figure 7.35).

References

1. Lander, J.J., *Prog. Solid State Chem.* **2**, 26 (1965).
2. Somorjai, G.A. (ed.), *The Structure and Chemistry of Solid Surfaces* (New York: Wiley, 1969).
3. Clark, B.C., Herman, R. and Wallis, R.F., *Phys. Rev.* **A139**, 860 (1965).
4. Maradudin, A.A. and Melngailis, J., *Phys. Rev.* **A133**, 1188 (1964).
5. Jones, E.R., McKinney, J.T. and Webb, M.B., *Phys. Rev.* **151**, 476 (1966).
6. Woodruff, D.P. and Seah, M.P., *Phys. Stat. Sol.* **1**, 429 (1970).
7. Morabito, J.M., Steiger, R.F. and Somorjai, G.A., *Phys. Rev.* **179**, 638 (1969).
8. Goodman, R.M., Farrell, H.H. and Somorjai, G.A., *J. Chem. Phys.* **48**, 1046 (1968).
9. MacRae, A.U., *Surface Sci.* **2**, 522 (1964).
10. Lyon, H.B. and Somorjai, G.A., *J. Chem. Phys.* **44**, 3707 (1965).
11. Mott, N.F. and Massey, H.S., *The Theory of Atomic Collisions* (Oxford: Oxford University Press, 1952).
12. Burhop, E.H.S., *Quantum Theory* (New York: Academic Press, 1961).
13. Lander, J.J. and Morrison, J., *J. Appl. Phys.* **34**, 3517 (1963).
14. Hirabayashi, K. and Takeishi, Y., *Surface Sci.* **4**, 150 (1966).
15. McRae, E.G., *J. Chem. Phys.* **45**, 3258 (1966).
16. Lax, M., *Rev. Mod. Phys.* **23**, 287 (1951).
17. Lax, M., *Phys. Rev.* **85**, 621 (1952).
18. McRae, E.G., *Surface Sci.* **25**, 491 (1971).
19. Kikuchi, S. and Nakagawa, S., *Sci. Pap. Inst. Phys. Chem. Res.* **21**, 256 (1933).
20. Miyake, S. and Hayakawa, K., *Acta Cryst.* **A26**, 60 (1970).
21. Kambe, K., *Z. Naturforsch.* **A22**, 322 (1967).
22. Kambe, K., *Z. Naturforsch.* **A23**, 1280 (1968).
23. Beeby, J.L., *J. Phys. C (Proc. Phys. Soc.)* **1**, 82 (1968).
24. Marcus, P.M. and Jepsen, D.W., *Phys. Rev. Letters* **20**, 925 (1968).
25. Ohtsuki, Y.H., *J. Phys. Soc. Japan* **24**, 1116 (1968).
26. Duke, C.B. and Tucker, C.W., *Phys. Rev. Letters* **23**, 1163 (1969).

27. Boudreaux, D.S. and Heine, V., *Surface Sci.* **8**, 426 (1967).
28. Ohtsuki, Y.H., *J. Phys. Soc. Japan* **24**, 481 (1968).
29. Hirabayashi, K., *J. Phys. Soc. Japan* **24**, 846 (1968).
30. Stern, R.M., Perry, J.J. and Boudreaux, D.S., *Rev. Mod. Phys.* **41**, 275 (1969).
31. Capert, G., *Surface Sci.* **13**, 361 (1969).
32. Stern, R.M., Taub, H. and Wagenfeld, H., *J. Phys. Soc. Japan* **28**, 723 (1970).
33. Pendry, J.B., *J. Phys. C* **4**, 2501 (1971).
34. Pendry, J.B., *J. Phys. C* **4**, 3095 (1971).
35. Pendry, J.B., *J. Phys. C* **5**, 2567 (1972).
36. Andersson, S. and Pendry, J.B., *J. Phys. C* **6**, 601 (1973).
37. Pendry, J.B., *Low Energy Electron Diffraction* (New York: Academic Press, 1974).
38. McCracken, G.M. and Freeman, N.J., *J Phys. B* **2**, 661 (1969).
39. Pronko, P.P., Appleton, B.R., Holland, O.W. and Wilson, S.R., *Phys. Rev. Letters* **43**, 779 (1979).
40. Pronko, P.P., Appleton, B.R., Holland, O.W. and Wilson, S.R., *Nucl. Inst. Methods* **170**, 227 (1980).
41. Barrett, J.H., Appleton, B.R. and Holland, O.W., *Phys. Rev.* **B22**, 4180 (1980).
42. Crawford, O.H., *Phys. Rev. Letters* **44**, 185 (1980).
43. Jakas, M.M. and Baragiola, R.A., *Phys. Rev. Letters* **44**, 424 (1980).
44. Oen, O.S., in *Proceedings of Atomic Collision in Solids*, Lyon, 1981.
45. Oen, O.S., in *Proceedings of Recent Theoretical Developments on Atomic Collision in Solids*, Strasbourg, 1981.
46. Mashkova, E.S. and Molchanov, V.A., *Rad. Effects* **23**, 215 (1974).
47. Balashova, L.L., Mashkova, E.S. and Molchanov, V.A., in *Proceedings VII International Conference on Atomic Collision in Solids*, Moscow, 1980, ed. by Bulgakov and Tulinov.
48. Andrä, H.J., Winter, H., Fröhling, R., Kirchner, N., Plöhn, H.J., Wittmann, W., Graser, W. and Varelas, C., *Nucl. Inst. Methods* **170**, 527 (1980).
49. Yamamura, Y., and Takeuchi, W., *Rad. Effects* **49**, 251 (1980).
50. Davies, J.A., Eriksson, D., Johansson, N.G.E. and Mitchell, I.V., *Phys. Rev.* **181**, 548 (1968).
51. Davies, J.A., Jackson, D.P., Matsunami, N., Norton, P.R. and Andersen, J.U., *Surface Sci.* **78**, 274 (1978).
52. Turkenburg, W.C., Soszka, W., Saris, F.W., Kersten, H.H. and Colenbrander, B.G., *Nucl. Inst. Methods* **132**, 587 (1976).
53. van der Veen, J.F., Smeenk, R.G., Tromp, R.M. and Saris, F.W., *Surface Sci.* **79**, 212 (1979).
54. Horiguchi, S., Koyama, K. and Ohtsuki, Y.H., *Phys. Stat. Sol.* **B87**, 757 (1978).
55. Hagstrum, H.D., *Phys. Rev.* **122**, 83 (1961).
56. Rau, C. and Sizmann, R., *Phys. Letters* **A43**, 317 (1973).
57. Yablinskii, Y.N. and Trubnikov, B.A., *J. Exp. Theor. Phys.* **25**, 1089 (1967).
58. Cobas, A. and Lamb, W.E., *Phys. Rev.* **65**, 327 (1944).
59. Hagstrum, H.D., *Phys. Rev.* **96**, 336 (1954).
60. Kitagawa, M. and Ohtsuki, Y.H., *Phys. Rev.* **B13**, 4682 (1976).
61. Bates, D.R., *Proc. Roy. Soc. London* **A247**, 294 (1958).
62. Zener, C., *Proc. Roy. Soc. London* **A137**, 696 (1932).
63. Kawai, R. and Ohtsuki, Y.H., (to the published).
64. Behrish, R., Eckstein, W., Meischner, P., Scherzer, B.M.U. and Verbeek, H., in *Atomic Collision in Solids* edited by S. Datz *et al.*, (Plenum, 1975), p. 315.
65. Zaidins, C.S., Ph.D. Thesis (California Inst. Technology, 1967).
66. Brandt, W. and Sizmann, R., *Phys. Lett.* **A37**, 115 (1971).
67. Zwiegel, J., Thesis (Technical University of Munich, 1979).

68. Rausch, E.O. and Thomas, E.W., in *Proceedings of Atomic Collision in Solids*, Moscow, 1977.
69. Matschke, F.E.P., *Charge State and Energy Distribution of Hydrogen Backscattered from Gold* (300*eV*–10*keV*), Reports from Max-Planck-Institut für Plasma-physik, Garching (1977).
70. Gemmingen, U. and Sizmann, R., *Surface Sci*, **114**, 445 (1982).
71. Gemmingen, U., Thesis (University of Munich, 1980).
72. Ritchie, R.H., *Phys. Rev.* **106**, 874 (1957).
73. Pabst, R., *Rad. Effects* **37**, 105 (1978).
74. Kliewer, K.L. and Fuchs, R., *Phys. Rev.* **153**, 498 (1967).
75. Raether, H., *Physics of Thin Films* **9**, 145 (1977).
76. Suzuki, K., Kitagawa, M. and Ohtsuki, Y.H., *Phys. Stat. Sol.* **B82**, 643 (1977).
77. Appelbaum, J.A. and Hamann, D.R., *Phys. Rev.* **B6**, 1122 (1972).
78. Lang, N.D. and Kohn, W., *Phys. Rev.* **B7**, 3541 (1973).
79. Ritchie, R.H., *Phys. Lett.* **A38**, 189 (1972).
80. Inkson, J.C. *Surface Sci.* **28**, 69 (1971).
81. Ray, R. and Mahan, G.D., *Phys. Lett.* **A42**, 301 (1973).
82. Harris, J. and Jones, R.O., *J. Phys. C* **6**, 3585 (1973).
83. Harris, J. and Jones, R.O., *J. Phys. C* **7**, 3751 (1974).
84. Hodges, C.H., *J. Phys.* C**8**, 1849 (1975).
85. Heinrichs, J., *Phys. Rev.* **B8**, 1346 (1973).
86. Muscat, J.P. and Newns, D.M., *Surface Sci.* **64**, 641 (1977).
87. Núñez, R., Echenique, P.M. and Ritchie, R.H. *J. Phys.* C **13**, 4229 (1980).
88. Hedin, L. and Lundquist, S., *Solid State Phys.* **23**, 33 (1969).
89. Ichikawa, M., Thesis (Waseda University, 1974).
90. Schmeits, M. and Lucas, A.A., *Surface Sci.* **64**, 176 (1977).
91. Lucas, A.A., *Phys. Rev.* **B20**, 4990 (1979).
92. Ohtsuki, Y.H., (Annual Meeting of Phys. Soc. of Japan, 1978).
93. Ohtsuki, Y.H., Koyama, K. and Yamamura, Y., *Phys. Rev.* **B20**, 5044 (1979).
94. Ritchie, R.H., *Phys. Rev.* **106**, 874 (1957).
95. Kawai, R., Ohtsuki, Y.H. and Itoh, N., *Surface Sci.* **114**, 137 (1982).
96. Berry, H.G., Curtis, L.T., Ellis, D.G. and Schectmann, R.M., *Phys. Rev. Lett.* **32**, 751 (1974).
97. Berry, H.G., Gabrielse, G. and Livingston, A.E., *Phys. Rev.* **A16**, 1915 (1977).
98. Andrä, H.J., *Phys. Lett.* **54**, 315 (1975).
99. Andrä, H.J. and Solver, J.D., *Phys. Rev. Lett.* **37**, 1212 (1976).
100. Eck, T.G., *Phys. Rev. Lett.* **33**, 1055 (1974).
101. Hermann, R.M., *Phys. Rev. Lett.* **35**, 1626 (1975).
102. Schröder, H. and Kupfer, E., *Z. Physik* **A279**, 13 (1976).
103. Ohtsuki, Y.H., Goto, W., Tazunoki, T. and Kawai, R., *Rad. Effects* (to be published).
104. Tolk, N.H., Tully, J.C., Kraus, J.S., Heiland, W., and Neff, S.H., *Phys. Rev. Lett.* **41**, 643 (1978).
105. Tolk, N.H., Tully, J.C., Kraus, J.S. Heiland, W. and Neff, S.H., *Phys. Rev. Lett.* **42**, 1475 (1979).
106. Ginzburg, W. and Frank, I., *Zh. Eksp. Theor. Fiz.* **16**, 15 (1946).
107. Pafomov, V.E., *J. Exp. Theor. Physics* **6**, 829 (1958).
108. Ritchie, R.H. and Eldrige, H.B., *Phys. Rev.* **126**, 1935 (1962).
109. Ashley, J.C., *Phys. Rev.* **155**, 208 (1967).
110. Kröger, E., *Z. Phys.* **235**, 403 (1970).
111. Hattendorf, H.D., *Phys. Stat. Sol.* **A42**, 489 (1977).
112. O'hori, T. and Ohtsuki, Y.H., *Phys. Rev.* **B15** (1982).
113. Lucas, A.A., Kartheuser, E. and Badre, R.G., *Phys. Rev.* **B2**, 2488 (1970).

INDEX

Abrahamson's parameters 8
Absorption coefficient:
  calculation of 58
  measurement of 71
Anomalous absorption 54
  coefficient 57
Anomalous transmission 57
Auger process 206
Axial dechannelling 148

Backscattering 200
Barkas effect 107
Bethe–Bloch formula 85
Bethe's dynamical theory 36
Blocking 15
Bohr velocity 3
Born approximation 7
Born–Mayer potential 8
Borrmann effect 57
Bragg rule 107
Bremsstrahlung 109
  coherent 110

Channelling 9
Channelling radiation 114, 115
Complex potentials 67
  measurement of 70
Continuum potential 10
Core electron excitations 61
Critical angle 14
Cumulant expansion 64

Damping effect 142
Debye temperature 50
Debye–Waller factor 49
Dechannelling:
  applications 153
  fraction 128
  phenomenological theory 125
Diffuse patterns 72
Diffusion coefficients:
  electron beams 149
  fast ions 129
    electron contribution 130
    phonon excitation 135
  slow ions 139
Diffusion power 126
Dynamical image potential 228
Dynamical surface potential 221
  derivation 224

Effective charge 166
Electron channelling 26
Electron diffraction 30
Energy reflection coefficient 201
Equilibrium charge 156
  of channelled ions 168
Equi-partition rule 92

Firsov potential 8
Firsov's model 92
Fokker–Planck equation 142

Half thickness 146
Hartree approximation 3
Hartree–Fock approximation 3
Heisenberg representation 34
Hyperchannelling 15

Inelastic scattering probability function 101

Kikuchi band 76
Kikuchi line 76
Kikuchi patterns 75

Lattice vibration 48
Landau–Zener method 215
Laue–Bragg condition 32
LEED:
  dynamical theory 193
    band theory 197
    multiple scattering theory 194
  kinematical theory 187
Lindhard's theory 10, 125
  modification of 18
  quantum corrections to 24

Multi-beam effects 44
Minimum yield 17
Molière potential 8
Multiple scattering model 33

Near-surface ion interaction 239
Negative region 81
Neutralization 206
Non-radiative surface plasmon 223
Nuclear stopping power 97

Okorokov effect 168, 169

Phonon excitations 63
Planar dechannelling 142

Radiative surface plasmon 223
Reflection coefficient 201
Resonance capture 212
Resonance error 40
Rutherford scattering 5
Rydberg energy 3

Scattering factor 31
Screening parameter 4
Screening theory 174
Secondary Bragg peaks 197
Self-consistent field method 3
Self-consistent theory 174
Shadowing 15
Shell correction 106
Skipping motion 233
Soliton states 181
Stopping power 83
  corrections to 106
  function 99
  nuclear 97
String potential 11
Structure factor 33
Surface channelling 203
Surface energy loss 235
Surface plasmon 221
Surface potential, induction 228
Surface resonance 199
Surface scattering 203
  photon emission 238
Surface structure analysis 204
Surface wake 228

Thomas–Fermi model 4
Thomas–Fermi potential 8
Thomas–Fermi–Dirac potential 8
Two-wave approximation 34
Thermal diffuse scattering 63
Transition radiation 239

Überall effect 110

Valence electron excitations 59

Wake 174
Wavelet potential 231

Yoshioka term 67, 69